CAMBRIDGE LIBRARY COLLECTION

Books of enduring scholarly value

Physical Sciences

From ancient times, humans have tried to understand the workings of the world around them. The roots of modern physical science go back to the very earliest mechanical devices such as levers and rollers, the mixing of paints and dyes, and the importance of the heavenly bodies in early religious observance and navigation. The physical sciences as we know them today began to emerge as independent academic subjects during the early modern period, in the work of Newton and other 'natural philosophers', and numerous sub-disciplines developed during the centuries that followed. This part of the Cambridge Library Collection is devoted to landmark publications in this area which will be of interest to historians of science concerned with individual scientists, particular discoveries, and advances in scientific method, or with the establishment and development of scientific institutions around the world.

Mathematical and Physical Papers

William Thomson, first Baron Kelvin (1824–1907), is best known for devising the Kelvin scale of absolute temperature and for his work on the first and second laws of thermodynamics, though throughout his 53-year career as a mathematical physicist and engineer at the University of Glasgow he investigated a wide range of scientific questions in areas ranging from geology to transatlantic telegraph cables. The extent of his work is revealed in the six volumes of his *Mathematical and Physical Papers*, published from 1882 until 1911, consisting of articles that appeared in scientific periodicals from 1841 onwards. Volume 6, published in 1911, includes articles from the period 1867–1907. The chapters in the first part of the work focus on voltaic theory and radioactivity, while later ones examine navigation and tides.

Cambridge University Press has long been a pioneer in the reissuing of out-of-print titles from its own backlist, producing digital reprints of books that are still sought after by scholars and students but could not be reprinted economically using traditional technology. The Cambridge Library Collection extends this activity to a wider range of books which are still of importance to researchers and professionals, either for the source material they contain, or as landmarks in the history of their academic discipline.

Drawing from the world-renowned collections in the Cambridge University Library, and guided by the advice of experts in each subject area, Cambridge University Press is using state-of-the-art scanning machines in its own Printing House to capture the content of each book selected for inclusion. The files are processed to give a consistently clear, crisp image, and the books finished to the high quality standard for which the Press is recognised around the world. The latest print-on-demand technology ensures that the books will remain available indefinitely, and that orders for single or multiple copies can quickly be supplied.

The Cambridge Library Collection will bring back to life books of enduring scholarly value (including out-of-copyright works originally issued by other publishers) across a wide range of disciplines in the humanities and social sciences and in science and technology.

Mathematical and Physical Papers

VOLUME 6

LORD KELVIN
EDITED BY JOSEPH LARMOR

CAMBRIDGE UNIVERSITY PRESS

Cambridge, New York, Melbourne, Madrid, Cape Town,
Singapore, São Paolo, Delhi, Tokyo, Mexico City

Published in the United States of America by Cambridge University Press, New York

www.cambridge.org
Information on this title: www.cambridge.org/9781108029032

This edition first published 1911
This digitally printed version 2011

ISBN 978-1-108-02903-2 Paperback

MATHEMATICAL

AND

PHYSICAL PAPERS

MATHEMATICAL
AND
PHYSICAL PAPERS

VOLUME VI

VOLTAIC THEORY, RADIOACTIVITY, ELECTRIONS NAVIGATION AND TIDES MISCELLANEOUS

BY THE RIGHT HONOURABLE

SIR WILLIAM THOMSON, BARON KELVIN

O.M., P.C., G.C.V.O., LL.D., D.C.L., SC.D., M.D., ...
PAST PRES. R.S., FOR. ASSOC. INSTITUTE OF FRANCE,
GRAND OFFICER OF THE LEGION OF HONOUR, KT PRUSSIAN ORDER *POUR LE MÉRITE*,
CHANCELLOR OF THE UNIVERSITY OF GLASGOW
FELLOW OF ST PETER'S COLLEGE, CAMBRIDGE

ARRANGED AND REVISED WITH BRIEF ANNOTATIONS BY

SIR JOSEPH LARMOR, D.Sc., LL.D., Sec. R.S.

LUCASIAN PROFESSOR OF MATHEMATICS IN THE UNIVERSITY OF CAMBRIDGE
AND FELLOW OF ST JOHN'S COLLEGE
REPRESENTATIVE OF THE UNIVERSITY IN PARLIAMENT

CAMBRIDGE:
AT THE UNIVERSITY PRESS
1911

PREFACE

THE present volume completes the record of Lord Kelvin's published scientific work. For convenience of reference a complete schedule of the volumes in which that work is contained is printed in front of this preface.

At the time of the discovery of radioactivity, Lord Kelvin had been engaged for some years in investigations on the electrification of air (this volume pp. 1–64), in continuation of the early work on atmospheric electricity which is collected mainly in the *Papers on Electrostatics and Magnètism.* The investigation of the conducting quality excited in gases by cathode and other rays formed a natural sequel to this work, and to the early investigations on the insulation of Electroscopes in which leakage through the air had not been suspected. Thus a large amount of attention was bestowed by Lord Kelvin, in conjunction with his pupils, on the phenomena of conduction excited in gases by Röntgen rays and by uranium. These researches soon developed into consideration of the voltaic equilibrium which is established between metals separated by a conducting gaseous medium, and into the general phenomena of contact electricity (pp. 110–149), and its thermodynamic relations (Vol. v. pp. 29–37). Some of the experimental detail in these connexions has now been superseded by improved methods and apparatus (cf. Sir J. J. Thomson's treatise on *The Conduction of Electricity in Gases*), but for the sake of historical clearness it has been found desirable to reprint it practically in full. The same section also includes theoretical speculations on the nature of the decomposition of the molecules by radioactive agencies, in continuation of papers of this kind already reprinted in the *Baltimore Lectures.*

A section on "Navigation and Tides" follows (pp. 244–305), which may be described as a continuation of the work reprinted in Vol. III. of *Popular Lectures and Addresses.*

The final section of miscellaneous matter includes an obituary notice of Archibald Smith, which is of high interest in connexion with the history of the correction of the compass for the iron of ships: it also contains appreciations of the work of Fleeming Jenkin, of Sir George Stokes, of Peter Guthrie Tait, and of James Watt; and it ends appropriately with reminiscences of Lord Kelvin's career at Glasgow University, contained in an address delivered on his installation as Chancellor.

The excellent work of the Cambridge University Press, and the courtesy of its officials, have as usual much promoted the adequate presentation of the material which constitutes the volume; while the careful collaboration of Dr GEORGE GREEN, in this as in previous volumes, has ensured the accuracy and completeness of the reprint.

J. L.

ST JOHN'S COLLEGE, CAMBRIDGE,
October 2, 1911.

CONTENTS

VOLTAIC THEORY, RADIOACTIVITY, ELECTRIONS.

NAVIGATION AND TIDES.

MISCELLANEOUS.

VOLTAIC THEORY, RADIO-ACTIVITY, ELECTRIONS

234. On Volta-convection by Flame.

[From *Brit. Assoc. Report*, Vol. xxxvii. 1867, pt. ii. pp. 17, 18; *Phil. Mag.* Vol. xxxv. Jan. 1868, pp. 64—66; *Annal. de Chimie*, Vol. xiv. 1868, pp. 487, 488. Reprinted in *Electricity and Magnetism*, art. xxiii. pp. 328, 329.]

235. Electrification of Air by Flame.

[From *Edinb. Roy. Soc. Proc.* Vol. xvi. [read July 15, 1889], pp. 262, 263.]

In continuation of experiments on the electricity of air within doors, which I made twenty-seven years ago, and which are described in §§ 296—300 of my *Electrostatics and Magnetism*, a series of observations was commenced under my instructions at the end of April last, within the Natural Philosophy Class Room and Laboratory, the Bute Hall, the University Tower, and other places inside and outside the buildings of Glasgow University, by Mr Magnus Maclean, official assistant to the Professor of Natural Philosophy, and Mr Goto of Tokio, Japan, for the purpose of endeavouring to find a relation between the electrification of air within a building and the atmospheric potential in its neighbourhood outside; and of finding causes which produced or changed the electrification of any given mass of air.

A large number of series of observations have been made by Mr Maclean and Mr Goto on the potentials of water-dropping collectors within the building, and at different points outside the building. Hitherto no definite relation has been discovered between the external potential and the potential at different points within large enclosures, such as the Bute Hall and smaller rooms of the University Buildings. The weather has been for the most part very settled and the external potential almost always positive in all positions from a few feet above the ground to the top of the University Tower; while the potential within doors here, in the new University Buildings, on the top of Gilmour Hill, just as in the Old College down in the densest part of Glasgow, was always negative except sometimes in the Natural Philosophy Lecture Room and Apparatus Room, where there were considerable disturbances, undoubtedly due to the electric light wiring. In one ordinarily unused room (the Physical Apparatus Museum) 31½ feet long by 24 feet broad and about 20 feet high, practically quite free from any sensible disturbance by electric light wires, or by electrical operations being performed in the Laboratory, a remarkable result has been observed within the past fortnight. The electric potential of a water-dropper having its nozzle at the centre of the room and about 7 feet above the floor, was always found about 2 volts negative at the commencement of the observations, and always increased to about 9 volts in the course of the first twenty minutes of a series of observations lasting generally forty minutes. During the last twenty minutes of the series the potential remained somewhat nearly constant at 9 volts. Within the room, two quadrant electrometers, each with an ordinary paraffin lamp and scale, were used; one of them for the outside water-dropper, and the other for the water-dropper within the room.

Towards ascertaining the cause of this change, an observation was made on the 4th July, between 10 and 11 A.M. The lamps were both extinguished, and one of them was lighted by a lucifer match every five minutes for the purpose of reading the electrometer deflection. It was found that in these circumstances there was not the increase of negative potential which had been found in every previous series of observations in the same place, and with all other circumstances the same, except the burning of the lamp. This single observation seemed to prove conclusively that the burning of the lamp produced a negative charge of the air of the room. Subsequent experiments made by Mr Goto, with the electrometer and its lamp and scale outside, and with paraffin lamps burning or not burning within the room, have confirmed this result, and are being continued to discover whether corresponding effects are produced by other kinds of flame, or by the presence of eight or nine people in the room. Mr Maclean and Mr Goto will also continue their observations on natural atmospheric electricity, in various localities, indoors and in the open air, and will, I hope, give a paper to the Royal Society of Edinburgh early next session on the subject.

236. On the Velocity of Crookes' Cathode Stream.

[From *Roy. Soc. Proc.* Vol. LII. 1893 [Dec. 8, 1892], pp. 331, 332; *Nature*, Vol. XLVII. Dec. 15, 1892, pp. 164, 165.]

In connection with his splendid discovery of the cathode stream (stream from the cathode in exhausted glass vessels subjected to electric force), Crookes found that when the whole of the stream, or a large part of the whole, is so directed as to fall on 2 or 3 sq. cm. of the containing vessel, this part of the glass becomes rapidly heated up to many degrees, as much as 200° or 300° C. sometimes, above the temperature of the surroundings.

Let v be the velocity, in centimetres per second, of the cathode stream, and ρ the quantity of matter of all the molecules in 1 c.c. of it. Supposing what Crookes' experiments seem to prove to be not far from the truth, that their impact on the glass is like that of inelastic bodies, and that it spends all their translational energy in heating the glass, the energy thus spent, per square centimetre of surface struck, per second of time, is $\frac{1}{2}\rho v^3$; of which the equivalent in gramme-water-centigrade thermal units is approximately $\frac{1}{2}\rho v^3/42{,}000{,}000$. The initial rate at which this will warm the glass, in degrees centigrade per second, is

$$\frac{\frac{1}{2}\rho v^3}{10^6 \times 42 \,.\, \sigma a} \quad \ldots\ldots\ldots\ldots\ldots\ldots\ldots\ldots(1),$$

where σ denotes the specific heat of the glass, and a the thickness of it at the place where the stream strikes it.

The limiting temperature to which this will raise the glass is

$$\frac{1}{E} \times \frac{\frac{1}{2}\rho v^3}{42{,}000{,}000} \ldots\ldots\ldots\ldots\ldots\ldots\ldots\ldots(2),$$

where E denotes the sum of the emissivities of the two surfaces of the glass in the actual circumstances.

It is probable that ρ differs considerably from the average density of the residual air in the enclosure. Let us take, however, for a conceivably possible example, $\rho = 10^{-8}$, which is what the mean density of the enclosed air would be if the vessel were exhausted to 8×10^{-8} of the ordinary atmospheric density.

To complete the example, take

$$v = 100{,}000 \text{ cm. per sec.}$$

(being about twice the average velocity of the molecules of ordinary air at ordinary temperature); and take

$$\sigma a = \tfrac{1}{8} \text{ cm.},$$

as it might be for an ordinary glass vacuum bulb; and take

$$E = \tfrac{1}{3000},$$

which may not be very far from the truth.

With these assumptions, we find, by (1) and (2) approximately, 1° per second for the initial rise, and 375° for the final temperature, which are not very unlike the results found in some of Crookes' experiments.

The pressure of the cathode stream of the velocity and density which we have assumed by way of example is ρv^2, or 100 dynes per square centimetre, or about 100 milligrams heaviness per square centimetre, which is ample for Crookes' wonderful mechanical results.

The very moderate velocity of 1 kilom. per second which we have assumed is much too small to show itself by the optical colour test. The fact that this test has been applied, and that no indication of velocity of the luminous molecules has been found, has, therefore, no validity as an objection against Crookes' doctrine of the cathode stream.

237. ON THE ELECTRIFICATION OF AIR. By LORD KELVIN and MAGNUS MACLEAN.

[From *Roy. Soc. Proc.* Vol. LVI. 1894 [May 31, 1894], pp. 84—94; *Errata*, Vol. LVII. 1895, p. xxxii; *Nature*, Vol. L. July 19, 1894, pp. 280—283; *Phil. Mag.* Vol. XXXVIII. Aug. 1894, pp. 225—235; *Écl. Élec.* Vol. L. Dec. 1, 1894, pp. 568—570.]

1. THAT air can be electrified either positively or negatively is obvious from the fact that an isolated spherule of pure water, electrified either positively or negatively, can be wholly evaporated in air*. Thirty-four years ago it was pointed out by one of us† as probable, that in ordinary natural atmospheric conditions the

* This demonstrates an affirmative answer to the question, Can a molecule of a gas be charged with electricity? (J. J. Thomson, *Recent Researches in Electricity and Magnetism*, § 36, p. 53), and shows that the experiments referred to as pointing to the opposite conclusion are to be explained otherwise.

Since this was written we find, in the *Electrical Review* of May 18, on page 571, in a lecture by Elihu Thomson, the following:—"It is known that as we leave the surface of the earth and rise in the air, there is an increase of positive potential with respect to the ground....It is not clearly proven that a pure gas, rarefied or not, can receive and convey a charge. If we imagine a charged drop of water suspended in air and evaporating, it follows that, unless the charge be carried off in the vapour, the potential of the drop would rise steadily as its surface diminished, and would become infinite as the drop disappeared, unless the charge were dissipated before the complete drying up of the drop by dispersion of the drop itself, or conveyance of electricity by its vapour. The charge would certainly require to pass somewhere, and might leave the air and vapour charged."

It is quite clear that "must" ought to be substituted for "might" in this last line. Thus the vagueness and doubts expressed in the first part of the quoted statement are annulled by the last three sentences of it.

† "Even in fair weather the intensity of the electric force in the air near the earth's surface is perpetually fluctuating. The speaker had often observed it, especially during calms or very light breezes from the east, varying from 40 Daniell's elements per foot to three or four times that amount during a few minutes, and returning again as rapidly to the lower amount. More frequently he had observed variations from about 30 to about 40, and back again, recurring in uncertain periods of perhaps about two minutes. These gradual variations cannot but be produced by electrified masses of air or cloud, floating by the locality of observation."—Lord Kelvin's *Electrostatics and Magnetism*, Art. XVI. § 282.

air for some considerable height above the earth's surface is electrified*, and that the incessant variations of electrostatic force which he had observed, minute after minute, during calms and light winds, and often under a cloudless sky, were due to motions of large quantities of positively or negatively electrified air in the immediate neighbourhood of the place of observation.

2. It was proved† by observations in the Old College of Glasgow University that the air was in general negatively electrified, not only indoors, within the old lecture-room‡ of Natural Philosophy, but also in the out-of-doors space of the College Court, open to the sky though closed around with high buildings, and between it and the top of the College Tower. The Old College was in a somewhat low situation, surrounded by a densely crowded part of a great city. In the new University Buildings, crowning a hill on the western boundary of Glasgow, similar phenomena, though with less general prevalence of negative electricity in the air, have been observed, both indoors, in the large Bute Hall, and in many other smaller rooms, and out-of-doors, in the court, which is somewhat similar to the courts of the Old College, but much larger. It is possible that the negative electricity found thirty years ago in the air of the Old College may have been due to its situation, surrounded by houses with their fires, and smoking factory-chimneys. In the New College much of the prevalence of negative electricity in air within doors has, however, been found to be due to electrification by the burning lamp§ used with the quadrant electrometer; and more recent observations, with electrification by flame absolutely

* "The out-of-doors air potential, as tested by a portable electrometer in an open place, or even by a water-dropping nozzle outside, two or three feet from the walls of the lecture-room, was generally on these occasions positive, and the earth's surface itself therefore, of course, negative—the common fair-weather condition—which I am forced to conclude is due to a paramount influence of positive electricity in higher regions of the air, notwithstanding the negative electricity of the air in the lower stratum near the earth's surface. On the two or three occasions when the in-door atmospheric electricity was found positive, and, therefore, the surface of the floor, walls, and ceiling negative, the potential outside was certainly positive, and the earth's surface out-of-doors negative, as usual in fine weather."—Lord Kelvin's *Electrostatics and Magnetism*, Art. xvi. § 300.

† *Ibid.* Q. 2, § 283. ‡ *Ibid.* §§ 296—300.

§ "Electrification of Air by Combustion," Magnus Maclean, M.A., F.R.S.E., and Makita Goto, Philosophical Society of Glasgow, November 20, 1889; "Electrification of Air by Water Jet," Magnus Maclean, M.A., F.R.S.E., and Makita Goto, *Philosophical Magazine*, August 1890.

excluded, throw doubt on the old conclusion, that both in town and country negative electrification is the prevailing condition of natural atmospheric air in the lower regions of the atmosphere.

3. The electric ventilation found in the Old College, and described in § 299 of *Electrostatics and Magnetism*, according to which air drawn through a chink, less than ½ inch wide, of a slightly open window or door, into a large room, showed the electrification which it had on the other side of the chink, whether that was the natural electrification of the open air, or positive or negative electrification produced by aid of a spirit-lamp and electric machine in an adjoining room, has been tried

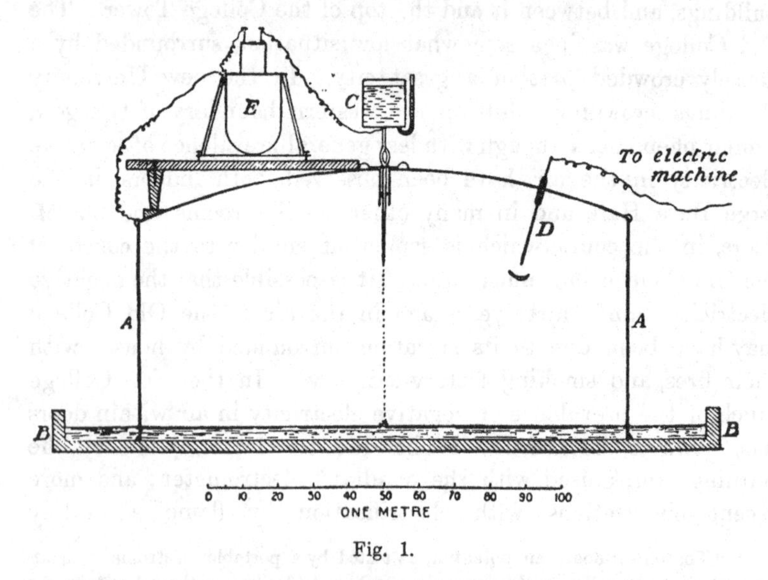

Fig. 1.

again in the New College with quite corresponding results. It has also been extended to the drawing in of electrified air through a tube to the enclosure represented in fig. 1 of the present paper; with the result that the water-dropping test indicated in the sketch amply sufficed to show the electrification, and verify that it was always the same as that of the air outside. When the tube was filled with loosely packed cotton-wool the electrification of the entering air was so nearly annulled as to be insensible to the test.

4. The object proposed for the experiments described in the present communication was to find if a small unchanged portion

of air could be electrified sufficiently to show its electrification by ordinary tests, and could keep its electrification for any considerable time; and to test whether or not dust in the air is essential to whatever of electrification might be observed in such circumstances, or is much concerned in it.

5. The arrangement for the experiments is shown in the diagram, fig. 1. *AA* is a large sheet-iron vat inverted on a large wooden tray *BB*, lined with lead. By filling the tray with

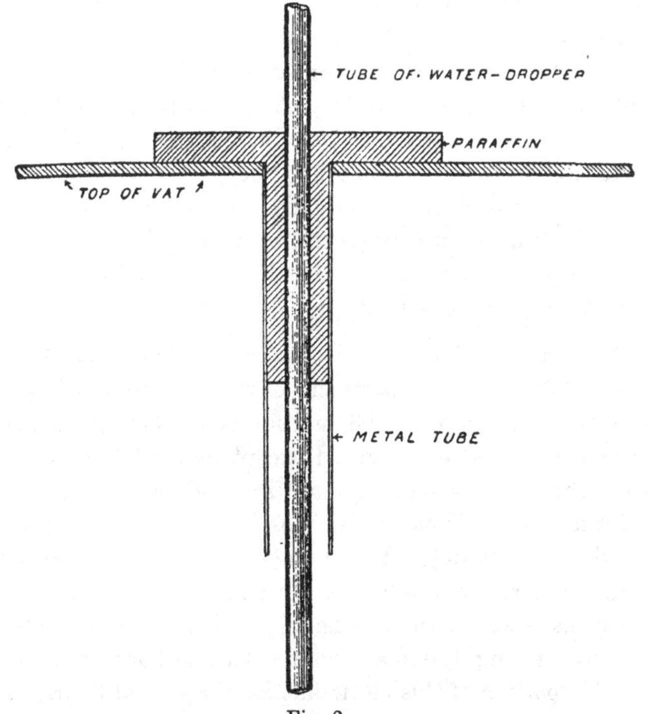

Fig. 2.

water the air is confined in the vat. There are two holes in the top of the vat: one for the water-dropper *C*, and one for the charging wire *D*. Both the water-dropper and the charging wire, ending with a pin-point as sharp as possible, are insulated by solid paraffin, which is surrounded by a metal tube, as shown in half size in fig. 2. To start with they were supported by pieces of vulcanite embedded in paraffin. But it was found that after the lapse of some days (possibly on account of ozone generated by the incessant brush-discharges) the insulation had utterly failed in both of them. The vulcanite pieces were then taken out, and solid paraffin, with the metal guard-tube round it to screen it from electrically influencing the water-dropper, was substituted. This has proved quite satisfactory: the water-dropper, with the flow of water stopped, holds a positive or a negative charge for hours.

6. A quadrant electrometer E (described in *Electrostatics and Magnetism*, §§ 346—353) was set up on the top of the vat near the water-dropper, as shown in fig. 1. It was used with lamp and semi-transparent scale to indicate the difference of potential between the water-dropper and the vat. The sensibility of the electrometer was 21 scale-divisions (half-millimetres) per volt; and as the scale was 90 centimetres long, difference of potentials up to 43 volts, positive or negative, could be read by adjusting the metallic zero to the middle of the scale. A frictional plate-electric machine was used, and by means of it, in connexion with the pin-point, the air inside the vat could be electrified either positively or negatively.

7. The vat was fixed in position in the Apparatus Room of the Natural Philosophy Department of the University of Glasgow on the 13th of December, 1893, and for more than three months the air inside was left undisturbed except by discharges from the pin-point through the electrifying wire, and by the spray from the water-dropper. Thus the air was becoming more and more freed of dust day by day. Yet at the end of the four months we found that the air was as easily electrified, either positively or negatively, as it was at the beginning; and that if we electrify it strongly by turning the machine for half-an-hour, it retains a considerable portion of this electrification for several hours.

8. Observations were taken almost daily since the 13th of December; but the following, taken on the 8th of February, the 12th of March, and the 23rd of April, will serve as specimens, the results being shown in each case by a curve. At all these dates the air must have been very free from dust. Both during the charging and during the observations the case of the electrometer and one pair of quadrants are kept metallically connected to the vat. During the charging the water-dropper and the other pair of quadrants were also kept in connexion with the vat. Immediately after the charging was stopped the charging-wire

was connected metallically to the outside of the vat, and left so with its sharp point unchanged in its position inside the vat during all the observations.

9. *Curve* 1. *February* 8, 1894.—The friction-plate machine was turned positive for half-an-hour. Ten minutes after the machine stopped the water-dropper was filled and joined to one pair of quadrants of the electrometer, while the other pair was joined to the case of the instrument. The first reading on the curve was taken four minutes afterwards, that is fourteen minutes after the machine stopped turning (18 volts).

Curve 2. *March* 3, 1894.—The friction-plate machine was turned positive for five minutes. The water-dropper was filled and joined to the electrometer immediately after the machine stopped turning. The spot was off the scale, and nine minutes elapsed before it appeared on the scale. The first reading on the curve was taken one minute afterwards, or ten minutes after the machine stopped turning (35·25 volts).

Curve 3. *March* 12, 1894.—A Voss induction-machine was joined to the charging wire, and run by an electric motor for 4 hours 19 minutes. A test was applied at the beginning of the run to make sure that it was charging negatively; and a similar test when it was disconnected from the charging wire in the vat showed it to be still charging negatively. The water-dropper was joined to the electrometer, and the spot appeared on the scale immediately. The first reading on the curve was taken half a minute after the machine was disconnected (30·65 volts).

Curve 4. *April* 23, 1894.—The friction-plate machine was turned positive for 30 seconds, with water-dropper running and joined to the electrometer. 20 seconds after the machine stopped the spot appeared on the scale, and the reading $1\frac{1}{2}$ minutes after the machine stopped turning is the first point on the curve (7·3 volts).

Curve 5. *April* 23, 1894.—The friction-plate machine was turned negative for 30 seconds, with the water-dropper running and joined to the electrometer. 10 seconds afterwards the spot appeared on the scale, and the reading 70 seconds after the machine stopped turning is the first point on the curve (7·6 volts).

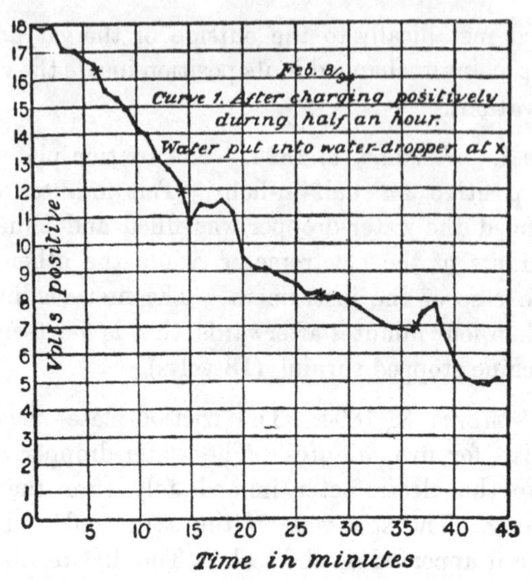
Feb. 8/94
Curve 1. After charging positively
during half an hour.
Water put into water-dropper at X
Volts positive.
Time in minutes
18
17
16
15
14
13
12
11
10
9
8
7
6
5
4
3
2
0
5
10
15
20
25
30
35
40
45

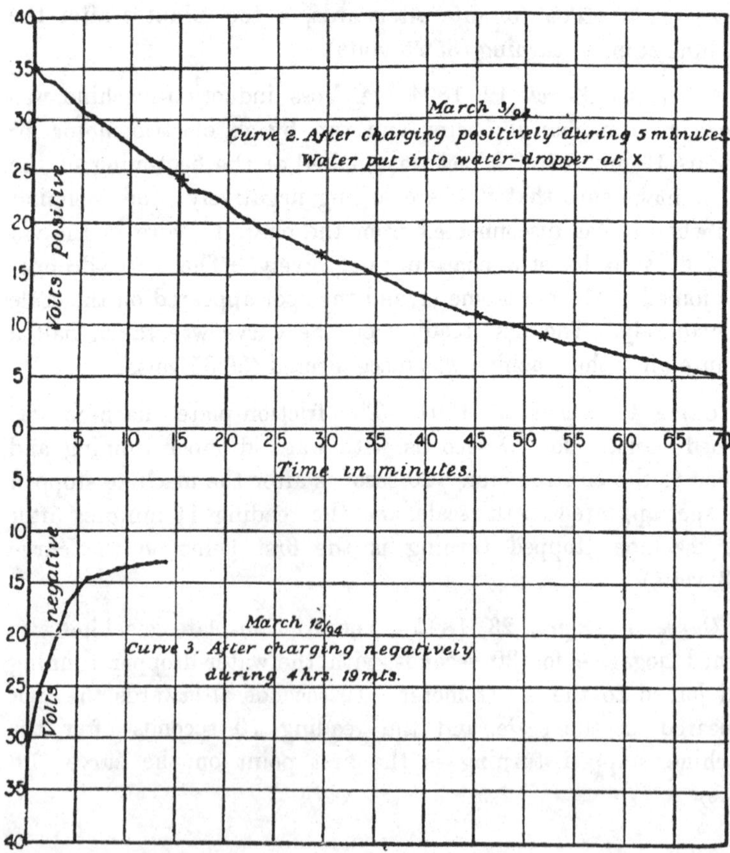
March 3/94
Curve 2. After charging positively during 5 minutes
Water put into water-dropper at X
Volts positive
Time in minutes.
March 12/94
Curve 3. After charging negatively
during 4 hrs. 19 mts.
Volts negative
40
35
30
25
20
15
10
5
0
5
10
15
20
25
30
35
40
45
50
55
60
65
70

The curves show, what we always found, that the air does not retain a negative electrification so long as it retains a positive. We also found, by giving equal numbers of turns to the machine, that the immediately resulting difference of potential between the water-dropper and the vat was greater for the negative than for the positive electrification; though the quantity received from the

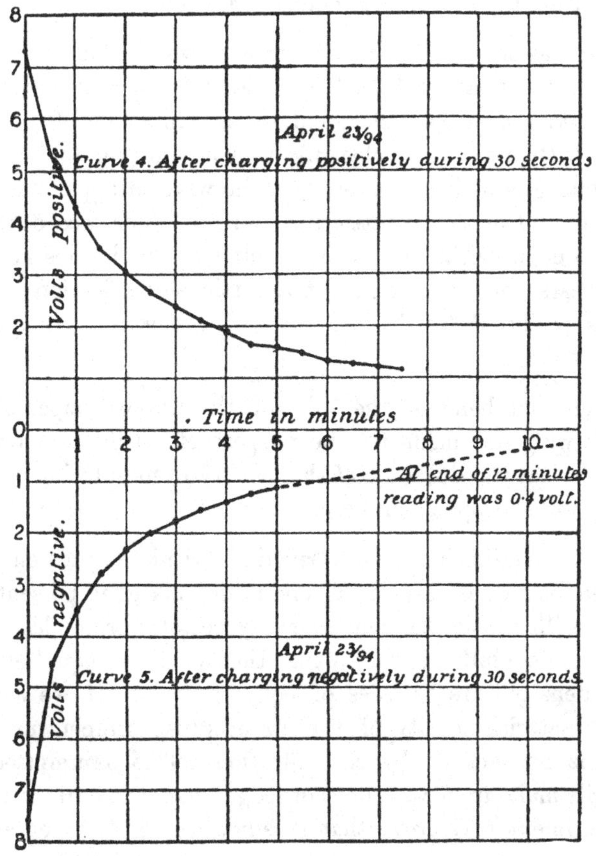

machine was probably less in the case of the negative electrification, because the negative conductor was less well insulated than the positive.

10. On the 21st of March two U-tubes were put in below the edge of the vat, one on either side, so that it might be possible to blow dusty, or smoky, or dustless air into the vat. To one tube was fitted a blowpipe-bellows, and by placing it on the top of a box in which brown paper and rosin were burning, the

vat was filled with smoky air. Again, several layers of cotton-wool were placed on the mouth of the bellows, so as to get dustless air into the vat. The bellows were worked for several hours on four successive days, and we found no appreciable difference (1) in the ease with which the air could be electrified by discharges from the wire connected to the electric machine, and (2) in the length of time the air retains its electrification.

But it was found that, as had been observed four years ago with the same apparatus*, with the water-dropper insulated and connected to the electrometer, and no electrification of any kind to begin with, a negative electrification amounting to four, five, or six volts gradually supervened if the water-dropper was kept running for 60 or 70 minutes, through air which was dusty, or natural, to begin with. It was also found, as in the observations of four years ago, that comparative little electrification of this kind was produced by the dropping of the water through air purified of dust.

The circular bend of the tube of the water-dropper shown in the drawing was made for the purpose of acting as a trap to prevent the natural dusty air of the locality from entering the vat when the water-dropper ran empty.

11. The equilibrium of electrified air within a space enclosed by a fixed bounding surface of conducting material presents an interesting illustration of elementary hydrostatic principles. The condition to be fulfilled is simply that the surfaces of equal electric "volume-density" are surfaces of equal potential, if we assume that the material density of the air at given temperature and pressure is not altered by electrification. This assumption we temporarily make from want of knowledge; but it is quite possible that experiment may prove that it is not accurately true; and it is to be hoped that experimental investigation will be made for answering this very interesting question.

12. For stable equilibrium it is further necessary that the electric density, if not uniform throughout, diminishes from the bounding surface inwards. Hence, if there is a portion of non-electrified air in the enclosure, it must be wholly surrounded by electrified air.

* Maclean and Goto, *Philosophical Magazine*, August 1890.

13. We may form some idea of the absolute value of the electric density, and of the electrostatic force in different parts of the enclosure, in the electrifications found in our experiments, by considering instead of our vat a spherical enclosure of diameter intermediate between the diameter and depth of the vat which we used. Consider, for example, a spherical space enclosed in metal of 100 centim. diameter, and let the nozzle of the water-dropper be so placed that the stream breaks into drops at the centre of the space. The potential shown by the electrometer connected with it, being the difference between the potentials of the air at the boundary and at the centre, will be the difference of the potentials at the centre due respectively to the total quantity of electricity distributed through the air and the equal and opposite quantity on the inner boundary of the enclosing metal; and we therefore have the formula

$$V = 4\pi \int_0^a \rho \left(\frac{r^2}{r} - \frac{r^2}{a}\right) dr,$$

where V denotes the potential indicated by the water-dropper, a the radius of the spherical hollow, and ρ the electric density of the air at distance r from the centre. Supposing now, for example, ρ to be constant from the surface to the centre (which may be nearly the case after long electrification as performed in our experiments), we find $V = \frac{2}{3}\pi\rho a^2$; whence $\rho = 3V/2\pi a^2$.

To particularise further, suppose the potential to have been 38 volts or 0·127 electrostatic C.G.S. (which is less than the greatest found in our experiments) and take $a = 50$ centim.: we find $\rho = 2{\cdot}4 \,.\, 10^{-5}$. The electrostatic force at distance r from the centre, being $\frac{4}{3}\pi\rho r$, is therefore equal to $10^{-4}r$. Hence a small body electrified with a quantity of electricity equal to that possessed by a cubic centimetre of the air, and placed midway ($r = 25$) between the surface and centre of the enclosure experiences a force equal to $2{\cdot}4 \,.\, 10^{-9} \,.\, 25$, or $6 \,.\, 10^{-8}$, or approximately $6 \,.\, 10^{-11}$ grammes weight. This is $4{\cdot}8 \,.\, 10^{-8}$ of the force of gravity on a cubic centimetre of air of density 1/800.

14. It is interesting to remark that negatively electrified air over negatively electrified ground, and with non-electrified air above it, in an absolute calm, would be in unstable equilibrium; and the negatively electrified air would therefore rise, probably in large masses, through the non-electrified air up to the higher

regions, where the positive electrification is supposed to reside. But with no stronger electrification than that which we have had within our experimental vat, the moving forces would be insufficient to produce instability comparable with that of air warmed by the ground and rising through colder air above.

15. During a thunderstorm the electrification of air, or of air and the watery spherules constituting cloud, need not be enormously stronger than that found in our experiments. This we see by considering that if a uniformly electrified globe of a metre diameter produces a difference of potential of 38 volts between its surface and centre, a globe of a kilometre diameter, electrified to the same electric density, reckoned according to the total electricity in any small volume (electricity of air and of spherules of water, if there are any in it), would produce a difference of potential of 38 million volts between its surface and centre. In a thunderstorm, flashes of lightning show us differences of potentials of millions of volts, but not perhaps of many times 38 million volts, between places of the atmosphere distant from one another by half a kilometre.

238. Preliminary Experiments to find if Subtraction of Water from Air Electrifies it. By Lord Kelvin, Magnus Maclean, and Alexander Galt.

[From *Brit. Assoc. Report*, 1894, pp. 554, 555; *Écl. Elec.* Vol. II. Jan. 12, 1895, p. 92.]

Experiments with this object were commenced by one of us in December 1868, but before any decisive result had been obtained, circumstances rendered a postponement of the investigation necessary.

A glass U-tube with vertical branches, each 18 in. long and about 1 in. bore, with the upper eight inches of one of the branches carefully coated outside and inside with clean shellac varnish, was held fixed by an uninsulated support attached to the upper end of this branch. The other branch was filled with little fragments of pumice soaked in strong pure sulphuric acid or in pure water; and a fine platinum wire, with one end touching the pumice, connected it to the insulated electrode of a quadrant electrometer. A metal cylinder, large enough to surround both branches of the U-tube without touching either, was placed so as to guard the tube from electric influences of surrounding bodies (of which the most disturbing is liable to be the woollen cloth sleeves of the experimenters or observers moving in the neighbourhood). This metal tube was kept in metallic connection with the outside metal case of the quadrant electrometer. The length of the exposed platinum wire between the U-tube and the electrometer was so short that it did not need a metal screen to guard it against irregular influences. An india-rubber tube (metal, metallically connected with the guard cylinder, would have been better) from an ordinary blowpipe bellows was connected to the uninsulated end of the U-tube. Air was blown through it steadily for nearly an hour. With the sulphuric pumice in the other branch the electrometer rose in the course of three-quarters of an hour to about nine volts positive. When the pumice was moistened with water, instead of sulphuric acid, no such effect was observed. The result of the first experiment proves decisively that the passage of the air through the U-tube gave positive electricity to the sulphuric acid, and therefore sent away the dried air with negative electricity. A corresponding experiment with fragments of chloride of calcium instead of sulphuric pumice gave a similar result. In repetition of the experiments, however, it has been noticed that the strong positive electrification of the U-tube seemed to commence somewhat suddenly when a gurgling sound, due to the bubbling of air through free liquid, whether sulphuric acid or chloride of calcium solution, in the bend of the U-tube, began to be heard. We intend to repeat the experiments with arrangements to prevent any bubbling of the air through liquid.

We have repeated our original experiment with pumice moistened with water in the insulated U-tube, and with an uninsulated U-tube filled with sulphuric pumice between the bellows and the insulated tube, so that the air entering it is artificially dried. With this arrangement the insulated U-tube was negatively electrified by the blowing of the air through it; but this electrification may have been due to the negative electrification of the dry entering air to be expected from the result of our first experiment. We intend to repeat the experiment with artificially dried *and dis-electrified* air blown through the U-tube containing pumice moistened with water.

239. ELECTRIFICATION OF AIR AND OTHER GASES BY BUBBLING THROUGH WATER AND OTHER LIQUIDS. By LORD KELVIN, MAGNUS MACLEAN, and ALEXANDER GALT.

[From *Roy. Soc. Proc.* Vol. LVII. 1895 [Feb. 21, 1895], pp. 335—346; *Nature*, Vol. LI. March 21, 1895, pp. 495—499; *Écl. Élec.* Vol. III. June 15, 1895, pp. 521—524.]

1. AT the meeting of the British Association in Oxford in August, 1894, a communication was given to Section A, entitled "Preliminary Experiments to find if Subtraction of Water from Air Electrifies it." These experiments were performed during July of 1894, and were a continuation of experiments which were commenced in the Physical Laboratory of the University of Glasgow in December of 1868 with the same object, but which were then, for various reasons, discontinued before any decisive result had been obtained.

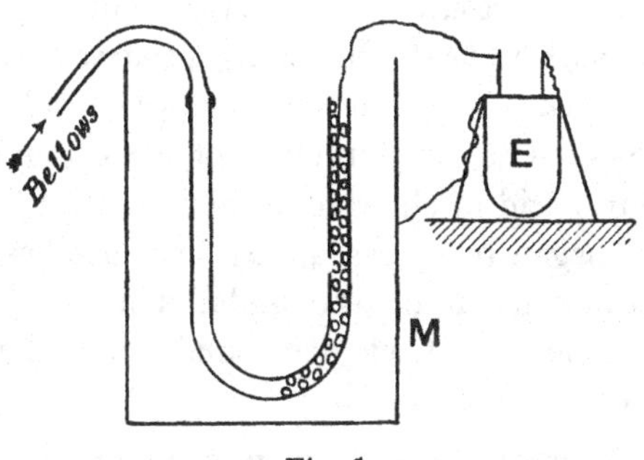

Fig. 1.

2. A glass U-tube with vertical branches (fig. 1), each 18 inches long and about 1-inch bore, with the upper 8 inches of one of the branches carefully coated outside and inside with clean shellac varnish, was held fixed by an uninsulated support attached to the upper end of this branch. The other branch was filled with little fragments of pumice soaked in strong sulphuric acid or

in water, and a fine platinum wire, with one end touching the pumice, connected it to the insulated electrode of a quadrant electrometer. A metal can, *M*, large enough to surround both branches of the U-tube without touching either, was placed so as to guard the tube from electric influences of surrounding bodies, the most disturbing of which is liable to be the woollen cloth sleeves of the experimenters or observers moving in the neighbourhood. This metal can was kept in metallic connection with the outside metal case of the quadrant electrometer. The length of the exposed platinum wire between the U-tube and the electrometer was so short that it did not need a metal screen to guard it against irregular influences. An india-rubber tube from an ordinary blow-pipe bellows was connected to the uninsulated end of the U-tube. Air was blown through it steadily for nearly an hour. With the pumice soaked in strong sulphuric acid in the other branch, the electrometer reading rose in the course of three-quarters of an hour to about 9 volts positive. *When the pumice was moistened with water, instead of sulphuric acid, no such effect was observed.* The result of the first experiment proves decisively that the passage of the air through the U-tube gave positive electricity to the sulphuric acid, and therefore sent away the dried air with negative electricity. A corresponding experiment with fragments of pure chloride of calcium instead of pumice in sulphuric acid, gave a similar result. In repetition of the experiments, however, it was noticed that the strong positive electrification of the U-tube seemed to commence somewhat suddenly when a gurgling sound—due to the bubbling of air through free liquid, whether sulphuric acid or chloride of calcium solution in the bend of the U-tube—began to be heard. It has since been ascertained that it was because no liquid accumulated in the bottom of the U-tube that no electric effect was found when the pumice was moistened with pure water.

3. Arrangements were made to prevent any bubbling of the air through liquid, by using a straight tube instead of a U-tube. In a large number of experiments with pumice, moistened with pure sulphuric acid, in the straight tube, and air blown through for about half an hour, no definite electrification was obtained. In this straight tube, as formerly with the U-tube, pumice moistened with pure water gave no electrification. Chloride of calcium in lumps, not specially dried, gave no effect in the

straight tube; but if previously heated to 180° or 200° C. and put into the straight tube when still hot, it gave an enormous positive electrification immediately on the commencement of blowing. Strong positive electrification was obtained a second time, by discharging the electrometer to zero, re-insulating, and re-commencing the blowing. But after discharging a second time, re-insulating, and re-commencing the blowing, no further electrification was found.

4. In continuation of these experiments, on the 25th of September the arrangement represented in fig. 2 was set up. An outer metallic guard-vessel, *M*, was kept connected by a wire to the case and to one pair of quadrants of a quadrant electrometer, *E*. Water in an inner glass or metal jar, *A*, was connected

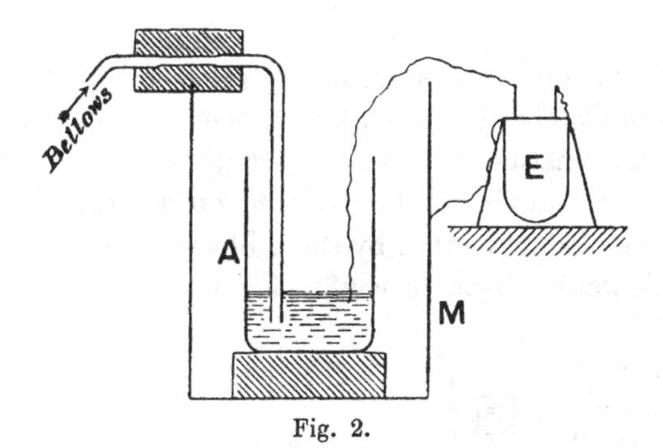

Fig. 2.

by a platinum wire to the other pair of quadrants of the electrometer. To have this inner jar well insulated, it was supported on a block of paraffin; and the upper end of a glass tube dipping into the water was fitted into one end of a tube of paraffin, to the other end of which was fitted a tube for ingress of air, from bellows, as shown in the figure. The insulation of this arrangement was found to be good. When air was blown through the water it was found that the jar containing the water became positively electrified.

5. To prevent splashing of water out of the jar, a paper cover was put on its mouth, or the jar was tilted, as shown in fig. 3, so that the bubbles broke against the inside of the jar. In three experiments thus made, the same electrification was still found, amounting to about 6 volts positive in a quarter of an hour.

6. As the jar was in every experiment positively electrified, the air, if unelectrified* when entering it, must have been negatively electrified when leaving it.

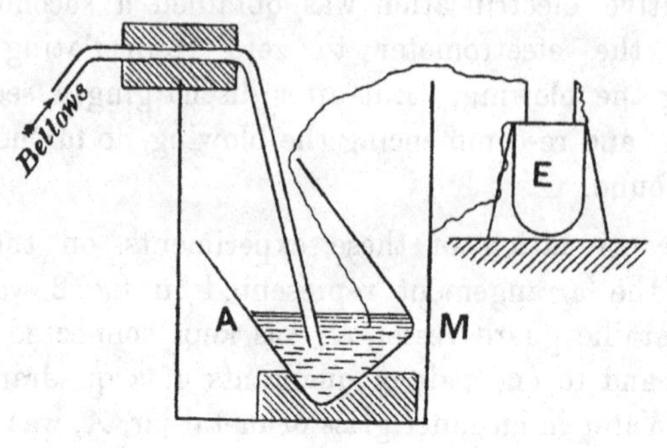

Fig. 3.

7. To test if the air was negatively electrified after bubbling, on the 11th of October the apparatus† shown in fig. 4 was set up. The apparatus consists of a large sheet iron vat, *VV*, 123 cm. in diameter and 70 cm. in height, inverted on a large wooden tray lined with lead, and supported by three blocks of wood. By filling the tray with water, the air is confined in the vat.

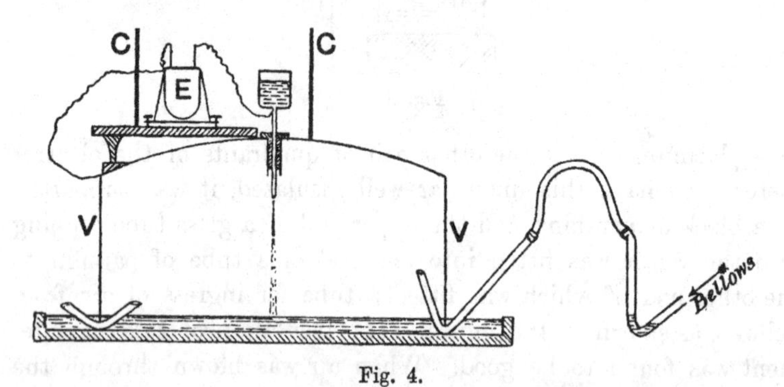

Fig. 4.

CC is a metal screen kept metallically connected with the case of the electrometer, and with the vat. It surrounds both the electrometer and the water-dropper, to prevent any external

* Air was similarly blown from bellows into the vat (see § 7) without any bubbling, and no electrification was observed.

† The vat, the water-dropper, and the electrometer are the same as in the apparatus described in the *Proceedings of the Royal Society*, Vol. LVI. year 1894, "Electrification of Air," by Lord Kelvin and Magnus Maclean.

varying electrifications from vitiating the proper results of our experiments.

This screening of the electrometer is absolutely necessary when it is used with high sensibility (70 scale divisions per volt in our experiments) in a laboratory or other place where various other electric experiments may simultaneously be going on. Four years ago the electrometer, the vat, and the water-dropper, were set up on the class-room table *without a metal screen.* When the deflection indicated about 4 volts negative (see § 8), the negative lead of Lord Kelvin's house electric-light circuit, which passes through the class-room, was joined to earth. This changed the deflection of the electrometer suddenly by 1 volt in the positive direction. When the positive lead was "earthed," the deflection was changed suddenly by 6 volts in the negative direction. Putting on sixteen 8 c.p. electric lamps, eight on each side of the class-room, changed the deflection by two-thirds of a volt in the negative direction.

8. In experimenting with the same apparatus* in 1890 it was found that the water jet gave negative electricity to the ordinary air of the laboratory enclosed in the vat. The present experiments fully confirm this result, showing a gradual negative electrification of the enclosed mass of air rising to about 5 volts in an hour, once every day for the first few days. For twenty-eight days after the vat was set up in October, 1894, fifteen observations of an hour each were taken, to find the effect of the water-dropper, with no other disturbing influence on the unchanged volume of air inside the vat. These experiments verify the conclusion (*Phil. Mag.* August, 1890) that the more the air inside the vat became free of dust, the less became the rate at which the air was negatively electrified by the water-dropper.

9. On the 15th October last the vat was lifted from the tray to remove some obstruction in the nozzle of the water-dropper, which was not then flowing freely. Curve (6) was obtained that afternoon. The air in the vat was the ordinary air of the laboratory, and the curve shows the effect of the water-dropper alone in electrifying the air negatively. For the next two days the water-

* *Phil. Mag.* August, 1890, "Electrification of Air by a Water Jet," by Maclean and Goto.

dropper was kept running continuously for about eight hours each day, to wash the dust out of the air, and on the 18th of October curve (7) was obtained. It shows a much less rate of negative electrification than curve (6). In the experiments of summer 1890 an aspirator was used to draw the air from the vat, and a tube full of cotton-wool was used to filter the air drawn into the vat.

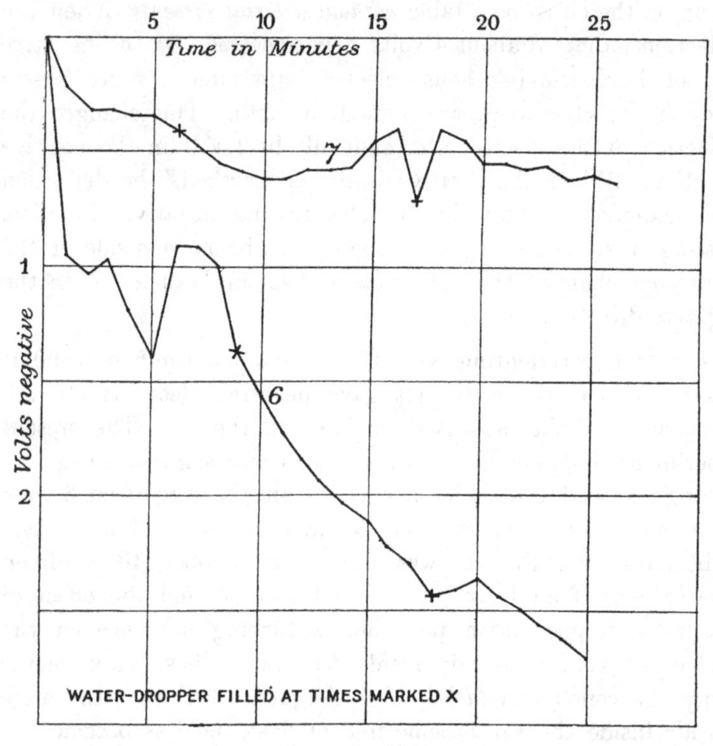

Curves 6 and 7.

Curves (1) to (5) are reproduced from the *Philosophical Magazine*, and they show that the more the air becomes free from dust the less is the rate at which the water-dropper electrifies. Thus curve (1) was obtained from the ordinary air of the laboratory, in the vat, and curve (2) after the aspirator was working for some time. In this curve the water-dropper itself was running for some time before the first observation was taken. The other curves were obtained after further continuous working of the aspirator.

After curve (4) was obtained the aspirator was worked continuously for twenty-five hours, and then curve (5) was obtained.

10. At the end of twenty-three days of October and November 1894 (§§ 8, 9 above), when the air inside the vat must have been fairly free from dust, and when the water-dropper of itself was giving little negative electrification, we bubbled air into it by a forked tube, one end of which was connected to a bellows, and three other open ends were below the water inside the vat. In five experiments thus made—two on November 7, two on November 8, and one on December 17—an average negative electrification of 5 volts in twelve minutes was obtained.

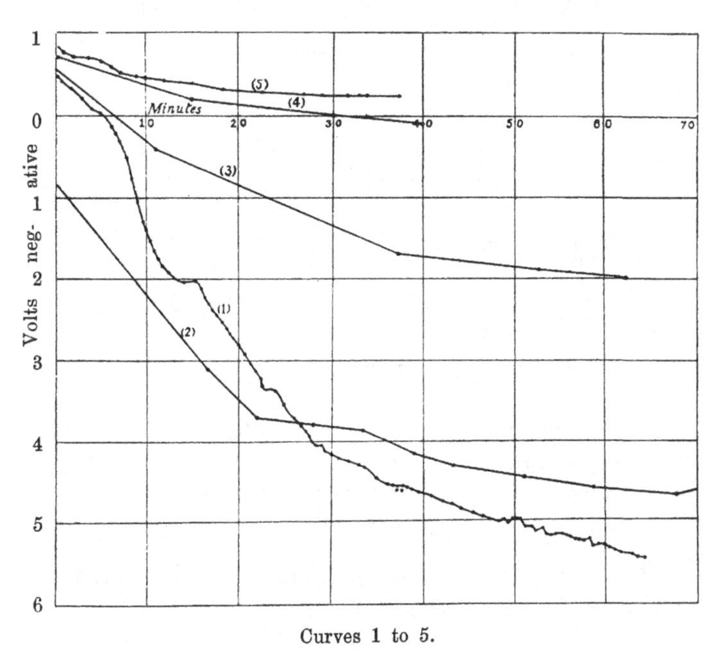

Curves 1 to 5.

11. We now arranged a U-tube with pure water in it (fig. 4) outside the vat. Air from the bellows bubbled through the water in this U, and was carried thence by a block-tin pipe into the vat without any further bubbling. Observations by the quadrant electrometer, while the water-dropper was running and the bellows worked, gave us measurements of the varying state of the electrification of the air in the vat. The average of fifteen experiments gave a negative electrification of the air in the vat of 8½ volts in

25 minutes. The rate at which the air was blown in in these experiments was such as to displace the entire volume of air in the vat in half an hour.

12. Curve (8) shows the rate of electrification of air, in one of the fifteen experiments, when thus bubbled through the water in the U-tube and then admitted into the vat.

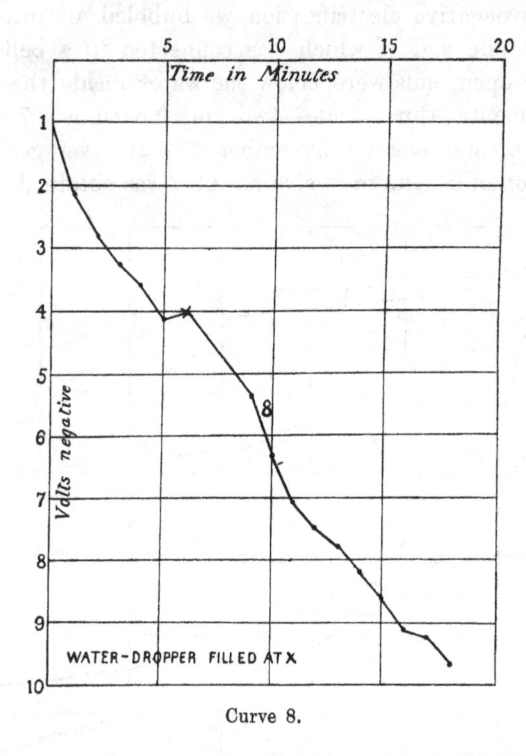

Curve 8.

13. Two U-tubes, in series, with water in each, did not seem to give a perceptibly cumulative effect.

14. The effect of one or more wire gauze strainers between the U-tube and the vat, or between the U-tube and the bellows, was next tested. The gauzes were placed between short lengths of lead tube, which were held together by a rubber tube slipped over them. The arrangement is shown by longitudinal and cross sections in fig. 5. Twelve wire gauzes, with or without cotton-wool between them, placed between the bellows and the U-tube, did not prevent the subsequent electrification by bubbling of the air thus filtered. But when placed between the U-tube and the

vat they almost entirely diselectrified the air, even without the cotton-wool, and still more decidedly when cotton-wool was loosely packed between the wire gauzes. A single wire-gauze strainer produced but little of diselectrifying effect.

15. The interpretation of these experiments is complicated, and the time required for each is lengthened, on account of the large mass of air in the vat to start with, whether uncharged or retaining electricity from previous experiments, and also on account of the effect of the water-dropper itself. Hence, in our later experiments, we fell back on the arrangement shown in fig. 2, by which we test the electrification of the liquid, and not directly that of the gas blown through it.

16. In our first experiments with this apparatus the amount of the electrification did not seem much affected when a paper

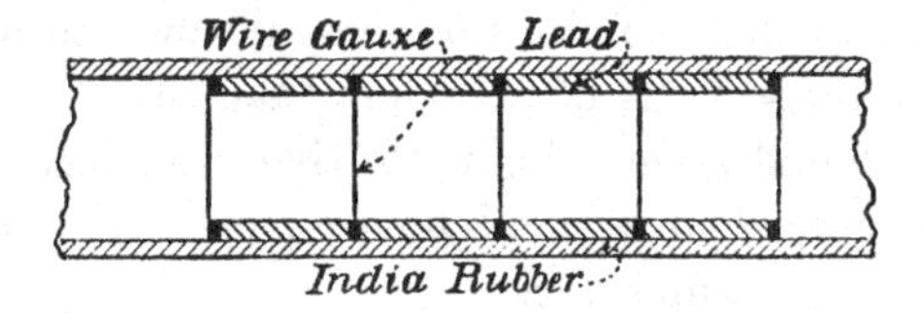

Fig. 5.

cover was put on the jar, or when we tilted the jar as shown in fig. 3. We now made a large number of tests with different covers and screens (chiefly of sheet copper or sheet zinc, or brass wire gauze) at different heights above the liquids, and we concluded that, if the screens are not within a centimetre and a half of the liquid surface, they do not directly affect the magnitude of the electrification obtained. In nearly all of the subsequent experiments a horizontal circular screen of thin sheet copper, leaving an air space of about 3 mm. all round between its edge and the inner surface of the jar, about 3 cm. above the liquid surface, was used to prevent spherules of the liquid from being tossed out of the jar by the bubbling.

17. In the following short summary of our results the duration of each experiment was 10 minutes. The effect of blowing air through water and other liquids is summarised in §§ 18 to 27, and of blowing other gases than air through water in §§ 28 to 31.

18. The jar contained 200 c.c. of the Glasgow town-supply water (from Loch Katrine). A mean of seventeen experiments showed an electrification of the jar to 4 volts positive when air was blown through it for 10 minutes.

19. A solution of zinc sulphate of different strengths was now used instead of the pure water. Three experiments, with 150 c.c. of water containing one drop of a saturated solution of the zinc sulphate, gave half the positive electrification that would, under similar circumstances, have been obtained from water only. With five drops no definite electrification was obtained. With greater proportions of the zinc sulphate solution up to saturation (twenty-four experiments altogether) the electrification was on the average slightly negative.

20. Twelve experiments were then made to test the effect of adding a solution of ammonia to the water. One drop reduced the electrification to one-half; two drops brought it down to one-quarter. With larger proportions of ammonia than this, up to a saturated solution, we found a very slight positive electrification, never amounting to more than a small fraction of a volt, and therefore negligible in the circumstances.

21. Seven experiments with sulphuric acid of different strengths all showed small *positive* electrification, the amount gradually decreasing from $\frac{1}{4}$ volt, in 10 minutes, with 0·5 per cent. acid in water to $\frac{1}{16}$ volt, in the same time, with acid of full strength.

Seven experiments with hydrochloric acid solution of different strengths all showed a small *negative* electrification, the amount gradually increasing from $\frac{1}{2}$ volt, in 10 minutes, with $\frac{1}{3}$ per cent. acid solution in water to $1\frac{1}{4}$ volts, in the same time, with acid solution of full strength.

Nine experiments with calcium chloride solution were made. A saturated solution and a solution diluted to 75 per cent. of full strength gave no result; but solutions of gradually diminished strength, from 50 per cent. down to $\frac{1}{10}$ per cent., showed a negative electrification from fully $\frac{1}{2}$ volt, in 10 minutes, down to $\frac{1}{13}$ volt.

Additions of very small quantities of washing soda to water greatly reduce the positive electrification obtained.

Loch Katrine water supersaturated with carbonic acid, and placed in the insulated jar, showed, when air was bubbled through it for ten minutes, a negative electrification of ¼ volt.

22. Ten drops of paraffin oil added to water reduced the electrification to about half of that obtained from water only. Thirty drops reduced it to about a tenth, which as it amounted to only 0·4 volt during the time of the experiment is negligible.

23. Ten drops of benzene reduced the electrification to half, and thirty drops to about a third of that taken by pure water.

24. A saturated solution of granulated phenol (carbolic acid) was made, and small portions of it added to the water in the jar. Several experiments showed no diminution in the electrification as long as the quantity of the phenol solution present in the water was under 10 per cent. With 25 per cent. the electrification was reduced to a third. With strengths greater than this up to saturation the electrification was reduced to one-sixth.

25. A saturated solution of common salt was prepared. Blowing air through 200 c.c. of water containing the quantities of the salt solution mentioned, gave us in 10 minutes the following electrifications:—

(*a*)	0·004 per cent. of saturated solution	of salt in water	2·4	volts positive
(*b*)	0·02	,,	,,	1·2	,,
(*c*)	0·1	,,	,,	0·6	,,
(*d*)	0·5	,,	,,	0·4	,,
(*e*)	2·0	,,	,,	0·15	,,
(*f*)	4·0	,,	,,	0·0	,,

26. Several experiments showed that with 200 c.c. of water containing not more than ten drops of absolute alcohol, practically the same amount of positive electrification (4 volts in ten minutes) is obtained as if pure water were used. With fifty drops less than 2 volts were got, and with 100 drops less than 1 volt. 25 and 50 per cent. alcohol in the water gave very small and hence negligible positive electrification.

27. One drop of saturated solution of copper sulphate in 200 c.c. of water showed 1 volt positive in 10 minutes. With ½ per cent. of it in the water, the electrification was reduced to a

fraction of a volt positive. With greater proportions of copper sulphate present, up to saturation, slightly negative electrifications were obtained, but never amounting to more than about one-tenth of a volt, and hence negligible.

28. On blowing carbonic acid gas, from a cylinder obtained from the Scotch and Irish Oxygen Company, through pure water in the glass jar, the water became electrified to $8\frac{3}{4}$ volts positive in ten minutes. Blowing the breath through water gave an electrification of 3 volts positive in the same time: this diminished result is doubtless due chiefly to the diminished rate of bubbling.

29. The blowing of oxygen from a cylinder, obtained from the Oxygen Company, through water, gave as a mean of four experiments a positive electrification to the water of half a volt in 10 minutes. When continued for 55 minutes, it gave the very decided result of 5 volts positive.

30. Hydrogen prepared from zinc and dilute sulphuric acid was passed into a large metal gas-holder; and was passed on from this to bubble through the water in the insulated jar. In two experiments this was done immediately after the preparation of the hydrogen; in another it was done after the hydrogen had remained 18 hours in the gas-holder. In each of the three experiments the water was electrified to 2 volts positive in 10 minutes.

When the hydrogen was allowed to pass direct through a tube from the Wolffe's bottle where it was generated, to bubble in the insulated jar, the magnitude of the effect obtained was very much larger. In one case a mixture of muriatic acid and sulphuric acid and water was used, and the reading went off the scale positive in 30 seconds (more than 10 volts). In other two experiments, with dilute sulphuric acid and zinc in the Wolffe's bottle, the electrifications obtained were 6 volts positive in 7 minutes, and 7·3 volts positive in 13 minutes, in the last of which the hydrogen was allowed to bubble through caustic potash contained in a small bottle between the Wolffe's bottle and the insulated jar.

The hydrogen was next generated in the insulated jar itself, the tube for ingress of air used in the ordinary experiments being taken away. 200 c.c. of pure water, along with some granulated zinc, was put into the jar. Then some pure sulphuric acid was added, and electrometer readings were taken. In two experiments with no screen in the jar (§ 16) the reading went off the scale *negative* (1) in 2 minutes and (2) in 4 minutes (more than 9 volts in each case). In another experiment, in other respects the same, but with a copper screen 7 cm. above the surface of the liquid, the electrification showed 2 volts *negative* in 2 minutes, then came back to zero in 5 minutes, and in the next 6 minutes went 4 volts *positive*. The jar and pair of quadrants connected with it were then metallically connected with the outer case of the electrometer for a few seconds, and reinsulated; in 5 minutes the reading went up to 2 volts *positive*. A little more sulphuric acid was added to the jar, which was disinsulated for a short time and reinsulated; the reading went up to 7 volts *positive* in 4 minutes. The jar was again disinsulated for a few seconds and reinsulated; the reading went up in 4½ minutes to 6½ volts *positive*.

31. Coal-gas, bubbled through water in the insulated jar, gave 1·4 volts positive in 10 minutes.

32. In the ordinary experiment of bubbling air through a small quantity of water in the bottom of the jar it was noticed that the electrification did not commence to be perceptible generally till about the end of the first minute; and that it went on augmenting perceptibly for a minute or more after the bubbling was stopped. The following experiment was therefore tried several times. One of us stood leaning over the jar, with the head about 10 ins. above it, and the mouth so partly closed that breathing was effected sideways; another blew the bellows; and another took the readings of the electrometer. After bubbling had been going on for some minutes, and the readings were rising gradually (4 volts per 10 minutes, as in § 18), blowing was stopped. As soon as the bubbling ceased, the first-mentioned observer, without moving his head or his body (see § 7, regarding the necessity to have the electrometer screened from outside influences) blew into the jar to displace the negatively electrified air in it. In every case the electrometer reading showed instantly a small rise in the positive direction.

In the carrying out of these experiments we have received

much valuable help from Walter Stewart, M.A., and Patrick Hamilton, B Sc.

33. The very interesting experiments described by Lenard, in his paper on the Electricity of Waterfalls*, and by Professor J. J. Thomson, on the Electricity of Drops†, show phenomena depending, no doubt, on the properties of matter to which we must look for explanation of the electrical effects of bubbling described in our present communication, and of the electrification of air by drops of water falling through it, to which we have referred as having been found in previous experiments which were commenced in 1890 for the investigation of the passage of electrified air through tubes‡.

* *Wiedemann's Annalen*, 1892, Vol. XLVI. pp. 584—636.

† *Phil. Mag.* April, 1894, Vol. XXXVII. pp. 341—358.

‡ "Electrification of Air by a Water Jet." By Magnus Maclean and Makita Goto, *Phil. Mag.* August, 1890, Vol. XXX. pp. 148—152.

240. ON THE DISELECTRIFICATION OF AIR. By LORD KELVIN, MAGNUS MACLEAN, and ALEXANDER GALT.

[From *Roy. Soc. Proc.* Vol. LVII. 1895 [March 21, 1895], pp. 436—439; *Nature*, Vol. LI. April 11, 1895, pp. 573, 574.]

1. THE experiment described in § 14 of our paper on the "Electrification of Air and other Gases by bubbling through Water and other Liquids" (*Roy. Soc. Proc.* February 21, 1895), proves that air, electrified negatively by bubbling through water and caused to pass through a metallic wire gauze strainer, gives up some, but not a large proportion, of its electricity to the metal. We have now made a fresh experimental arrangement for the purpose of investigating diselectrification of air which has been electrified, whether positively or negatively, by other means than bubbling through water: with apparatus represented in figs. 1 and 2, which is simplified from that of our former paper by the omission of the apparatus for electrification by bubbling, and for collecting large quantities of electrified air.

2. In fig. 1, *A*, *B* represent the two terminals of a Voss electric machine connected, one of them to a metal can, *CC′* (a small biscuit canister of tinned iron), and the other to a fine needle, of which the point *n* is in the centre of the can. The wire making the connection to the needle passes through the centre of a hole in the side of the can, stopped by a paraffin plug. Air is blown from bellows through a pipe, *E*, near the bottom of the can, and allowed to escape from near the top through an electric filter, *F*, called the tested filter, from which it passes through a long block-tin pipe, *GG*, about $3\frac{3}{4}$ metres long and 1 cm. internal diameter, and thence through a short tunnel in a block of paraffin, *K*. From this, lastly, it passes through a second electric filter, *R*, into the open air. This second filter, which we sometimes call the *testing filter*, sometimes the *electric receiver*, is kept in metallic connection with the insulated terminal, *I*, of a quadrant electrometer, *Q*. The metal can and the block-tin

pipe are metallically connected to the outer case and uninsulated terminal, T, of the quadrant electrometer.

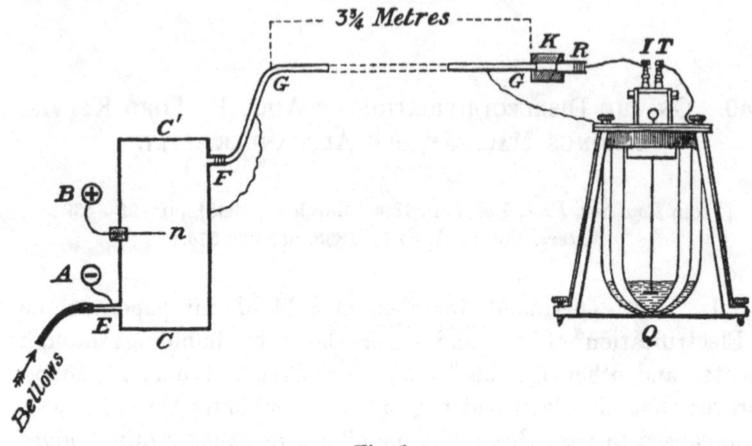

Fig. 1.

3. The testing filter or electric receiver consists of twelve discs of brass-wire cloth fixed across the mouth of a short metal pipe supported on the end of the paraffin tunnel in the manner

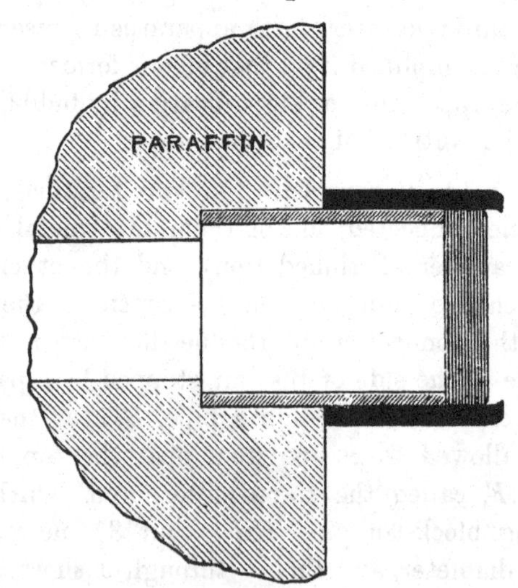

Fig. 2.

represented in fig. 2, on a scale of twice the size of the filter which we have actually used, or of true size for a filter on a tube of 2 cm. diameter, which for some purposes may be better. One of eleven similar discs, of size adapted to a tube of 2 cm. diameter, and an outermost disc with projecting lugs, are shown, true size, and with the gauge of the wire-cloth which we have actually used, shown true size, in fig. 3. The eleven little circular discs of wire-cloth are held in position by bending over them the four lugs belonging to the outermost disc, and all are kept compactly together by a short piece of india-rubber tube stretched over them outside as shown in fig. 2.

4. We commenced with a few experiments to test the efficiency of the testing filter, R, with no tested filter at F, and merely continuous block-tin pipe, FGG, from the can to the paraffin tunnel. First, working the bellows with no electrification

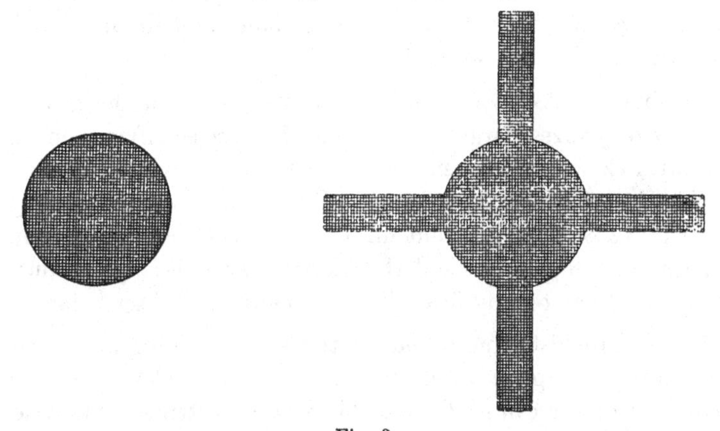

Fig. 3.

Twenty-five wires to the centimetre. Diameter of each wire is 0·16 mm. Hence each aperture is 0·24 mm. square.

of the needle point, we found no sensible electric effect on the electrometer, which proved that, whether from natural electrification of the air of the laboratory, or by the action of the bellows, or by the passage of the air through the long metal pipe, no electrification sensible to our test was produced. After that we kept the needle point, n, electrified, either positively or negatively, for five or six minutes at a time by turning the little Voss machine, and we found large effects rising to about $3\frac{1}{2}$ volts in five minutes, positive or negative, according as n was positive or negative.

5. The apparatus is now ready to test the efficacy of filters or other appliances of different kinds placed at F for the purpose

3—2

of diselectrifying air which has been electrified, whether positively or negatively, by the electrified needle point *n*. We begin with a filter of 12 wire-gauze discs, placed at *F* and kept in metallic connection with the tin pipe outside. This nearly halved the electricity shown by the electrometer. We then tried 24, 48, 72, 96, 120 wire-gauze discs, successively, placed in groups of 24, and separated from one another by short lengths of 2 cm. of lead tube, in the line of the flow of the air between *F* and *G* (fig. 1), all kept in metallic connection with the block-tin pipe and the outer case of the electrometer. We were surprised with the smallness of the additions to the diselectrifying efficiency of the 12 strainers first tried; for example, the filter of 120 wire gauzes only reduced the electrical indication to a little less than one-half of what it was with the 12 which we first tried.

We found that cotton-wool between the spaces in the groups of 24 wire gauzes largely increased the diselectrifying effect. Thus, with 72 wire gauzes and cotton-wool we succeeded in reducing the electrical effect to about one-twelfth of what it was with only a filter of 12 wire gauzes; but hitherto we have not succeeded in rendering imperceptibly small the electricity yielded by the outflowing air to the testing filter *R* in our method of observation.

6. We intend trying various methods of obtaining more and more nearly complete diselectrification of the electrified air flowing out of the can at *F*; and this for air electrified otherwise than by the needle point, as shown in the diagram: for instance, by an electrified flame in place of the needle point; or again by bubbling through water or other liquids. Meantime, the mere fact that the electricity, whether positive or negative, given to air by an electrified needle point, can be conveyed through 3 or 4 metres of small metal tube (1 cm. diameter), and shown on a quadrant electrometer by a receiving filter, is not without interest. We may add now that, with the receiving filter removed and merely a fine platinum wire put in the mouth of the paraffin tunnel, we have found that enough of electricity is taken from the outflowing air to be amply shown by the quadrant electrometer; which renders even more surprising the fact that the diselectrifying power of 120 strainers of fine wire-gauze should be so small as we have found it.

241. On the Electrification of Air.

[From *Glasg. Phil. Soc. Proc.* Vol. xxvi. [March 27, 1895], pp. 233—243; *Nature*, Vol. lii. May 16, 1895, pp. 67—70.]

1. Continuous observation of natural atmospheric electricity has given ample proof that cloudless air at moderate heights above the earth's surface, in all weathers, is electrified with very far from homogeneous distribution of electric density. Observing, at many times from May till September, 1859, with my portable electrometer on a flat open sea-beach of Brodick Bay in the Island of Arran, in ordinary fair weather at all hours of the day, I found the difference of potentials, between the earth and an insulated burning match at a height of 9 feet above it (2 feet from the uninsulated metal case of the instrument, held over the head of the observer), to vary from 200 to 400 Daniell's elements, or as we may now say volts, and often during light breezes from the east and north-east, it went up to 3000 or 4000 volts. In that place, and in fair weather, I never found the potential other than positive (never negative, never even down to zero), if for brevity we call the earth's potential at the place zero. In perfectly clear weather under a sky sometimes cloudless, more generally somewhat clouded, I often observed the potential at the 9 feet height to vary from about 300 volts gradually to three or four times that amount, and gradually back again to nearly the same lower value in the course of about two minutes*. I inferred that these gradual variations must have been produced by electrified masses of air moving past the place of observation. I did not remark then, but I now see, that the electricity in these moving masses of air must, in all probability have been chiefly positive to cause the variations which I observed, as I shall explain to you a little later.

* *Electrostatics and Magnetism* (Sir William Thomson), Art. xvi. §§ 281, 282.

2. Soon after that time a recording atmospheric electrometer* which I devised, to show by a photographic curve the continuous variation of electric potential at a fixed point, was established at the Kew Meteorological Observatory, and has been kept in regular action from the commencement of the year 1861 till the present time. It showed incessant variations quite of the same character, though not often as large, as those which I had observed on the sea-beach of Arran.

Through the kindness of the Astronomer Royal, I am able to place before you this evening the photographic curves for the year 1893, produced by a similar recording electrometer which has been in action for many years at the Royal Observatory, Greenwich. They show, as you see, not infrequently, during several hours of the day or night, negative potential and rapid transitions from large positive to large negative. Those were certainly times of broken weather, with at least showers of rain, or snow, or hail. But throughout a very large proportion of the whole time the curve quite answers to the description of what I observed on the Arran sea-beach thirty-six years ago, except that the variations which it shows are not often of so large amount in proportion to the mean or to the minimums.

3. Thinking over the subject now, we see that the gradual variations, minute after minute through so wide a range as the 3 or 4 to 1, which I frequently observed, and not infrequently rising to twenty times the ordinary minimum, must have been due to *positively* electrified masses of air, within a few hundred feet of the place of observation, wafted along with the gentle winds of 5 or 10 or 15 feet per second which were blowing at the time. If any comparably large quantities of negatively electrified air had been similarly carried past, it is quite certain that the minimum observed potential, instead of being in every case positive, would have been frequently large negative.

4. Two fundamental questions in respect to the atmospheric electricity of fair weather force themselves on our attention:—(1) What is the cause of the prevalent positive potential in the air near the earth, the earth's potential being called zero? (2) How comes the lower air to be electrified to different electric densities whether positive or negative in different parts? Obser-

* *Electrostatics and Magnetism*, Art. XVI. §§ 271, 292.

vations and laboratory experiments made within the last six or eight years, and particularly two remarkable discoveries made by Lenard, which I am going to describe to you, have contributed largely to answering the second of these questions.

5. In an article "On the Electrification of Air by a Water-jet," by Magnus Maclean and Makita Goto*, experiments were described showing air to be negatively electrified by a jet of water shot vertically down through it from a fine nozzle into a basin of water about 60 centimetres below it. It seemed natural to suppose that the observed electrification was produced by the rush of the fine drops through the air; but Lenard conclusively proved, by elaborate and searching experiments, that it was in reality due chiefly, if not wholly, to the violent commotions of the drops impinging on the water surface of the receiving basin, and he found that the negative electrification of the air was greater when they were allowed to fall on a hard slab of any material thoroughly wetted by water, than when they fell on a yielding surface of water several centimetres deep. He had been engaged in studying the great negative potential which had been found in air in the neighbourhood of waterfalls, and which had generally been attributed to the inductive action of the ordinary fine weather electric force, giving negative electricity to each drop of water-spray before it breaks away from conducting communication with the earth. Before he knew Maclean and Goto's paper, he had found strong reason for believing that that theory was not correct, and that the true explanation of the electrification of the air must be found in some physical action not hitherto discovered. A less thorough inquirer might have been satisfied with the simple explanation of the electricity of waterfalls naturally suggested by Maclean and Goto's result, and might have rested in the belief that it was due to an electrifying effect produced by the rush of the broken water through the air; but Lenard made an independent experimental investigation in the Physical Laboratories of Heidelberg and Bonn, by which he learned that the seat of the negative electrification of the air electrified is the lacerated water at the foot of the fall, or at any rocks against which the water impinges, and not the multitudinous interfaces between air and water, falling freely in drops through it.

* *Philosophical Magazine*, 1890, second half-year.

6. It still seems worthy of searching inquiry to find electrification of air by water falling in drops through it, even though we now know that, if there is any such electrification, it is not the main cause of the great negative electrification of air which has been found in the neighbourhood of waterfalls. For this purpose an experiment has been very recently made by Mr Maclean, Mr Galt, and myself, in the course of an investigation regarding electrification and diselectrification of air with which we have been occupied for more than a year. The apparatus

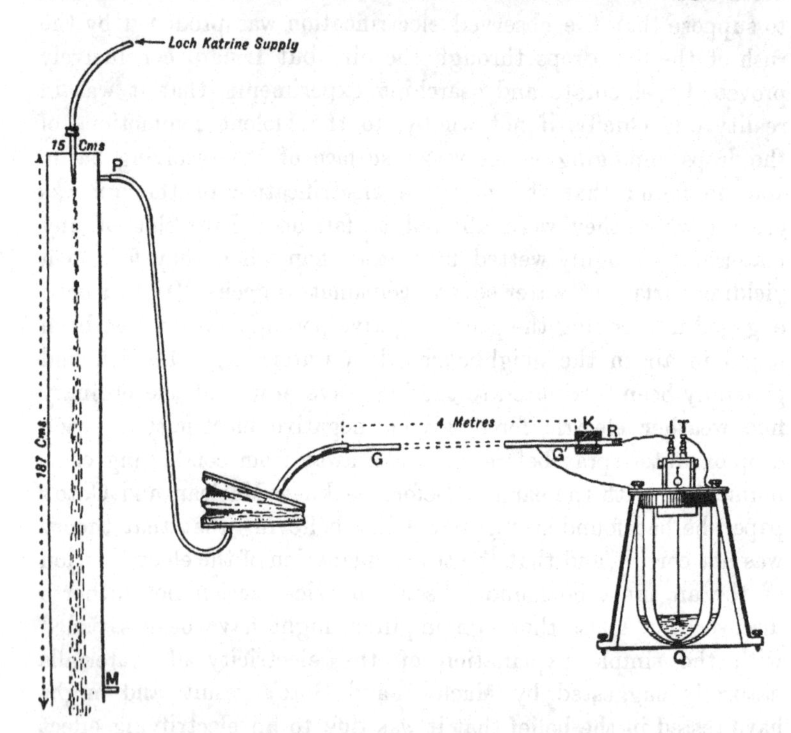

which we used is before you. It consists of a quadrant electrometer connected with an insulated electric filter* applied to test the electrification of air drawn from different parts of a tinned iron funnel, 187 centimetres long and 15 centimetres diameter, fixed in a vertical position with its lower end open and its upper end closed, except a glass nozzle, of 1·6 mm. aperture, admitting a jet of Glasgow supply water (from Loch Katrine) shot vertically

* Kelvin, Maclean, Galt, "On the Diselectrification of Air," *Proc. Roy. Soc.* March 21, 1895, *supra*, p. 33.

down along its axis. The electric filter (*R* in the drawing), a simplified and improved form of that described in the *Proceedings* of the Royal Society for March 21, consists of twelve circles of fine wire gauze rammed as close as possible together in the middle of a piece of block tin pipe of 1 cm. bore and 2 cm. length. One end of it is stuck into one end of a perforation through a block of paraffin, *K*, which supports it. The other end (*G'*) of this perforation is connected by block tin pipe (which in the apparatus actually employed was 4 metres long, but might have been shorter), and indiarubber tubing through bellows to one or other of two short outlet pipes (*M* and *P*) projecting from the large funnel.

7. We first applied the india-rubber pipe to draw air from the funnel at the *upper* outlet, *P*, and made many experiments to test the electricity given by it to the receiving filter, *R*, under various conditions as to the water-jet; the bellows being worked as uniformly as the operator could. When the water fell fairly through the funnel with no drops striking it, and through 90 cm. of free air below its mouth, a small negative electrification of *R* was in every case observed (which we thought might possibly be attributable to electrification of the air where the water was caught in a basin about 90 cm. below the mouth of the funnel). But when the funnel was slanted so that the whole shower of drops from the jet, or even a small part of it, struck the inside of the funnel, the negative electrification of *R* was largely increased. So it was also when the shower, after falling freely down the middle of the funnel, impinged on a metal plate in metallic communication with the funnel, held close under its mouth, or 10 or 20 cm. below it. For example, in a series of experiments made last Monday (March 25), we found ·28 of a volt in 15 minutes with no obstruction to the shower; and 4·18 volts in five minutes, with a metal plate held three or four centimetres below the mouth of the funnel; the air being drawn from the upper outlet (*P*). Immediately after, with *P* closed, the air drawn from the lower outlet (*M*), but all other circumstances the same, we found ·20 of a volt in five minutes with no obstruction; and 6·78 volts in five minutes with the metal plate held below the mouth as before.

8. These results, and others which we have found, with

many variations of detail, confirm, by direct test of air drawn away from the neighbourhood of the waterfall through a narrow pipe to a distant electrometer, Lenard's conclusion that a preponderatingly strong negative electrification is given to the air at every place of violent impact of a drop against a water-surface, or against a wet solid. But they do not prove that there is *no* electrification of air by drops of water falling through it. We always found, in every trial, decisive proof of negative electrification; though of comparatively small amount when there was no obstruction to the shower between the mouth of the funnel and the catching basin 90 cm. below it. We intend to continue the investigation, with the shower falling freely far enough down from the mouth of the funnel to make quite sure that the air which we draw off from any part of the funnel is not sensibly affected by impact of the drops on anything below.

9. The other discovery* of Lenard, of which I told you, is that the negative electrification of air, in his experiments with pure water, is diminished greatly by very small quantities of common salt dissolved in it, that it is brought to nothing by ·011 per cent.; that positive electrification is produced in the air when there is more than 011 per cent. of salt in the water, reaching a maximum with about 5 per cent. of salt, when the positive electrical effect is about equal to the negative effect observed with pure water, and falling to 14 per cent. of this amount when there is 25 per cent. of salt in the solution. Hence sea-water, containing as it does about 3 per cent. of common salt, may be expected to give almost as strong positive electrification to air as pure water would give of negative in similar circumstances as to commotion. Lenard infers that breaking waves of the sea must give positive electricity to the air over them; he finds, in fact, a recorded observation by Exner, on the coast of Ceylon, showing the normal positive electric potential of the air to be notably increased by a storm at sea. I believe Lenard's discovery fully explains also some very interesting observations of atmospheric electricity of my own, which I described in a letter to Dr Joule, which he published in the *Proceedings* of the Literary and Philosophical Society of Manchester for

* "Ueber die Electricität der Wasserfälle." Table xvii. p. 628. *Annalen der Physik und Chemie*, 1892, Vol. XLVI.

October 18, 1859*. "The atmospheric effect ranged from 30° to about 420° [of a heterostatic torsion electrometer of 'the divided-ring' species] during the four days which I had to test it; that is to say, the electromotive force per foot of air, measured horizontally from the side of the house, was from 9 to above 126 zinc-copper water cells. The weather was almost perfectly settled, either calm, or with slight east wind, and in general an easterly haze in the air. The electrometer twice within half an hour went above 420°, there being at the time a fresh temporary breeze from the east. What I had previously observed regarding the effect of east wind was amply confirmed. Invariably the electrometer showed very high positive in fine weather, before and during east wind. It generally rose very much shortly before a slight puff of wind from that quarter, and continued high till the breeze would begin to abate. I never once observed the electrometer going up unusually high during fair weather without east wind following immediately. One evening in August I did not perceive the east wind at all, when warned by the electrometer to expect it; but I took the precaution of bringing my boat up to a safe part of the beach, and immediately found by waves coming in that the wind must be blowing a short distance out at sea, although it did not get so far as the shore......On two different mornings the ratio of the house to a station about sixty yards distant on the road beside the sea was ·97 and ·96 respectively. On the afternoon of the 11th instant, during a fresh temporary breeze of east wind, blowing up a little spray as far as the road station, most of which would fall short of the house, the ratio was 1·08 in favour of the house electrometer—both standing at the time very high—the house about 350°. I have little doubt but that this was owing to the negative electricity carried by the spray from the sea, which would diminish relatively the indications of the road electrometer."

10. The negative electricity spoken of in this last sentence, as "carried by the spray from the sea," was certainly due to the inductive effect of the ordinary electrostatic force in the air close above the water, by which every drop or splash breaking away from the surface must become negatively electrified; but this only partially explains the difference which I observed between

* Republished in *Electrostatics and Magnetism*, "Atmospheric Electricity," Art. XVI. § 262.

the road station and the house station. We now know, by the second of Lenard's two discoveries, to which I have alluded, that every drop of the salt water spray, falling on the ground or rocks wetted by it, must have given positive electricity to the adjoining air. The air, thus positively electrified, was carried towards and over the house by the on-shore east wind which was blowing. Thus, while the road electrometer under the spray showed less electrostatic force than would have been found in the air over it and above the spray, the house electrometer showed greater electrostatic force because of the positively electrified air blown over the house from the wet ground struck by the spray.

11. The strong positive electricity, which, as described in my letter to Joule, I always found in Arran with east wind, seemed at first to be an attribute of wind from that quarter. But I soon found that in other localities east wind did not give any very notable augmentation, nor perhaps any augmentation at all, of the ordinary fair weather positive electric force, and for a long time I have had the impression that what I observed in this respect, on the sea-beach of Brodick Bay in Arran, was really due to the twelve nautical miles of sea between it and the Ayrshire coast east-north-east of it; and now it seems to me more probable than ever that this is the explanation when we know from Lenard that the countless breaking waves, such as even a gentle east wind produces over the sea between Ardrossan and Brodick, must every one of them give some positive electricity to the air wherever a spherule of spray falls upon unbroken water. It becomes now a more and more interesting subject for observation (which I hope may be taken up by naturalists having the opportunity) to find whether or not the ordinary fine weather positive electric force at the sea coast in various localities is increased by gentle or by strong winds from the sea, whether north, south, east or west of the land.

12. From Lenard's investigation we now know that every drop of rain falling on the ground or on the sea*, and every drop of fresh water spray of a breaking wave, falling on a fresh water lake, sends negative electricity from the water surface to

* "Ueber die Electricität der Wasserfälle," *Annalen der Physik und Chemie*, 1892, Vol. XLVI. p. 631.

the air; and we know that every drop of salt water, falling on the sea from breaking waves, sends positive electricity into the air from the water surface. Lenard remarks that more than two-thirds of the earth's surface is sea, and suggests that breaking sea-waves may give contributions of positive electricity to the air which may possibly preponderate over the negative electricity given to it from other sources, and may thus be the determining cause of the normal fair weather positive of natural atmospheric electricity. It seems to me highly probable that this preponderance is real for atmospheric electricity at sea. In average weather, all the year round, sailors in very small vessels are more wet by sea-spray than by rain, and I think it is almost certain that more positive electricity is given to the air by breaking waves than negative electricity by rain. It seems also probable that the positive electricity from the waves is much more carried up by strong winds to considerable heights above the sea, than the negative electricity given to the air by rain falling on the sea; the greater part of which may be quickly lost into the sea, and but a small part carried up to great heights. But it seems to me almost certain that the exceedingly rapid recovery of the normal fair weather positive, after the smaller positive or the negative atmospheric electricity of broken weather, which was first found by Beccaria in Italy 120 years ago, and which has been amply verified in Scotland and England*, could not be accounted for by positively electrified air coming from the sea. Even at Beccaria's Observatory, at Garzegna di Mondovi in Piedmont, or at Kew, or Greenwich, or Glasgow, we should often have to wait a very long time for reinstatement of the normal positive after broken weather, if it could only come in virtue of positively electrified air blowing over the place from the sea; and several days, at least, would have to pass before this result could possibly be obtained in the centre of Europe.

13. It has indeed always seemed to me probable that the rain itself is the real restorer of the normal fair weather positive. Rain or snow, condensing out of the air high up in the clouds, must itself, I believe, become negatively electrified as it grows, and must leave positive electricity in the air from which it falls. Thus rain falling from negatively electrified air would leave it

* *Electrostatics and Magnetism*, Art. XVI. § 287.

less negatively electrified, or non-electrified or positively electrified; rain falling from non-electrified air would leave it positively electrified; and rain falling from positively electrified air would leave it with more of positive electricity than it had before it lost water from its composition. Several times within the last thirty years I have made imperfect and unsuccessful attempts to verify this hypothesis by laboratory experiments, and it still remains unproved. But I am much interested just now to find some degree of observational confirmation of it in Elster and Geitel's large and careful investigation of the electricity produced in an insulated basin by rain or snow falling into it, which they described in a communication published in the *Sitzungsberichte* of the Vienna Academy of Sciences, of May 1890. They find generally a large electrical effect, whether positive or negative, by rain or snow falling into the basin for even so short a time as a quarter of a minute, with however, on the whole, a preponderance of negative electrification.

14. But my subject this evening is not merely natural atmospheric electricity, although this is certainly by far the most interesting to mankind of all hitherto known effects of the electrification of air. I shall conclude by telling you very briefly, and without detail, something of new experimental results regarding electrification and diselectrification of air, found within the last few months in our laboratory here by Mr Maclean, Mr Galt, and myself. We hope before the end of the present session of the Royal Society to be able to communicate a sufficiently full account of our work.

15. Air blown from an uninsulated tube, so as to rise in bubbles through pure water in an uninsulated vessel, and carried through an insulated pipe to the electric receiving filter, of which I have already told you, gives negative electricity to the filter. With a small quantity of salt dissolved in the water, or sea water substituted for fresh water, it gives positive electricity to the air. There can be no doubt but these results are due to the same physical cause as Lenard's negative and positive electrification of air by the impact of drops of fresh water or of salt water on a surface of water or wet solid.

16. A small quantity of fresh water or salt water shaken up vehemently with air in a corked bottle electrifies the air, fresh

water negatively, salt water positively. A "Winchester quart" bottle (of which the cubic contents is about two litres and a half), with one-fourth of a litre of fresh or salt water poured into it, and closed by an india-rubber cork, serves very well for the experiment. After shaking it vehemently till the whole water is filled with fine bubbles of air, we leave it till all the bubbles have risen and the liquid is at rest, then take out the cork, put in a metal or india-rubber pipe, and by double-acting bellows, draw off the air and send it through the electric filter. We find the electric effect, negative or positive according as the water is fresh or salt, shown very decidedly by the quadrant electrometer: and this, even if we have kept the bottle corked for two or three minutes after the liquid has come to rest before we take out the cork and draw off the air.

17. An insulated spirit lamp or hydrogen lamp being connected with the positive or with the negative terminal of a little Voss electric machine, its fumes (products of combustion mixed with air) sent through a block-tin pipe, four metres long, and one centimetre bore, ending with a short insulating tunnel of paraffin and the electric filter, give strong positive or strong negative electricity to the filter.

18. Using the little biscuit-canister and electrified needle, as described in "our communication"* to the Royal Society "On the Diselectrification of Air," but altered to have two insulated needles with varied distances of from half a centimetre to two or three centimetres between them, we find that when the two needles are kept at equal differences of potential positive and negative, from the enclosing metal canister, little or no electrification is shown by the electric filter; and when the differences of potential from the surrounding metal are unequal, electrification, of the same sign as that of the needle whose difference of potential is the greater, is found on the filter.

When a ball and needle-point are used, the effect found depends chiefly on the difference of potentials between the needle-point and the surrounding canister, and is comparatively little affected by opposite electrification of the ball. When two balls are used,

* *Proceedings of the Royal Society*, March 21, 1895. [*Supra* p. 33.]

and sparks in abundance pass between them, but little electricity is deposited by the sparks in the air, even when one of the balls is kept at the same potential as the surrounding metal. [The communication was illustrated by a repetition of some of the experiments shown on the occasion of a Friday evening lecture* on Atmospheric Electricity at the Royal Institution on May 18, 1860, in which one half of the air of the lecture-room was electrified positively, and the other half negatively, by two insulated spirit lamps mounted on the positive and negative conductors of an electric machine.]

* *Electrostatics and Magnetism*, Art. XVI. §§ 285, 286.

242. On the Electrification and Diselectrification of Air and Other Gases. By Lord Kelvin, Magnus Maclean, and Alexander Galt.

[From *Brit. Assoc. Report*, 1895, pp. 630—633; *Nature*, Vol. LII. Oct. 17, 1895, pp. 608—610.]

1. Experiments were made for the purpose of finding an approximation to the amount of electrification communicated to air by one or more electrified needle points. The apparatus consisted of a metallic can 48 cm. high and 21 cm. in diameter, supported by paraffin blocks, and connected to one pair of quadrants of a quadrant electrometer. It had a hole at the top to admit the electrifying wire, which was 5·31 metres long, hanging vertically within a metallic guard tube. This guard tube was always metallically connected to the other pair of quadrants of the electrometer and to its case, and to a metallic screen surrounding it. This prevented any external influences from sensibly affecting the electrometer, such as the working of the electric machine which stood on a shelf 5 metres above it.

2. The experiment is conducted as follows:—One terminal of an electric machine is connected with the guard tube and the other with the electrifying wire which is let down so that the needle is in the centre of the can. The can is temporarily connected to the case of the electrometer. The electric machine is then worked for some minutes, so as to electrify the air in the can. As soon as the machine is stopped the electrifying wire is lifted clear out of the can. The can and the quadrants in metallic connection with it are disconnected from the case of the electrometer, and the electrified air is very rapidly drawn away from the can by a blowpipe bellows arranged to exhaust. This releases the opposite kind of electricity from the inside of the can, and allows it to place itself in equilibrium on the outside of the can

and on the insulated quadrants of the electrometer in metallic connection with it.

3. We tried different lengths of time of electrification, and different numbers of needles and tinsel, but we found that one needle and four minutes of electrification gave nearly maximum effect. The greatest deflection observed was 936 scale divisions. To find, from this reading, the electric density of the air in the can, we took a metallic disc, of 2 cm. radius, attached to a long varnished glass rod, and placed it at a distance of 1·45 cm. from another and larger metallic disc. This small air condenser was charged from the electric light conductors in the laboratory to a difference of potential amounting to 100 volts. The insulated disc thus charged was removed and laid upon the roof of the large insulated can. This addition to the metal in connection with it does not sensibly influence its electrostatic capacity. The deflexion observed was 122 scale divisions. The capacity of the condenser is approximately $\pi \times 2^2/(4\pi \times 1{\cdot}45) = 1/1{\cdot}45$. The quantity of electricity with which it was charged was $1/1{\cdot}45 \times 100/300 = 1/4{\cdot}35$ electrostatic unit. Hence the quantity to give 936 scale divisions was $1/4{\cdot}35 \times 936/122 = 1{\cdot}7637$.

The bellows was worked vigorously for two and a half minutes, and in that time all the electrified air would be exhausted. The capacity of the can was 16,632 cubic centimetres, which gives, for the quantity of electricity per cubic centimetre, $1{\cdot}7637/16632 = 1{\cdot}06 \times 10^{-4}$. The electrification of the air in this case was positive: it was about as great as the greatest we got, whether positive or negative, in common air when we electrified it by discharge from needle points. This is about four times the electric density which we roughly estimated as about the greatest given to the air in the inside of a large metal vat, electrified by a needle point and then left to itself, and tested by the potential of a water-dropper with its nozzle in the centre of the vat, in experiments made two years ago, and described in a communication to the Royal Society of date May 1894*.

4. In subsequent experiments electrifying common air in a large gas-holder over water by an insulated gas flame burning within it with a wire in the interior of the flame kept electrified

* "On the Electrification of Air," by Lord Kelvin and Magnus Maclean, *supra*, p. 6.

by an electric machine to about 6,000 volts, whether positively or negatively, we found as much as $1{\cdot}5 \times 10^{-4}$ for the electric density of the air. Electrifying carbonic acid in the same gas-holder, *whether positively or negatively*, by needle points, we obtained an electric density of $2{\cdot}2 \times 10^{-4}$.

5. We found about the same electric density ($2{\cdot}2 \times 10^{-4}$) of *negative* electricity in carbonic acid gas drawn from an iron cylinder lying horizontally, and allowed to pass by a U-tube into the gas-holder without bubbling through the water. This electrification was due probably not to carbonic acid gas rushing through the stopcock of the cylinder, but to bubbling from the liquid carbonic acid in its interior, or to the formation of carbonic acid snow in the passages and its subsequent evaporation. When carbonic acid gas was drawn slowly from the liquid carbonic acid in the iron cylinder placed upright, and allowed to pass, without bubbling, through the U-tube into the gas-holder over water, no electrification was found in the gas unless electricity was communicated to it from needle points.

6. The electrifications of air and carbonic acid described in Sections 4 and 5 were tested, and their electric densities measured, by drawing by an air pump a measured quantity of the gas* from the gas-holder through an indiarubber tube to a receiver of known efficiency and of known capacity in connection with the electrometer. We have not yet measured how much electricity was lost in the passage through the indiarubber tube. It was not probably nothing; and the electric density of the gas before leaving the gas-holder was no doubt greater, though perhaps not much greater, than what it had when it reached the electric receiver.

7. The efficiency of the electric receivers used was approximately determined by putting two of them in series, with a paraffin tunnel between them, and measuring by means of two quadrant electrometers the quantity of electricity which each took from a measured quantity of air drawn through them. By performing this experiment several times, with the order of the two

* The gas-holder was 38 cm. high and 81 cm. in circumference. Ten strokes of the pump raised the water inside to a height of 8·1 cm., so that the volume of air drawn through the receivers in the experiments was 428 cubic cm. per stroke of the pump. This agrees with the measured effective volume of the two cylinders of the pump.

receivers alternately reversed, we had data for calculating the proportion of the electricity taken by each receiver from the air entering it, on the assumption that the proportion taken by each receiver was the same in each case. This assumption was approximately justified by the results.

8. Thus we found for the efficiencies of two different receivers respectively 0·77 and 0·31 with air electrified positively or negatively by needle points; and 0·82 and 0·42 with carbonic acid gas electrified negatively by being drawn from an iron cylinder placed on its side. Each of these receivers consisted of block tin pipe 4 cm. long and 1 cm. diameter, with five plugs of cotton wool kept in position by six discs of fine wire gauze. The great difference in their efficiency was, no doubt, due to the quantities of cotton wool being different, or differently compressed in the two.

9. We have commenced, and we hope to continue, an investigation of the efficiency of electric receivers of various kinds, such as block tin, brass, and platinum tubes from 2 to 4 cm. long and from 1 mm. to 1 cm. internal diameter, all of smooth bore and without any cotton wool or wire gauze filters in them; also a polished metal solid insulated with a paraffin tunnel. This investigation, made with various quantities of air drawn through per second has already given us some interesting and surprising results, which we hope to describe after we have learned more by further experimenting.

10. In addition to our experiments on electric filters we have made many other experiments to find other means for the dis-electrification of air. It might be supposed that drawing air in bubbles through water should be very effective for this purpose, but we find that this is far from being the case. We had previously found that non-electrified air drawn in bubbles through pure water becomes negatively electrified, and through salt water positively. We now find that positively electrified air drawn through pure water, and negatively electrified air through salt water, has its electrification diminished but not annulled if the primitive electrification is sufficiently strong. Negatively electrified air drawn in bubbles through pure water, and positively electrified air drawn through salt water, has its electrification augmented.

11. To test the effects of heat we drew air through combustion tubes of German glass about 180 cm. long and 2½ or

$1\frac{1}{2}$ cm. bore, the heat being applied externally to about 120 cm. of the length. We found that when the temperature was raised to nearly a dull red heat, air, whether positively or negatively electrified, lost little or nothing of its electrification by being drawn through the tube. When the temperature was raised to a dull red heat, and to a bright red, high enough to soften the glass, losses up to as much as four-fifths of the whole electrification were sometimes observed, but never complete diselectrification. The results, however, were very irregular. Non-electrified air never became sensibly electrified by being drawn through the hot glass tubes in our experiments; but it gained strong positive electrification when pieces of copper foil, and negative electrification when pieces of carbon, were placed in the tube, and when the temperature was sufficient to powerfully oxidise the copper or to burn away the charcoal.

12. Through the kindness of Mr E. Matthey, we have been able to experiment with a platinum tube 1 metre long and 1 mm. bore. It was heated either by a gas flame or an electric current. When the tube was cold, and non-electrified air drawn through it, we found no signs of electrification by our receiver and electrometer. But when the tube was made red or white hot, either by gas burners applied externally or by an electric current through the metal of the tube, the previously non-electrified air drawn through it was found to be electrified strongly positive. To get complete command of the temperature we passed a measured electric current through 20 cm. of the platinum tube. On increasing the current till the tube began to be at a scarcely visible dull red heat we found but little electrification of the air. When the tube was a little warmer, so as to be quite visibly red hot, large electrification became manifest. Thus 60 strokes of the air-pump gave 45 scale divisions on the electrometer when the tube was dull red, and 395 scale divisions (7 volts) when it was a bright red (produced by a current of 36 ampères). With stronger currents, raising the tube to white-hot temperature, the electrification seemed to be considerably less.

243. On the Generation of Longitudinal Waves in Ether.

[From *Roy. Soc. Proc.* Vol. LIX. [Feb. 13, 1896], pp. 270—273; *Nature*, Vol. LIII. March 12, 1896, pp. 450, 451; *Écl. Élec.* Vol. VI. March 14, 1896, pp. 493, 494.]

In a short note published in last week's *Nature*, of which a copy is appended, I suggested an arrangement of four insulated and electrified spherical conductors with their centres in one line, giving rise to ethereal waves in the surrounding atmosphere, of which the disturbance in the line of centres is essentially longitudinal. But at any finite distance from this line there must also be laminar or distortional waves of the kind expressed in Maxwell's equations. The object of my present communication is to show an arrangement by which a large space of air is traversed by pressural disturbance, or by waves essentially longitudinal, or by condensational-rarefactional vibrations; but a very small proportion, practically evanescent, of laminar waves.

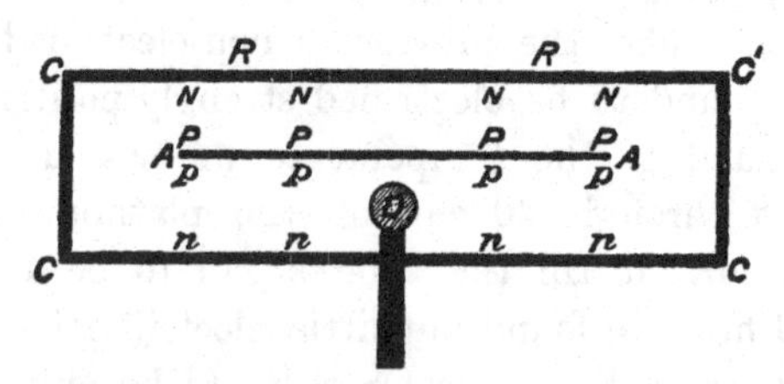

Let *AA* be a plane circular metal plate insulated within a metal case *CCC′C′*, as indicated in the drawing. Let *D* be a discharger which can be pushed in so as to make contact with *A*.

Let *A* be charged to begin with, positively for instance as indicated by the letters *PPpp*; *NNnn* showing negative electricity induced by it. Let now the discharger be pushed in till a spark passes. The result, as regards the space between *AA* and the roof *RR* over it, will be either an instantaneous transmission of commencement of diminution of electrostatic force, or a set of electric waves of almost purely longitudinal displacement, according as ether is incompressible or compressible.

Hence, if the theory of longitudinal waves, suggested by Röntgen as the explanation of his discovery (for the consideration of which he has given strong reasons), be true, it would seem probable that a sensitive photographic plate in the space between *AA* and *RR* should be acted on, as sensitive plates are, by Röntgen rays. Either a Wimshurst electrical machine or an induction-coil, adapted to keep incessantly charging *AA* with great rapidity so as to cause an exceedingly rapid succession of sparks between *D* and *A*, might give a practical result. In trying for it, the light of the sparks at *D* must be carefully screened to prevent general illumination of the interior of the case and ordinary photographic action on the sensitive plate.

The arrangement may be varied by making the roof of sheet aluminium, perhaps about a millimetre thick, and placing the sensitive photographic plate, or phosphorescent substance, on the outside of this roof, or in any convenient position above it. When a photographic plate is used there must, of course, be an outer cover of metal or of wood, to shut out all ordinary light from above. This arrangement will allow the spark gap at *D* to be made wider and wider, until in preference the sparks pass between *AA* and the aluminium roof above it. The transparency of the aluminium for Röntgen light will allow the photographic plate to be marked, if enough of this kind of light is produced in the space between the roof and *AA*, whether with or without sparks.

The new photography has hitherto, so far as generally known, been performed only by light obtained from electric action in vacuum; but that vacuum is not essential, for the generation of the Röntgen light might seem to be demonstrated by an experiment by Lord Blythswood, which he described at a meeting of the Glasgow Philosophical Society last Wednesday (Feb. 5). As a result he exhibited a glass photographic dry plate with splendidly clear marking which had been produced on it when placed inside its dark slide, wrapped round many times in black velvet cloth, and held in front of the space between the main electrodes of his powerful Wimshurst electrical machine, but not in the direct line of the discharge. He also exhibited photographic results obtained from the same arrangement with only the difference that the dark slide, wrapped in black velvet, was held in the direct line of the discharge. In this case the photographic result was due, perhaps wholly, and certainly in part, to electric sparks or brushes inside the enclosing box, which was, as usual, made of mahogany with metal hinges and interior metal mountings. It is not improbable that the results of the first experiments described by Lord Blythswood may also be wholly due to sparking within the wooden case. I have suggested to him to repeat his experiments with a thoroughly well closed aluminium box, instead of the ordinary photographic dark slide which he used, and without any black cloth wrapped round outside. The complete metallic enclosure will be a perfect guarantee against any sparks or brushes inside.

If the arrangement which I now suggest, with no sparks or brushes between *AA* and the roof, gives a satisfactory photographic result, or if it shows a visible glow on phosphorescent material placed anywhere in the space between *AA* and the roof above it, or above the aluminium roof, it would prove the truth of Röntgen's hypothesis. But failure to obtain any such results would not disprove this hypothesis. The electric action, even with the place of the spark so close to the field of the action sought for as it is at *D*, in the suggested arrangement, may not be sudden enough or violent enough to produce enough of longitudinal waves, or of condensational-rarefactional vibrations, to act sensibly on a photographic plate, or to produce a visible glow on a phosphorescent substance.

(Extract from *Nature*, referred to above.)

"*Velocity of Propagation of Electrostatic Force.*

"Dr Bottomley's note published in *Nature*, of January 23, quotes an extract from my Baltimore Lectures of October, 1884, in which this subject is spoken of, with an illustration consisting of two metal spheres at a great distance asunder, having periodically varying opposite electrifications maintained in them by a wire connecting them through an alternate current dynamo.

"For an illustration absolutely freed from connecting wire and all complications, consider four metal spheres, *A*, *B*, *c*, *d*, with their centres all in one straight line;—their relative magnitudes and positions being such as shown in the accompanying diagram. Let each of the four be initially electrified, *A* and *c* positively, *B* and *d* negatively. Let the charges on *c* and *d* be so strong that a spark is only just prevented from passing between them by the influence of B and A. Let A be gradually brought nearer to B till a spark passes between them. Will the consequent spark between c and d take place at the same instant or a little later? It is not easy to see how this question could be answered experimentally; but remembering the wonderful ingenuity shown by Hertz in finding how to answer questions related to it, we need not, perhaps, despair to see *it* also answered by experiment.

"The elastic solid theory restricted to the supposition of incompressibility (which is expressed by Maxwell's formulas) makes the difference of times between the two sparks infinitely small.

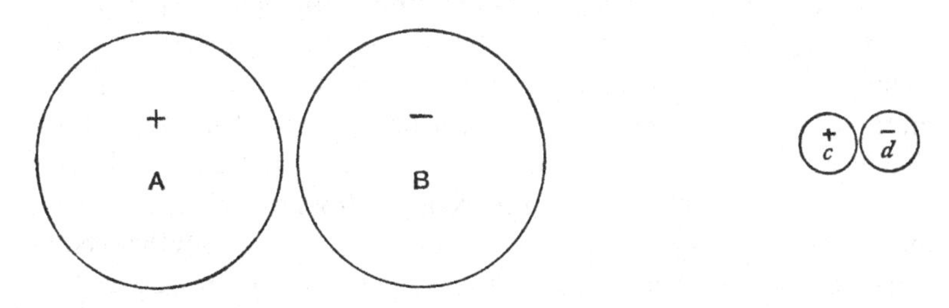

The unrestricted elastic solid theory gives for the difference of times the amount calculated according to the velocity of the condensational-rarefactional wave.

"But I feel that it is an abuse of words to speak of the 'elastic solid theory of electricity and magnetism' when no one hitherto has shown how to find in an elastic solid anything analogous to the attraction between rubbed sealing-wax and a little fragment of paper; or between a loadstone or steel magnet and a piece of iron; or between two wires conveying electric currents. Elastic solid, however, we must have, or a definite mechanical analogue of it, for the undulatory theory of light and of magnetic waves and of electric waves. And consideration of the definite knowledge we have of the properties of a real elastic solid, which we have learned from observation and experiment, aided by mathematics, is exceedingly valuable in suggesting and guiding ideas towards a general theory which shall include light (Old and New), old and new knowledge of electricity, and the whole of electro-magnetism."

244. On Lippmann's Colour Photography with Obliquely Incident Light.

[From *Nature*, Vol. LIV. May 7, 1896, pp. 12, 13.]

In the discussion which followed Prof. Lippmann's splendidly interesting communication to the Royal Society, April 23, on colour photography, I suggested the possibility of applying his method to the Röntgen X-light; but at the same time remarked that it might be found impracticable on account of the smallness of the specular reflection of the X-light from polished surfaces, unless at obliquities little short of 90°. Lord Blythswood's experiments, communicated to the Royal Society on March 19, seemed to prove decisively something of true specular reflection of X-light, incident on a plain mirror of speculum metal at 45°. Experiments, which he has since made by means of a concave mirror of speculum metal, have demonstrated beyond all doubt that there is regular reflection at nearly normal incidence; but they have also proved that the amount of regularly reflected light is exceedingly small in proportion to diffuse light caused to emanate from the mirror, by the incidence of X-light upon it. Experiments by Joly, of Dublin, have, I believe, proved somewhat abundant specular reflection of the X-light, at incidences little short of 90°, on surfaces of bodies transparent to ordinary light. And the extremely small refractivity of the photographic gelatine film for X-light, will allow incidences little short of 90° upon the metal mirror, to be used instead of the normal incidences which Prof. Lippmann has hitherto used. But for very oblique incidences the mercury mirror, with its surface fitted to the not rigorously plane surface of the photographic film, would be unsuitable; and the plan, which Lord Rayleigh described in the discussion, of forming the film on a solid metallic mirror, might be substituted for it.

All things considered, it seems not improbable that Lippmann's process may be applied successfully to X-rays at nearly grazing incidences on metallic mirrors, and possibly even on non-metallic mirrors.

Suppose now, for instance, the directions of the incident and reflected rays to be inclined to the mirror at angles of ·1 of a radian (5·7°). The distance between the planes of stratification in the photograph would be ten times that which would be produced by the same light at normal incidence. Thus if, for example, the wave-length of the particular X-light used is 5×10^{-6} cms. (or one-tenth of that of green light), the photograph would show tints of from green to violet when viewed normally, or at less or more oblique angles, by Lippmann's ordinary arrangements.

It is quite possible, however, that when we know something of the composition of Röntgen light, we may find such great differences of wave-lengths* in it, and so much difficulty to obtain approximately homogeneous X-light by sifting through metal plates (as we sift ordinary visible light by coloured glasses), or by other means if other means can be found, that the experiment which I have suggested may fail on account of want of homogeneousness of the incident light.

But here, suggested to me by thinking of oblique incidence for the photographic light, is an illustrative experiment which (with variations of detail to facilitate realisation) cannot fail if Prof. Lippmann will think it worth while to try it. Place a point source of homogeneous violet light (wave-length 4×10^{-5} cms.) so near to the centre of the mirror and sensitive film that rays shall be received at all angles of incidence from zero up to 56° (being the angle of which the secant is 1·788). The thickness of each stratum will vary in different parts of the photograph in simple proportion to the secant of the angle of incidence, and in

* It is to be hoped however that, very soon, we shall have definite knowledge of wave-lengths of Röntgen X-light by diffraction fringes actually seen instead of estimates of their smallness from diffraction fringes not seen. I should explain that I am writing on the supposition which seems to me, after much correspondence with Sir George Stokes, to be exceedingly probable that Röntgen light is merely ordinary transverse-vibrational light of very short period. That its period is less than one-fifth that of green light seems well proved by the skilful experiments described by Perrin in *Comptes Rendus*, Jan. 27, 1896, p. 187; and by Sagnac, *Comptes Rendus*, Mar. 30 [1896], p. 783.

the centre it will be equal to the half wave-length. It will therefore vary from 2×10^{-5} in the centre to $3{\cdot}6 \times 10^{-5}$ at the circle of 56° incidence. This photograph, viewed or thrown on a screen as nearly as may be normally, according to Prof. Lippmann's ordinary procedure, will be seen as a complete spectrum in concentric circles, with violet in the centre, and red, of wave-length $7{\cdot}15 \times 10^{-5}$, at the circle of 56° incidence; but, if viewed by an eye placed at the position of the source of the violet light which photographed it, it will, according to the principles explained by Dr Lippmann in his paper, be seen of uniform violet light throughout its whole area.

245. ON MEASUREMENTS OF ELECTRIC CURRENTS THROUGH AIR AT DIFFERENT DENSITIES DOWN TO ONE FIVE-MILLIONTH OF THE DENSITY OF ORDINARY AIR. By LORD KELVIN, J. T. BOTTOMLEY, and MAGNUS MACLEAN.

[From *British Association Report*, 1896, pp. 710, 711.]

246. ON THE COMMUNICATION OF ELECTRICITY FROM ELECTRIFIED STEAM TO AIR. By LORD KELVIN, MAGNUS MACLEAN, and ALEXANDER GALT.

[From *Brit. Assoc. Report*, 1896, p. 721 [title only]; *Electrician*, Vol. XXXVIII. 1897, p. 115; *Nature*, Vol. LIV. Oct. 29, 1896, pp. 622, 623.]

247. Experiments on the Electrical Phenomena Produced in Gases by Röntgen Rays, by Ultra-violet Light, and by Uranium. By Lord Kelvin, J. C. Beattie, and M. S. de Smolan.

I. *Electrification of Air by Röntgen Rays.*

[From *Edinb. Roy. Soc. Proc.* Vol. xxi. Dec. 21, 1896, pp. 393—397; *Nature*, Vol. lv. Dec. 31, 1896, pp. 199, 200.]

1. To test whether or not the Röntgen rays have any electrifying effect on air, the following arrangement was made.

A lead cylinder 76 cms. long, 23 cms. diameter, was constructed; and both ends were closed with paraffined cardboard, transparent to the Röntgen rays. Outside the end distant from the electrometer (see diagram 1) a Röntgen lamp* was placed. In the other

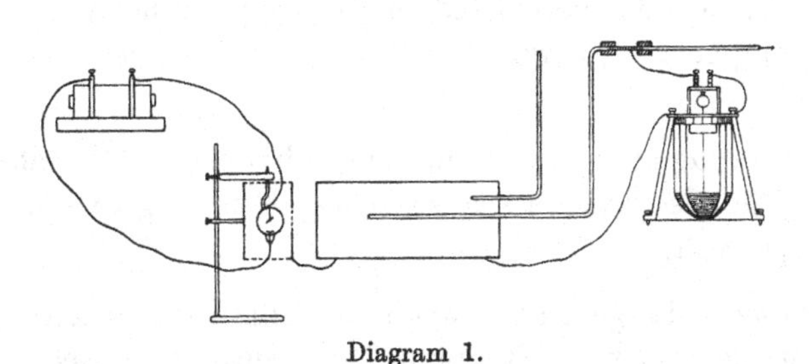

Diagram 1.

end two holes were made, one in the middle, through which passed a glass tube (referred to below as suction pipe) of sufficient length to allow the end in the lead cylinder to be put into any desired place in the cylinder. By means of this, air was drawn through an electric filter† by an air pump. The other hole, at a

* The Röntgen lamp was a vacuum vessel with an oblique platinum plate (Jackson pattern).

† Kelvin, Maclean, Galt, *Proc. Roy. Soc.*, London, March 21, 1895. [*Supra* p. 33.]

little distance from the centre, contained a second glass tube by which air was drawn through india-rubber tubing from the open-air quadrangle outside the laboratory.

In one series of experiments the end of the suction pipe was kept in the axial line of the lead cylinder at various points 10 cms. apart, beginning with a point close to the end distant from the Röntgen lamp.

In every case the air drawn through the filter was found to be negatively electrified when no screen or an aluminium screen was interposed between the Röntgen lamp and the near end of the lead cylinder. The air was found not electrified at all, or very slightly negative, when a lead screen was interposed.

When the Röntgen lamp was removed or stopped, and air was still pumped through the filter, no deflection was observed on the electrometer. This proved that the air of the quadrangle was not electrified sufficiently to show any deflection when thus tested by filter and electrometer.

Similar results were obtained with the end of the suction pipe placed so as to touch the floor of the lead cylinder, or the roof, or the sides. Whether the air was pumped away from a place in the cylinder permeated, or from a place not permeated, by the Röntgen rays, it was in all cases found to be negatively electrified.

The following are some of the results obtained on December 16 and 17. The electrometer was so arranged as to give 140 scale divisions per volt.

Conditions.—Large lead cylinder metallically connected with sheath of electrometer. Röntgen lamp surrounded by a lead sheath, which latter was also connected to electrometer-sheath. There was a window in this lamp-sheath 2·5 cms. broad and 5 cms. high. This window could be screened by aluminium or by lead. These screens were always connected metallically to sheaths. During all the experiments a Bunsen lamp (not shown in the diagram) was kept constantly burning, with its flame about 30 cm. below the Röntgen lamp.

Results.—Röntgen lamp in action; air drawn from lowest point of end of lead cylinder next to the R. lamp.

December 16:—

3.55 p.m.	−61 scale divisions in 2 mins. with				aluminium screen
”	−63	”	”	”	no screen
”	−14	”	”	”	lead screen
4.20 p.m.	Air drawn from point on lowest line of lead cylinder 26 cms. distant from R. L. end				
”	−14 scale divisions in 2 mins. with				lead screen
”	−78	”	”	”	no screen
”	−24	”	”	”	lead screen
”	−83	”	”	”	aluminium screen
”	−13	”	”	”	lead screen

December 17:—

10.47 a.m.

R. L. acting, and air drawn through filter					End of suction pipe kept in axial line of cylinder (cms.)
−44	in 2 mins.	with	alumin.	screen	68 from R. L. end
0	”	”	lead	”	68 ” ”
−28	”	”	no	”	58 ” ”
−24	”	”	no	”	48 ” ”
0	”	”	lead	”	48 ” ”
−23	”	”	alumin.	”	48 ” ”
−26	”	”	alumin.	”	38 ” ”
−9	”	”	lead	”	38 ” ”
−7	”	”	lead	”	28 ” ”
−26	”	”	alumin.	”	28 ” ”
−36	”	”	alumin.	”	18 ” ”
−21	”	”	alumin.	”	8 ” ”

2. We had previously made experiments with a sheet-iron funnel 1 metre long, 14·5 cms. diameter; and with a glass tube 150 cms. long, 3·5 cms. diameter; and with an aluminium tube 60 cms. long, 4·5 cms. diameter. Air was pumped from different parts while the Röntgen rays were shining along the tube from one end, which was closed by paraffined paper stretched across it. In every case the air was found to be negatively electrified.

In those earlier experiments the air drawn away was replaced by air coming in from the laboratory at the open end of the tube. We found evidence of disturbance due to electrification of air of the laboratory by brush discharges from electrodes between the induction coil and Röntgen lamp, and perhaps from circuit-break spark of induction coil. These sources of disturbance are eliminated by our later arrangement of lead cylinder covered with cardboard at both ends, as described above, and air drawn into it from open-air outside the laboratory.

3. We have also found a very decided electrification of air—sometimes negative, sometimes positive—when the Röntgen rays are directed across a glass tube or an aluminium tube, through which air was drawn from the quadrangle outside the laboratory, to the filter.

A primary object of our experiments was to test whether air electrified positively or negatively lost its charge by the passage of Röntgen rays through it. We soon obtained an affirmative answer to this question, both for negative and positive electricity. We found that positively electrified air lost its positive electricity, and in some cases acquired negative electricity, under the influence of Röntgen rays; and we were thus led to investigate the effect of Röntgen rays on air unelectrified to begin with.

3A. The arrangement described in § 1 was again used to test whether or not air was electrified by ultra-violet rays. The ultra-violet rays were produced by an arc lamp. This lamp was placed about a cm. distant from the closed end of the large lead cylinder. The rays passed into the cylinder through a quartz window. The air in the cylinder from the immediate neighbourhood of this window was drawn through an electric filter. No effect was produced on the electrometer. An exactly corresponding arrangement—§ 1—with Röntgen rays gave negative electrification of the air.

II. *On Apparent and Real Diselectrification of Solid Dielectrics Produced by Röntgen Rays and by Flame.*

[From *Edinb. Roy. Soc. Proc.* Vol. XXI. Feb. 15, 1897, pp. 397—403; *Nature*, Vol. LV. March 18, 1897, pp. 472—474.]

4. The fact that air is made conductive by flame, by ultra-violet light, by Röntgen rays, and by the presence of bodies at a white heat, has been shown by many experimenters. We propose in this communication to give some results bearing on this conductivity of air, based chiefly on experiments of our own.

5. We have examined more particularly the behaviour of paraffin and of glass.

In our first experiments with paraffin we used a brass ball of about an inch diameter, connected to the insulated terminal of an electrometer by a thin copper wire soldered to the ball. The ball and the wire were both coated to the depth of about $\frac{1}{8}$th of an inch with paraffin. The ball was then laid on a block of paraffin in a lead box with an aluminium window, both of which were in metallic connection with the case of the electrometer. By this means we avoided all inductive effects.

The electrometer was so arranged as to read 140 scale divisions per volt.

After testing the insulation the paraffin ball was charged positively and the rays played on it. After two minutes the electrometer reading was steady at 0·5 of the initial reading. The electrometer was then discharged by metallic connection, and again charged positively. Its reading remained steady after three minutes at 0·63 of the initial charge. In the third and fourth experiments the readings after three minutes were ·81 and ·90 of the initial charges respectively.

The ball was next charged negatively. When the rays were played on it a steady reading was obtained after four minutes at 18 of the initial charge. In the second, third, and fourth experiments the steady readings after four minutes were ·45, ·70, and ·78 of the initial charges respectively.

6. The paraffin was then removed and the brass tail polished with emery paper; whether the charge was positive or negative, it fell in about five seconds to one definite position, 50 scale divisions on the positive side of the metallic zero, when the Röntgen rays were played on the charged ball.

7. These experimental results demonstrate that the Rontgen rays *did not produce sensible conductance* between the brass ball, when it was coated with paraffin, and the surrounding metal sheath; and that *they did produce it* when there was only air and no paraffin between them. From experiments by J. J. Thomson, Righi, Minchin, Benoist and Hurmuzescu, Borgmann and Gerchun, and Röntgen*, we know that air is rendered temporarily conduc-

* J. J. Thomson, *Proceedings R. S. L.*, February 13, 1896; Righi, *Comptes Rendus*, February 17, 1896; Benoist and Hurmuzescu, *Comptes Rendus*, February 3, March 17, April 27, 1896; Borgmann and Gerchun, *Electrician*, February 14, 1896; Röntgen, *Würzburger Phys. Med. Gesellschaft*, March 9, 1896; Minchin, *Electrician*, March 27, 1896.

tive by Röntgen rays, and Röntgen's comparison of the effect of the rays with that of a flame shows that our experimental results are explained by the augmentation of the electrostatic capacity (quasi-condenser) of the brass ball by the outside surface of its coat of paraffin being put into conductive communication with the surrounding lead sheath and the connected metals.

8. In our second experiments we have endeavoured to eliminate the influence of the varying capacity of this quasi-condenser. For this purpose, we placed a strip of metal connected to the insulated terminal of the electrometer inside an aluminium cylinder; the space between the metal and the cylinder was first filled with air, afterwards with paraffin. The aluminium was connected to the case of the electrometer, and inductive disturbances were avoided by surrounding the copper wire connecting the metal to the insulated terminal with a lead sheath in metallic connection with the electrometer sheath (see diagram 2).

In our first experiments with this apparatus we had air, instead of the main mass of paraffin, separating the insulated metal from the surrounding aluminium tube, as shown in the diagram, and we had only small discs of paraffin serving as insulating supports for the ends of the metal, and not played on by the Röntgen rays. When the metal thus supported was charged, whether positively or negatively, the Röntgen rays diselectrified it in about five seconds; not, however, to the metallic zero of the electrometer, but to a "rays-zero" depending on the nature of the insulated metal surrounding it.

With paraffin between the aluminium cylinder and the insulated metal within, as shown in the diagram, the following results were obtained:—

December 30, 1896. 5.30 p.m.—Interior metal charged negatively. Total charge, 356.

Röntgen lamp in action and no screen, 39 scale divisions discharged in 5 mins.
R. L. not acting . . . . . 25 " " 5 "
R. L. again acting and no screen . 17 " " 5 "

5.45.—Interior metal charged positively. Total charge, 244.

R. L. in action and lead screen . 1 scale division discharged in 3 mins.
R. L. in action and no screen . 6 " " 3 "
R. L. not acting . . . . . 0 " " 3 "

December 31, 1896. 10.54 a.m.—Interior metal charged positively. Total charge, 163.

R. L. not acting	2 scale divisions discharged in		3 mins.
R. L. acting and no screen . .	1	,,	,,	3 ,,
11.0.—R. L. stopped . . .	1·5	,,	,,	2 ,,
R. L. again acting, no screen . .	3	,,	,,	2 ,,
R. L. stopped	2·5	,,	,,	3 ,,

11.12.—Interior metal charged negatively. Total charge, 342.

R. L. not acting	10 scale divisions discharged in		3 mins.
R. L. acting, no screen	21	,,	,,	3 ,,
11.18.—R. L. stopped . . .	11·5	,,	,,	3 ,,
R. L. acting, no screen	16·5	,,	,,	3 ,,

These results are quite in accordance with those found in similar experiments by Röntgen; and they show that if paraffin is made conductive, it is only to so small an extent that it is scarcely perceptible by the method we have used.

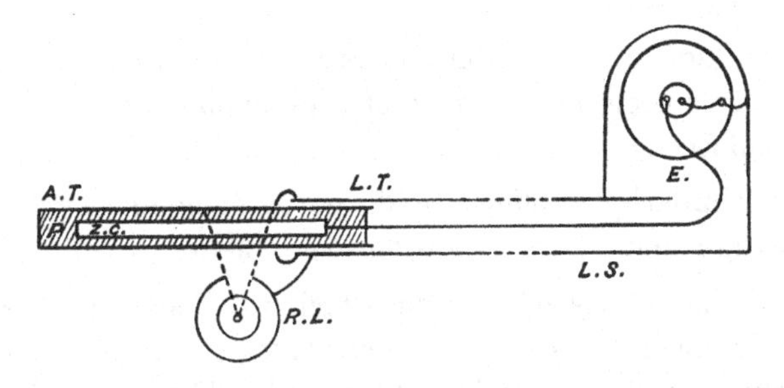

Diagram 2. A.T., Aluminium Tube; L.T., Lead Tube; R.L., Röntgen lamp; L.S., Lead sheaths; E., Electrometer; P., Paraffin; Z.C., Zinc cylinder.

9. To make a similar series of experiments with glass, we used a piece of glass tubing 9·5 mm. in diameter, length 70 cms., and 1 cm. external diameter. The inside of this tube was coated with a deposit of silver, which was placed in metallic connection with the insulated terminal of the electrometer. The outside of the glass was covered with wet blotting-paper connected to sheaths.

With this arrangement we obtained the following results:—

February 8, 1897.—Insulated terminal of electrometer charged to −333 scale divisions from the metallic zero.

4.23.—Röntgen lamp, acting	0·5 sc. div. lost in		3 mins.
,, ,, not acting	1·0	,,	,,	5 ,,

Charge to +164 scale divisions from the metallic zero.

4.36.—Röntgen lamp, not acting	13	sc. div. lost in	7 mins.
" " acting	8·5	" "	5 "
" " not acting	6·0	" "	6 "
" " acting	3·5	" "	5 "
" " not acting	3·5	" "	5 "

[Sensibility of electrometer, 140 scale divisions per volt.]

We next removed a part of the wet blotting-paper from the outside of the glass, and, after having charged the insulated interior metal deposited on the inside of the glass, we heated the exposed part with a spirit flame, in this way making the glass a conductor. Thus with a charge of +280 scale divisions from the metallic zero, the loss in 30 seconds, during which time the glass was heated in the spirit flame, was 90 scale divisions; in the next minute, with no further heating, the loss was 20 scale divisions. Reapplication of heat gave complete discharge in 2½ minutes. Thus we see that our method is amply sensitive to the conductance produced in glass by heating.

We conclude that the Röntgen rays do not produce any conductance perceptible in the mode of experimenting which we have hitherto followed.

10. A similarity in effects produced by flame and by Röntgen rays is brought out by the following experiments.

Two similar sticks of paraffin, which we shall call *A* and *B* respectively, each of about 4 sq. cm. cross section, were coated throughout half their lengths with tinfoil. These tinfoils ought to be each metallically connected to sheaths.

To obtain a sufficiently delicate test for their electric state, a metal disc of 3 cms. diameter was fixed horizontally to the insulated terminal of the electrometer.

The two pieces of paraffin were first diselectrified by being held separately in the flame of a spirit-lamp. Their non-tinfoiled ends were then pressed together, and their electric state again tested after separation. It was found that they were still free from electric charge. After this *B* was charged by being held over the pointed electrode of an inductive electric machine. The quantity of electricity given to it in this way was roughly measured by noting the electrometer reading when the paraffin was held at a distance of 4 cms. above the metal disc connected to the insulated terminal of the electrometer.

The free ends of *A* and *B* were again held together, and, after separation, both pieces were tested separately. The charged one, *B*, had suffered no appreciable loss, and the other, *A*, induced an electrometer reading of a few scale divisions in the same direction, when held as near as possible to the metal disc without touching it. This showed that an exceedingly minute quantity of electricity had passed from *B* to *A* when they were in contact.

A was then diselectrified by being held alone in the flame. The ends of *A* and *B* were again put together, and in this position were passed through the flame. They were tested with their ends still pressed together, and it was found that when held as near as possible to the metallic disc without touching it, no reading was produced on the electrometer. After this they were separated and tested separately; and it was found that *B*, when held over the disc, gave a large reading in the same direction as before it had been passed through the flame, and *A* (which was previously non-electrified) gave a reading of about the same amount in the opposite direction.

The same results were obtained when Röntgen rays were substituted for the flame.

The explanation clearly is this: the flame or the Röntgen rays put the outer paraffin surfaces of *A* and *B* temporarily in conductive communication with the tinfoils, but left the end of *B*, pressed as it was against the end of *A*, with its charge undisturbed. This charge induced an equal quantity of the opposite electricity on the outer surfaces of the paraffin of *A* and *B* between the tinfoils; half on *A*, half on *B*.

When the application of flame or rays was stopped, this electrification of the outer paraffin surfaces became fixed. *B*, presented to the electrometer, showed the effect of the charge initially given to its end, and an induced opposite charge of half its amount on the sides between the end and the tinfoil. *A* showed on the electrometer only the effect of its half of the whole opposite charge induced on the sides by the charge on *B*'s end.

We have here another proof that paraffin is not rendered largely conductive by the Röntgen rays. Had it been made so, then the charge given to the end would have leaked through the body of the paraffin to the outside, and have been carried away either by the tinfoil or by the conductive air surrounding the non-tinfoiled parts.

To show that the induced charges were fixed on the sides, the two sticks, A and B, were next coated with tinfoil throughout their whole length, only one end of each being uncovered. The uncoated end of B was then charged and pressed against that of A, and the two were held either in the flame of a spirit-lamp or in the Röntgen rays. When taken out of the flame or the Röntgen rays, and then separated and tested separately, it was found that B had retained its charge practically undiminished, and that A had acquired a very slight charge of the opposite kind.

11. Instead of placing the two ends of the paraffin in immediate contact, four pieces of metal of $\frac{1}{10}$ of a mm. thickness were placed one at each corner of one of the ends, so that when the sticks of paraffin were placed end to end there was now an air space of $\frac{1}{10}$ of a mm. between the paraffin ends. When B was charged and A not charged, and the two put end to end, and then exposed to flame or to Röntgen rays, it was found that B's end still retained its charge, and A's end acquired a very slight opposite charge.

With an air space of $\frac{1}{5}$ of a mm. the same results were obtained.

With the air space increased to 1 mm. the charge on B was less after the two had been passed through the flame or the rays.

12. Similar experiments were made with rods of glass and of ebonite, with similar results.

III. *On the Influence of Röntgen Rays in respect to Electric Conduction through Air, Paraffin, and Glass.*

[From *Edinb. Roy. Soc. Proc.* Vol. XXI. March 1, 1897, pp. 403—406; *Nature,* Vol. LV. March 25, 1897, pp. 498, 499; *Phil. Mag.* Vol. XLV. March 1898, pp. 277, 278.]

13. We have in §§ 5 to 10 described experiments respecting electric conduction through air, paraffin, or glass, when Röntgen rays fall on metal surrounded by air, paraffin, or glass, and positively or negatively electrified to potentials of two or three volts. We found that although air is rendered conductive, paraffin and glass are not rendered sensibly conductive when the differences of

potential concerned are not more than two or three volts per centimetre of air, or per centimetre of paraffin, or per half-millimetre of glass.

We have now to describe an extension of the investigation to much higher voltages, in which we use an arrangement of two (quasi) Leyden jars, *A* and *B*, with their inside coatings connected together. The outside coating of *A* was connected to sheaths, the outside of *B* to the insulated terminal of the electrometer. In all the experiments to be described, *B* remained the same.

It consisted of a cylindrical lead can, 25 cms. long, 4 cms. diameter. A metal bar about 1 cm. diameter, 25 cms. long, was supported centrally on paraffin filling the whole space between it and the containing lead. This metal bar was connected by a wire to the internal coating of *A*. To protect this wire from inductive effects, it was surrounded by a tube of lead connected to sheaths.

The Leyden *A*, which was placed opposite the Röntgen lamp, was different according as we were experimenting on the discharge through air through paraffin, or through glass.

To get a definite difference of potential, the two pairs of quadrants of the electrometer were first placed in metallic connection. Then one terminal of a battery or of an electrostatic inductive machine was connected to the internal coatings of the jars, and the other terminal to sheaths. The difference of potential produced was measured by a multicellular voltmeter in the case of differences under 500 volts, and on a vertical single vane voltmeter for higher differences.

When the desired difference of potential had been established, the metallic connection of the battery or electric machine with the internal coatings of *A* and *B* was broken, and this charged body left to itself. To find the loss due to imperfect insulation, the pair of quadrants in metallic connection with the outside coating of *B* was insulated in the ordinary way, and the deviation of the electrometer reading from the metallic zero per half-minute was observed. To find the loss when the rays were acting, the two pairs of quadrants were again placed in metallic connection, the Röntgen lamp set a-going, then the pair of quadrants connected to the outside coating of *B* was insulated from the other pair, and the deviation from metallic zero again observed per half-minute.

14. In the experiments with air, the Leyden A consisted of an aluminium cylinder, 16 cms. long, 3 cms. in diameter. This cylinder projected beyond the lead tube, and was connected to sheaths. The insulated metal inside it, which was a flat strip of aluminium, about 10 cms. long and $1\frac{1}{2}$ cms. wide, cut from the same sheet as the surrounding aluminium tube, was supported at one end by a small piece of paraffin so placed as to be out of reach of the action of the Röntgen lamp. The rays from the lamp were allowed to pass from a lead cylinder surrounding it by a small hole about ·3 of a square cm. in area. They fell on the aluminium sheath transparent to them, and rendered the air between it and the insulated aluminium within conductive.

We tried various differences of potential, ranging from a few volts to 2200 volts. In one series of experiments we charged the insulated metal to −97·5 volts, and then disconnected the battery electrodes. The lamp was then set a-going, and the electrometer deviation taken each half-minute for a minute and a half with one pair of quadrants insulated. The rays were then stopped, the quadrants metallically connected, and metallic zero again found. Then the reading during another period of one and a half minutes, with the rays acting, was observed, and so on until no deviation from the metallic zero of the electrometer was found with one pair of quadrants insulated, and the rays falling on the aluminium outside coating of the Leyden A. The sensibly complete discharge thus observed took place in about a quarter of an hour. We found that the rate of deviation from the metallic zero was the same as the difference of potential fell from −97·5 volts to about −4 volts. With differences of potential of −930, −1750, and −2000 volts the rate of deviation was not appreciably greater than with ±20 volts.

This confirms and extends, through a very wide range of voltage, the interesting and important discovery announced by J. J. Thomson and M'Clelland, in their paper in the Cambridge Philosophical Society *Proceedings* of March 1896, to the effect that the conduction of electricity through air under influence of the Röntgen rays is almost independent of the electric pressure when it exceeds a few volts per centimetre.

15. In the experiments on paraffin, the outside coating of the Leyden *A* consisted of an aluminium cylinder 27 cms. long, 4 cms. diameter, connected to sheaths. A metal bar about 1·73 cms. in diameter, and 30 cms. long, supported centrally on paraffin filling the whole space between it and the aluminium sheath, constituted the inside coating. With this arrangement we made experiments with differences of potential of ± 94, ± 119, ± 238, − 2000, + 2500, and − 2400 volts. At none of these potentials did we find any perceptible increase of conductance produced by the Röntgen rays above the natural conductance of the paraffin when undisturbed by them.

16. In the experiments with glass, the Leyden *A* consisted of a glass tube silvered on the inside. The inside silvering was placed in metallic connection with the inside coating of *B*. That part of the glass tube which projected beyond the lead sheath was covered with wet blotting-paper connected to the sheaths. We observed the behaviour of glass under the Röntgen rays at differences of potential of + 800, + 1500, + 2000 volts. We found no indication of increased conductance due to the rays at these voltages.

We are forced to conclude that the experiments described by J. J. Thomson and M'Clelland do not prove any conductance to be induced in paraffin or glass by the Röntgen rays. It seems to us probable that the results described in their paper—pages 7 and 8—are to be explained by electrifications induced on surfaces of glass or of paraffin in contact with air rendered temporarily conductive by the Röntgen rays. (See § 7.)

IV. *On the Conductive Effect Produced in Air by Röntgen Rays and by Ultra-Violet Light.*

[From *Edinb. Roy. Soc. Proc.* Vol. XXI. Feb. 1, 1897, pp. 406—417; *Nature*, Vol. LV. Feb. 11, 1897, pp. 343—347.]

17. We propose next to describe results of experiments on the electrical effects of Röntgen rays and of ultra-violet light when shone on metals, or through air between two metals mutually insulated; and electrified to begin with, by previously producing a difference of potentials between platinum electrodes of an electrometer metallically connected with them. In some of our experiments this potential-difference was zero, and the initial ±

electrifications of the opposed surfaces depended solely on difference of volta-electric quality between their opposed surfaces.

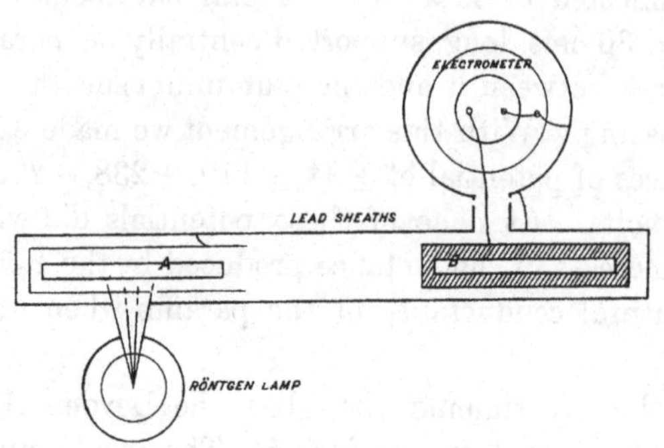

Diagram 3.

18. To investigate the effects of Röntgen rays, a hollow cylinder of unpolished aluminium connected to the electrometer sheaths was used. Along the axis of this a metallic bar was placed, supported by its ends on small blocks of paraffin so situated as not to be shone on by the Röntgen rays. This insulated metal was connected by a copper wire to the insulated terminal of the electrometer. To protect it from inductive effects it was enclosed in a lead tube connected to the other terminal and to sheaths (see diagram 4).

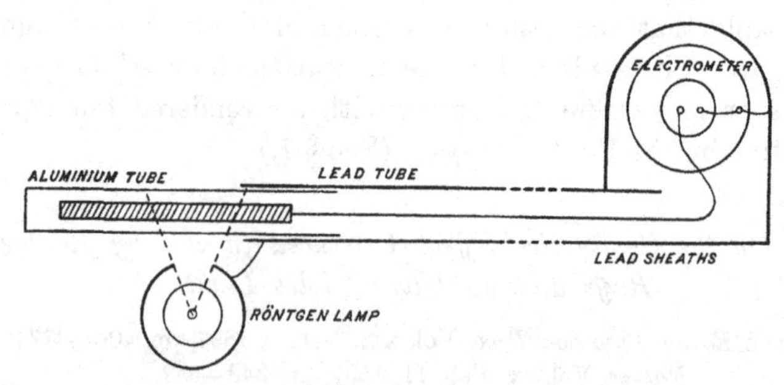

Diagram 4.

The Röntgen lamp was placed in a lead cylinder connected to sheaths. The rays passed into the tube of aluminium through a window in the lead cylinder, which could be screened or unscreened at will, as described in § 1.

The course of the experiment was the same with each insulated metal. The metal was charged first positively, then negatively;

the Röntgen rays were then shone on it through the aluminium cylinder surrounding it, and the electrometer readings taken at fixed intervals, until a steady reading on the electrometer was obtained. The point at which the electrometer readily remained steady with the rays acting we shall call the *rays-zero*.

Finally, the insulated metal was discharged by metallic connection in the electrometer, and re-insulated; the rays were again shone on it until the rays-zero was again reached.

The following figures, taken from the laboratory book, show the effect obtained in this way when the insulated metal was amalgamated zinc.

The zero with the electrometer quadrants in metallic connection we shall afterwards speak of as the *metallic zero*.

December 31, 1896. 5.56 p.m.—Readings with one pair of electrometer quadrants insulated, and with Röntgen lamp acting.

					Time
−72	scale divisions	from	metallic zero	after	5 secs.
−87	"	"	"	"	10 "
−91	"	"	"	"	15 "
−92	"	"	"	"	30 "
−93	"	"	"	"	2 mins.

Afterwards steady.

Thus the difference between the rays-zero and the metallic zero is in this case − 93 scale divisions, or − 0·66 of a volt.

[Sensibility of electrometer 140 sc. divs. per volt.]

This deviation from the metallic zero was not stopped by placing an aluminium screen over the window of the lead cylinder; on the other hand, it was stopped if a lead screen was used. If a positive or a negative charge was given to the insulated metal, and the Röntgen rays were shone through the aluminium cylinder surrounding it, the discharge went on till the rays-zero was reached; only then was the electrometer reading steady.

In the following table, Column II. gives the potential differences of the rays-zero from the metallic zero for twelve different metals insulated within the unpolished aluminium cylinder as described above. Column III. gives the differences for two of the same metals in the interior, but with the surrounding aluminium cylinder altered by polishing its inner surface with emery paper.

I Insulated metal	II	III
Magnesium tape . .	−0·671 of a volt	
Amalgamated zinc . .	−0·66 ,,	
Polished aluminium . .	−0·465 ,,	
Polished zinc . . .	−0·343 ,,	
Unpolished aluminium .	−0·349 ,, . .	+0·35 of a volt
Polished lead . . .	−0·257 ,,	
Polished copper . .	+0·129 ,,	
Polished iron nail . .	+0·182 ,,	
Palladium wire . .	+0·255 ,,	
Gold wire	+0·264 ,, . .	+0·930 of a volt
Carbon	+0·429 ,,	

It is to be noted that the preceding experiments tell us insufficiently as to what would happen had we shone the rays on an insulated metal surrounded by an absolutely identical metallic surface connected to sheaths. Another experiment towards answering this question will be described in a later part of our paper.

The preceding results of the action of Röntgen rays are very similar to, and wholly in accordance with, the results found by Mr Erskine Murray, and described by him in a communication to the Royal Society of London, March 19, 1896.

19. They are analogous to those found for ultra-violet light by Righi (*Rend. R. Acc. dei Lincei*, 1888, 1889); Hallwachs (*Wiedemann's Annalen*, 34, 1888); Elster and Geitel (*Wiedemann's Annalen*, 38, 41, 1888); Branly (*Comptes rendus*, 1888, 1890), and others.

We have also made some experiments with ultra-violet light, in which this similarity is further brought out. The method we have employed is that of Righi.

A cage of brass wire gauze was made and connected to sheaths. Inside it the insulated metal was placed on a block of paraffin, and connected to the insulated terminal of the electrometer by a thin copper wire protected against inductive effects. The light from an arc lamp was then shone through the gauze, so as to fall on the insulated metal perpendicular to its surface (see diagram 5).

The experiments were of the same nature as those with the Röntgen rays, except that wire gauze letting through the ultra-violet light was substituted for the non-perforated aluminium cylinder transparent to the Röntgen rays. The insulated metal disc was 2 cms. distant from the gauze of brass wire. The steady electrometer readings after the two pairs of quadrants were insulated and the ultra-violet light shining (which we shall hereafter refer to as the *ultra-violet-light-zero*) were observed.

The insulated metal was afterwards charged positively, and then negatively. The rate of discharge was observed until the ultra-violet-light-zero was reached.

With polished zinc as the insulated metal the following results were obtained.

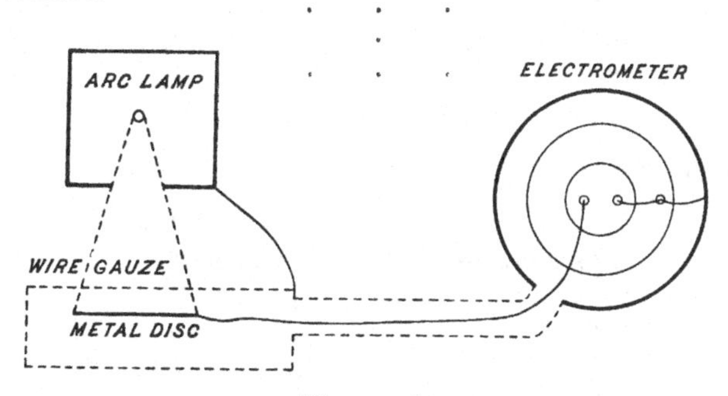

Diagram 5.

The insulation was first tested. When no ultra-violet light was used it was found that the electrometer reading remained the same whether the two pairs of quadrants were in metallic connection or not. With the ultra-violet light shining the reading with the quadrants in metallic connection was the same as before, the readings with the quadrants disconnected were :—

January 14. 3 h. 41 m. p.m.

		Time
− 25 sc. divs.	from metallic zero after	15 secs.
− 45 ”	” ”	30 ”
− 59 ”	” ”	45 ”
− 67 ”	” ”	1 min.
− 80 ”	” ”	1½ ”
− 89 ”	” ”	2 ”
− 99 ”	” ”	3 ”
− 101 ”	” ”	4 ”

Afterwards steady.

[Sensibility of electrometer 146 sc. divs. per volt.]

The difference thus found, between the metallic zero and the ultra-violet-light-zero, is − 101 or − 0·72 of a volt.

3 h. 47 m. Zinc charged positively to 219 scale divisions from the metallic zero.

Reading from metallic zero with ultra-violet light shining:—

	Time
+124	after 15 secs.
+ 64	„ 30 „
+ 23	„ 45 „
− 13	„ 1 min.
− 55	„ 1½ „
− 79	„ 2 „
− 93	„ 2½ „
−100	„ 3½ „
−103	„ 4 „

Afterwards steady.

3 h. 55 m. Zinc charged negatively to 238 scale divisions from metallic zero:—

	Time
−177 sc. divs. from metallic zero after	15 secs.
−149 „ „ „	30 „
−132 „ „ „	45 „
−124 „ „ „	1 min.
−113 „ „ „	2 „
−111 „ „ „	3 „

Afterwards steady.

The following table shows the steady potential differences in the electrometer due to the conductive effect of ultra-violet light in our apparatus between the brass wire gauze and plates of various other metals.

Insulated metal:—

Polished zinc	−0·75 of a volt
Polished aluminium	−0·66 „
German silver	−0·19 „
Gilded brass	+0·04 „
Polished copper	+0·12 „
Oxidised copper	+1·02 „

The copper was oxidised by being held in a Bunsen flame.

In the case of polished zinc, polished aluminium, polished copper, and oxidised copper, both positive and negative charges were discharged at the same rate, if we reckon the charge of the insulated metal from its ultra-violet-light-zero. The rates of reaching the ultra-violet-light-zero were not observed for gilded brass and german silver.

It must again be noticed that our experiments do not tell us what would happen if an insulated metal, shone on by ultra-violet light, were surrounded by a metal of precisely the same quality of surface connected to sheaths.

20. So far we have mentioned only experiments in which the rays, whether Röntgen or ultra-violet, fell perpendicularly on the insulated metal. We have also made some experiments with the rays going parallel to the metal surfaces.

For this purpose a cardboard box 46 cms. long, 19 cms. square (see diagram 6), lined, in the first instance, with tinfoil, connected to sheaths, was used. Inside this box an insulated disc of oxidised copper of 10 cms. diameter was supported in such a way as to allow of its being fixed at different distances from the tinfoil-coated end-wall of the box facing it.

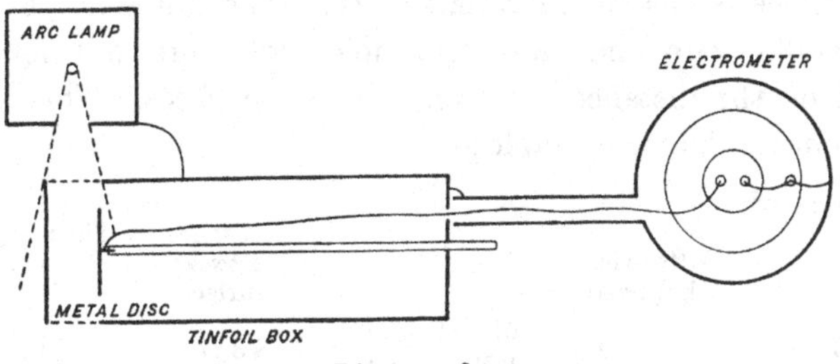

Diagram 6.

The distance between the disc and the tinfoil was at first 4 cms. The arc lamp was distant about 20 cms. from the box. The light from it shone through a slit in the tinfoil covering the side of the box perpendicular to the surface of the oxidised copper. The slit was 4 cms. long, 1 cm. broad. Its length was first placed parallel to the copper surface, so that the light admitted by it shone in the space between the two metals in such a way as not to illuminate either directly. It was found (1) that the ultra-violet-light-zero did not deviate from the metallic zero when the sheet of light passed between the two metals; (2) that a negative charge given to the insulated oxidised copper was not discharged; and (3) that a positive charge was removed very slowly—about 4 scale divisions per minute from a charge of 197 scale divisions from the metallic zero.

When the length of the slit was placed perpendicular to the surface, so that a small portion of both metals, as well as the intervening air, was illuminated, it was found that the reading deviated + 1 scale division per minute from the metallic zero. The oxidised copper was charged positively and negatively. Discharge took place at about 4 scale divisions per minute from a charge of + 202 scale divisions; and 3 scale divisions per minute from a charge of − 246 scale divisions: the charge reckoned from the metallic zero in each case.

The slit was then so arranged as to allow the light to shine on the oxidised copper alone. In this case the deflection went towards an ultra-violet-light-zero at about + 6 scale divisions per minute; and both positive and negative charges were discharged, the negative much more quickly than the positive.

21. The ultra-violet light was now shone between the oxidised copper and the disinsulated tinfoil wall opposite to it, parallel to their surfaces so as to illuminate both. The difference between the metallic zero and the ultra-violet-light-zero was found to depend on the distance between the two surfaces. This will be seen from the following table:—

January 28.

	Ultra-violet-light-zero		Distance between surfaces	Time required to come to steady reading
12.20 p.m.	+150	sc. divs. from metallic zero	4·3 cms.	4 mins.
2.0 „	+134	„ „	3·0 „	9 „
2.10 „	+121	„ „	2·0 „	5 „
2.20 „	+102	„ „	1·0 „	5 „
2.40 „	+ 86	„ „	0·6 „	5 „
2.50 „	+169	„ „	4·0 „	10 „
3.0 „	+161	„ „	5·0 „	5 „
3.20 „	+199	„ „	7·0 „	5 „

[Sensibility of electrometer 140 sc. divs. per volt.]

The fact that in experiments (2) and (6) a longer time was required before a steady reading was obtained, probably depended on the way the light fell on the surface and on variations in intensity of the light.

In this table we see that the steady electrometer reading (which we have called the ultra-violet-light-zero) is largely influenced by the distance between the plates, being greater the greater the distance. This is a very remarkable result. It was first discovered by Righi, and very clearly described in papers of his to which we have referred. It may be contrasted with the non-difference of electrometer readings for different distances between the plates in a volta-zinc-copper and single fluid cell.

22. [*Added February* 6.—We have also made an exactly similar series of experiments with Röntgen rays. The same insulated oxidised copper plate was placed inside the same tinfoil box, and the Röntgen rays shone in between the two metals so as to shine on both. The following results were obtained with the oxidised copper at different distances:—

February 5. 11.30 a.m.

Rays-zero	Distance between surfaces
+23·5 sc. divs. from metallic zero	1·2 cms.
+25·0 ,, ,, ,,	2·2 ,,
+23·0 ,, ,, ,,	3·8 ,,
+23·0 ,, ,, ,,	6·0 ,,

We next removed the oxidised copper plate, and substituted a polished zinc disc. With it we obtained the following results:—

Rays-zero	Distance between surfaces
−82 sc. divs. from metallic zero	1 cm.
−79 ,, ,, ,,	1·5 ,,
−81 ,, ,, ,,	3·0 ,,
−90 ,, ,, ,,	7·0 ,,
−90 ,, ,, ,,	7·5 ,,

The steady reading of the rays-zero was very nearly reached in each case in about 15 secs., but the observation was continued for one or two minutes till we found the reading steady.

Thus we see that, as previously found by Mr Erskine Murray, the rays-zero is independent, or nearly independent, of the distance between the opposed metallic surfaces.]

23. Towards realising the case of an insulated metal surrounded by metal of identical surface-quality connected to sheaths, we covered over the oxidised copper with tinfoil. The tinfoil wall facing it was very rough, and not so well polished. The insulated tinfoil was 4 cms. distant from the end of the box to which its surface was parallel.

When the ultra-violet light fell on the insulated metal alone through a slit, the ultra-violet-light-zero was + 53 scale divisions from the metallic zero. A charge given to it, whether positive or negative, was discharged slowly. After making these experiments, we again observed the difference of zeros, and found that now the ultra-violet light reading was at the end of the first four minutes + 2 scale divisions from the metallic zero; at the end of the next four minutes it was − 8 scale divisions from it.

When the ultra-violet light fell on the disinsulated metal and not on the insulated, the insulated when charged retained its charge.

With the light shining on both through a window 7 cms. broad, 13 cms. high, both positive and negative charges given to the insulated metal were discharged, and the ultra-violet light deviated from the metallic zero by − 152 scale divisions.

This difference was reduced to about − 30 scale divisions when the experiments were repeated after the apparatus had been left to itself for a night.

24. To make similar experiments with the Röntgen rays, it was found necessary to cover the window near the lamp with tinfoil gauze connected to sheaths, and the window on the opposite side was covered with non-perforated tinfoil. In this way direct electrostatic induction was avoided. We had also a thin sheet aluminium window between the tinfoil gauze and the Röntgen lamp.

When the Röntgen rays fell on both insulated and disinsulated metal the rays-zero was − 5 scale divisions from the metallic zero, and both positive and negative charges fell to this zero in a few seconds.

With the rays shining only on the insulated metal the same small difference of zeros was obtained, and both positive and negative charges fell to the rays-zero, though much more slowly than before, in about four minutes.

With the Röntgen rays shining on the insulated tinfoil through the disinsulated tinfoil gauze, the rays-zero was − 9 scale divisions from the metallic zero, and both positive and negative charges were removed in about a minute.

On substituting an aluminium gauze for the tinfoil gauze, and sending rays through it on the insulated tinfoil, the rays-zero was + 25 scale divisions from the metallic zero.

[*Added February* 6.—With a polished zinc disc as the insulated metal, and with the same windows to the tinfoil box, the Röntgen rays were shed in between the insulated zinc and the opposite wall of tinfoil from a slit in a lead screen outside. This slit was 4 cms. long by 1 cm. broad. The distance between the two metals was 7 cms. The rays illuminated only part of the air space between the two, and also a part of the tinfoil covering the two windows.

The following are some of the results obtained:—

[Sensibility of electrometer 140 sc. divs. per volt.]

February 5, 1897. Zinc charged negatively to 285 scale divisions from the metallic zero.

Reading from metallic zero with Röntgen lamp acting:—

	Time
−276 scale divisions . .	after 1 min.
−265 " . .	" 2 "
−255 " . .	" 3 "
−243 " . .	" 4 "
−227 " . .	" 5 "
−214 " . .	" 6 "
−184 " . .	" 8 "

Discharge still continued.

The zinc was then discharged by metallic connection. The readings, with the Röntgen light shining, and the two pairs of electrometer quadrants again disconnected, were:—

	Time
− 4 sc. divs. from metallic zero after	½ min.
−13 " " "	1½ "
−41 " " "	2½ "
−53·5 " " "	3½ "
−61 " " "	4½ "
−67 " " "	5½ "
−70·5 " " "	6½ "
−71·0 " " "	7 "

The difference between the rays-zero and the metallic zero is thus found to be −71 sc. divs., or −0·5 of a volt. Immediately after this experiment, we removed the lead window and allowed the Röntgen light to shine on both metals, still 7 cms. apart. We then found the difference of zeros to be − 89 sc. divs., or − 0·64 of a volt; but instead of seven minutes, scarcely a quarter of a minute was taken to reach the rays-zero after the metallic connection was broken. These results are substantially in accordance with Erskine Murray's §§ 9 of his paper already referred to.]

V. *Experiments on Electric Properties of Uranium.*

[From *Edinb. Roy. Soc. Proc.* Vol. XXI. April 4, 1897, pp. 417—428; *Nature*, Vol. LVI. May 6, 1897, p. 20.]

25. Potential differences of uranium-conductance-zero from metallic zero for different metals in air.

We have used two different methods to measure the potential difference between two mutually insulated metals when the air between them is rendered conductive by the presence of uranium. The more convenient method is to take uranium as one of the mutually insulated metals. To do this we fixed a metallic disc, 3 cms. diameter, to the insulated terminal of a quadrant electrometer. Opposite this metallic disc, and separated from it by air, we placed a disc of uranium, 5·5 cms. diameter, connected to the other terminal of the electrometer. With this arrangement a steady reading, the metallic zero, was obtained when the quadrants of the electrometer were in metallic connection. After contact between the quadrants was broken at the electrometer a deviation from the metallic zero took place gradually to a point, the uranium-conductance-zero we shall call it, depending on the volta difference between the two opposed surfaces of metals, more or less tarnished as they generally are. On the other hand, if the insulated metal had a charge given to it of such an amount as to cause the electrometer reading to deviate from the metallic zero beyond the uranium-conductance-zero, the reading quickly fell to this conductance-zero and there remained steady. When no charge was given to the insulated metal the steady conductance-zero was reached in about half a minute. The following table gives the potential differences found in this way:—

Metals	Potential Difference. Volts
Polished aluminium (1) immediately after being polished	−1·13
Polished aluminium (1), next day	−1·90
Polished aluminium (2)	−1·00
Amalgamated zinc	−0·80
Polished zinc	−0·71
Unpolished zinc	−0·55
Polished lead	−0·54
Tinfoil	−0·49
Unpolished aluminium (1)	−0·41
Polished copper	−0·17
Unpolished copper	+0·07
Silver coin	+0·05
Carbon	+0·20
Oxidised copper (*a*)	+0·42
Oxidised copper (*b*)	+0·90

With a third specimen of oxidised copper a potential difference of +0·35 of a volt was obtained. This specimen was afterwards connected to sheaths; a piece of polished aluminium was placed opposite it and connected to the insulated terminal of the electrometer. The uranium disc, insulated on paraffin, was then placed between them, and the deviation observed was equivalent to a potential difference of −1·53 volts; that is, we obtained an effect equivalent to the sum of the effects we obtained when the metals were separately insulated in air opposite uranium.

26. Instead of placing the uranium directly opposite the insulated metal in air we also observed the conductance-zero by mutually insulating two metals in air, one of which was transparent to the uranium influence.

For this purpose we made a tinfoil box, with tinfoil sufficiently thin to be transparent to the uranium influence. The tinfoil forming the box was connected to sheaths. Inside it another metal was insulated on a glass stem, and placed so as to be parallel to one end of the tinfoil box. This metal was connected to the insulated terminal of the electrometer. The uranium was placed outside the box, about half a centimetre distant from the end to which the insulated metal was parallel. The same conductance-zero was obtained with the uranium insulated, or with it connected to sheaths. The time required to reach the uranium-conductance-zero with this arrangement was usually four or five minutes, and a charge given to the insulated metal large enough to produce a deviation beyond the conductance-zero was discharged till this zero was reached. A charge, causing the electrometer to deviate in the opposite direction, was discharged to the metallic zero and thence on to the uranium-conductance-zero, where it remained steady.

With polished aluminium as the insulated metal, the potential difference obtained was $-0{\cdot}7$ of a volt.

27. Effect of various screens on the rate of reaching the zero. With the second arrangement, described in § 25, it was possible to obtain a relative idea of the transparency to the uranium effect of screens of various materials. For example, when a sheet of lead, about 2 mms. in thickness, was placed between the uranium and the tinfoil, no deviation from the metallic zero was obtained. In other words, lead is not transparent to the uranium influence. Glass 3 mms. thick did not entirely stop the deviation; it reduced the deviation in the first minute, however, to $\frac{1}{6}$ of the amount obtained with only air between the uranium and the outside wall of tinfoil. A copper screen, 0·24 mm. in thickness, reduced the rate to $\frac{1}{3}$; two copper screens, total thickness 0·48 mm., reduced it to $\frac{1}{12}$; three copper screens reduced it to $\frac{1}{40}$. A mica screen did not reduce the rate at all. A zinc screen, 0·235 mm. thick, reduced it to $\frac{1}{2}$. Two zinc screens, total thickness 0·47 mm., reduced it to $\frac{1}{7}$. Paraffin, 3 mms. thick, when placed between the two mutually insulated metals, stopped the deviation to the conductance-zero.

28. Conductance-zero at different distances.

In the experiments described in the preceding section, the distances between the two mutually insulated metals was 2 cms. To observe the conductance-zero at different distances an aluminium box connected to sheaths was substituted in place of the tinfoil one, and oxidised copper insulated on a glass stem inside it. As before, with the tinfoil box, the uranium was placed outside the aluminium box, about 5 mms. from the end, to which the oxidised copper was kept parallel. The distance of the oxidised copper could be varied by moving the glass rod to which it was attached. The results obtained were as follows:—

Distance in cms.	Potential differences in volts
1·5	+0·97
4·0	+0·98
0·5	+0·96
8·0	+1·03
2·0	+0·95

29. Comparison of uranium-conductance-zero with water-arc-zero.

In the first arrangement for measuring the conductance-zero in § 25 we had discs of uranium and aluminium separated by air. We varied this by placing the uranium so close to the insulated aluminium that it could be brought into electric connection with it by a drop of water. The deviation obtained in the electrometer by this means was always in the same direction as the uranium-conductance-zero between the surfaces when dry, and was usually smaller in magnitude. For instance, with aluminium-air-uranium the deviation was + 0·8 of a volt, with aluminium-water-uranium it was + 0·43 of a volt*.

30. Uranium-conductance-zero in different gases and at different atmospheric pressures.

To investigate the behaviour of uranium in different gases and at different atmospheric pressures another piece of uranium 3 cms. long, 1 cm. broad, and $\frac{1}{2}$ cm. thick, was mounted firmly in a glass bulb 6 cms. long, 3 cms. diameter on a platinum electrode fused into one end of the bulb. The uranium in the glass bulb was surrounded throughout two-thirds of its length by a zinc cylinder $1\frac{1}{2}$ cms. in diameter. This zinc cylinder was kept in position by a stiff platinum electrode fused into the other end of the glass (see diagram 7). Two glass tubes were fixed on to the bulb, one at each end; by means of these any desired gas could be introduced, or any desired vacuum could be obtained.

* On the other hand, when the uranium surface was covered with water to the depth of about a millimetre, and an air space left between the wet uranium surface and the opposed insulated metal, so that we had a uranium-water-air-metal arc, the rate of deviation from the metallic zero was reduced so much as to be scarcely appreciable.

The gas used was first stored in a reservoir over water. It was then bubbled through strong sulphuric acid and drawn over caustic potash, calcium chloride, and phosphoric anhydride into the glass bulb. The bulb was first exhausted to an atmospheric pressure of about 6 mms.; then the gas to be used was passed into it. This was repeated about twenty times. Finally it was strongly heated so as to drive off any adhering layers of gas, and then allowed to cool in an atmosphere of the gas at 760 mms. pressure. One of the tubes was then sealed up; the other was closed by a good fitting and well-greased glass stopcock.

The vacuums up to 2 mms. pressure were obtained by means of a double-barrelled air-pump. Higher vacuums were obtained by means of a Töpler pump.

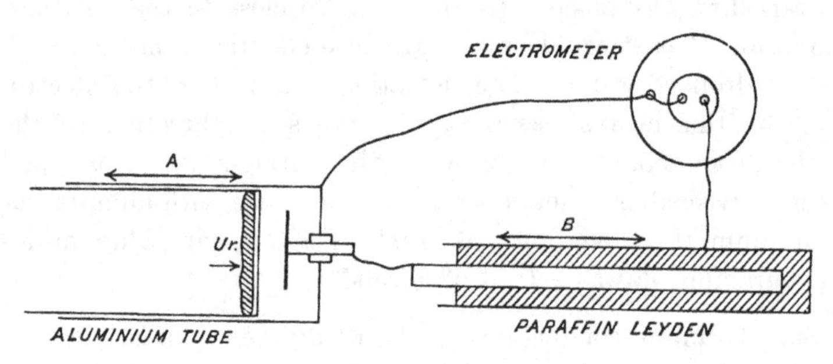

Diagram 7.

To observe the conductance-zero the uranium was connected to the insulated terminal of the electrometer, and the zinc cylinder to sheaths. In the following table, the results obtained in air, hydrogen, and oxygen are given:—

Pressure in mms.	Difference of potential between the uranium-conductance-zero and metallic zero		
	Hydrogen	Oxygen	Air
760	+·17 of a volt (in about a min.)	+·105 of a volt (in about a min.)	+·11 of a volt (in about a min.)
193	+·12 of a volt (in about a min.)		
66	+·05 of a volt (6 min.)	+·11 of a volt (3 min.)	
8	+·04 of a volt (8 min.)		
2	. . . .	+·10 of a volt in 27 min.	
$<\frac{1}{1000}$	+·05 of a volt in 28 min.		

The uranium-conductance-zero between mutually insulated uranium and zinc differs much less from the metallic zero than in our previous experiments. This is probably due to the oxidation of the zinc of the zinc cylinder. The conductance-zero, however, it will be noticed, is approximately the same in all three gases.

31. Leakage in air at ordinary pressure at different voltages.

We used in our first experiments the two Leydens method described in § 13. The Leyden *B*, whose external coating was connected to the insulated terminal of the electrometer, and its internal coating to the internal coating of *A*, was the paraffin Leyden described in § 13. The Leyden *A* was a cylinder of aluminium, with one end closed with aluminium. This formed the external coating. The internal coating was a disc of aluminium insulated in paraffin. The uranium was placed inside a cardboard cylinder, with one end open and the other covered with aluminium so as to touch the aluminium (see diagram 8).

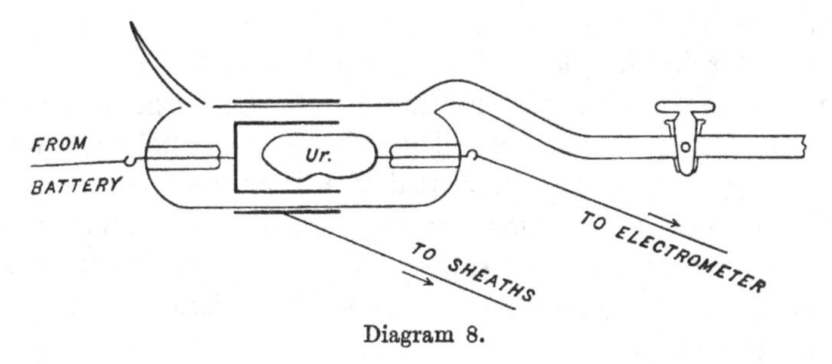

Diagram 8.

This cardboard cylinder could be moved backwards and forwards in the aluminium cylinder, so that the distance between the insulated disc in the latter and the aluminium end of the former could be varied. The uranium influence thus acted through the aluminium end of the cardboard box, and made the air between the end and the insulated aluminium disc conductive. The leakage was in this way made slow enough to be easily observed on the electrometer. The rate of leak was not perceptibly increased when the piece of uranium was heated or when the sunlight fell on it. The aluminium end of the cardboard box and the outside coating of the aluminium cylinder were connected to sheaths. The insulated aluminium disc was connected to the inside coating of *B*. The inside coatings were charged to a known potential, and then left to themselves.

The air space between the insulated aluminium disc and the aluminium end of the cardboard box was 2 centimetres. The voltages used were therefore voltages per two centimetres of air space. With this arrangement the leakage per minute at different voltages was:—

Voltage	Leakage per minute in scale divisions
6	56
10	65·5
44	113
88	128
176	156
750	219
1250	229
2000	260
3000	276

[Sensibility of electrometer 24 sc. divs. per volt of subsidence of difference of potential between coatings of *A*.]

We also measured the leakage at different voltages with the zinc cylinder in the glass bulb described in § 30, charged to a definite potential, and the uranium connected to the insulated terminal of the electrometer. The voltages up to 90 volts were obtained by connecting one terminal of a battery to the zinc, and keeping it connected during the experiments, while the other terminal was connected to sheaths. For higher voltages the zinc was charged to the given potential, and then disconnected from the charging body.

Voltage per 2 mms.	Leakage per minute in scale divisions
2	92
4	100
22	120
92	129
132	138
200	130
300	137
415	136

[Sensibility of electrometer 140 sc. divs. per min.]

The appended curves (diagram 9) were drawn by taking the leakage per minute as ordinate, the voltage as abscissa. Curve *a* gives the results obtained with the two Leydens arrangement reduced to voltages per 2 mms. between the internal and external coatings of *A*. Curve *b* gives the results obtained with the smaller piece of uranium in the glass bulb.

32. Leakage in other gases at ordinary pressure.

We have also observed the rate of leaks in hydrogen, oxygen, and carbonic acid at ordinary pressure at different voltages. The glass bulb referred to in §§ 30 and 31 was used for this purpose. The voltages were obtained by connecting the zinc to one terminal of a battery, and the other terminal to sheaths. The uranium was connected to the insulated terminal of the electrometer. While the connection between the battery and the zinc was being made the uranium was put in metallic connection

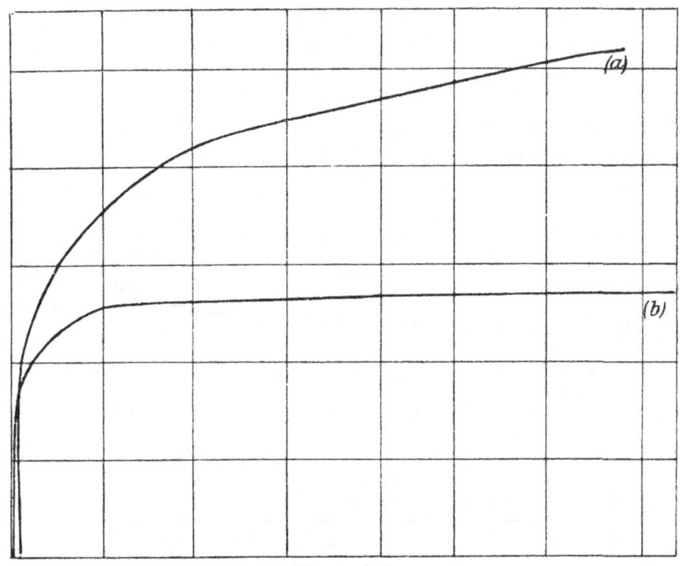

Diagram 9.

with the case of the electrometer; afterwards it was disinsulated, and the deviation in the electrometer observed per minute for a number of minutes. The following results were obtained for these gases:—

HYDROGEN

Voltage per 2 mms.	Leakage per minute in scale divisions
2	32
4	37
22	39
34	38
100	39
135	38

OXYGEN

Voltage per 2 mms.	Leakage per minute in scale divisions
4	125
96	157

CARBONIC ACID

Voltage per 2 mms.	Leakage per minute in scale divisions
4	94
95	167
238	183
255	180
2900	Discharge by sparking

[Sensibility of electrometer 140 sc. divs. per volt.]

The results given for these three gases are comparable to the second series of results given in § 7 for air.

We see that the rate of leak is greater in oxygen than in air; no comparative figures need be given, as these would vary according to the voltage chosen.

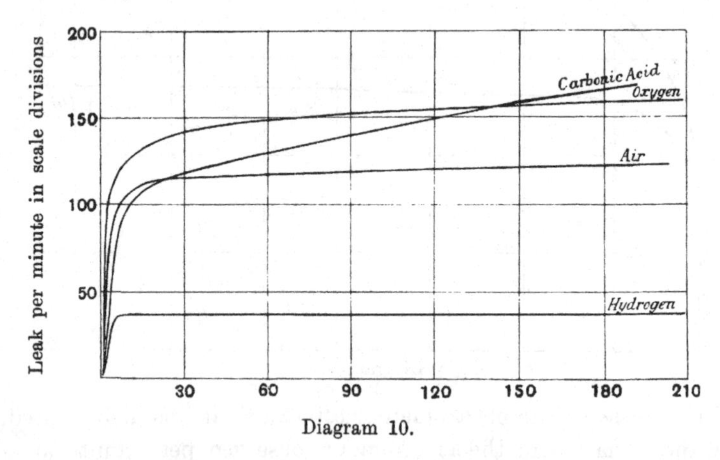

Diagram 10.

The leakage in hydrogen is less than that in air; in carbonic acid it is less for 4 volts, but greater for 90 volts, than it is in air; for the latter voltage the leakage in carbonic acid is greater even than the corresponding leakage for oxygen. The appended curves show the peculiarities of the leakage in the different gases (diagram 10).

33. Leakage in different gases at different atmospheric pressures.

The method of filling the glass bulb with any given gas, and the way in which the different vacuums were obtained, has been described in § 30.

The following tables give the results obtained with the gases we have up till now experimented on.

Air.

α Atmospheric Pressure in mms.	β Leakage per min. for 4 Volts	γ Leakage for 96 Volts	$\frac{\beta}{\alpha}$	$\frac{\gamma}{\alpha}$
760	100	131	·132	·172
240	44	46	·183	·192
190	40	39	·210	·205
121	24	26	·197	·214
64	12	13·5	·187	·212
58	11	10·0	·189	·172
23	4·4	3·75	·191	·163
3·6	1·2	1·2	·339	·339

It will be seen from the last two columns of the table that the rate of leak at 4 volts and at 96 volts is nearly proportional to the atmospheric pressure. The results obtained at 3·6 mms. are not very reliable. With lower pressures no appreciable leakage at these two voltages was observed.

Hydrogen.

α Atmospheric Pressure	β Leakage per Minute at 4 Volts per 2 mms.	$\frac{\beta}{\alpha}$	$\frac{\beta}{\sqrt{\alpha}}$
760	37	·0487	1·34
197	11	·056	·77
66	4	·061	·49
8	1·5	·187	·53

With lower pressures no leakage was observed. The leakage is at higher pressures somewhat approximately proportional to the pressure, at lower ones to the square root of the pressure.

Oxygen.

α Atmospheric Pressure in mms.	β Leakage per Minute for 4 Volts per 2 mms.	$\frac{\beta}{\alpha}$	$\frac{\beta}{\sqrt{\alpha}}$
760	125	·16	4·5
205	48·5	·236	3·38
64	15·0	·234	1·87
2	2·0	1·0	1·414

Carbonic Acid.

Atmospheric Pressure in mms.	Leakage per Minute for 4 Volts per 2 mms.	Leakage per Minute for 100 Volts per 2 mms.
760	94	167
62	18	21
2	...	Not observable

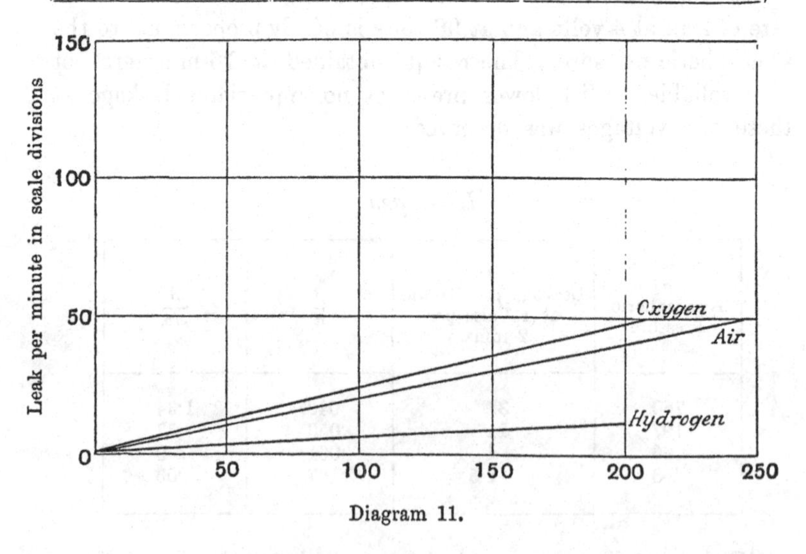

Diagram 11.

The curves for air, oxygen, and hydrogen given in diagram 11 were obtained by taking the atmospheric pressure in mms. as abscissa, and the leakage per minute for 4 volts as ordinate.

34. Voltage necessary to produce spark at different atmospheric pressures.

We found that, at ordinary atmospheric pressure, sparking took place in air at 4800 volts. At 232 mms. pressure the potential necessary to produce a spark fell to between 1500 and 2000 volts. At 127 mms. it had fallen to between 1100 and 1300 volts. At 54 mms. it was 700 volts; at 7 mms. 420 volts; at 2 mms. about 400 volts. At about $\frac{1}{1000}$ mm. the voltage had risen again to 2000 volts.

ELECTRIC EQUILIBRIUM BETWEEN URANIUM AND AN INSULATED METAL IN ITS NEIGHBOURHOOD.

[From *Edinb. Roy. Soc. Proc.* Vol. XXII. [read March 1, 1897], pp. 131—133; accidentally omitted from *Edinb. Roy. Soc. Proc.* Vol. XXI. p. 417, but published in *Nature*, Vol. LV. March 11, 1897, pp. 447, 448; reprinted in *Math. and Phys. Papers*, Vol. V. pp. 36, 37.]

THE wonderful fact that uranium held in the neighbourhood of an electrified body diselectrifies it, was first discovered by H. Becquerel. Through the kindness of M. Moissan we have had a disc of this metal, about five centimetres in diameter and a half-centimetre in thickness placed at our disposal.

We made a few preliminary observations on its diselectrifying property. We observed first the rate of discharge when a body was charged to different potentials. We found that the quantity lost per half-minute was very far from increasing in simple proportion to the voltage, from 5 volts up to 2100 volts; the electrified body being at a distance of about 2 cms. from the uranium discs. [*Added March* 9.—We have to-day seen Prof. Becquerel's paper in *Comptes rendus* for March 1. It gives us great pleasure to find that the results we have obtained on discharge by uranium at different voltages have been obtained in another way by the discoverer of the effect. A very interesting account will be found in the paper above cited, which was read to the French Academy of Sciences on the same evening, curiously enough, as ours was read before the Royal Society of Edinburgh.]

These first experiments were made with no screen placed between the uranium and the charged body. We afterwards found that there was also a discharging effect, though much slower, when the uranium was wrapped in tinfoil. The effect was still observable when an aluminium screen was placed between the uranium, wrapped in tinfoil, and the charged body.

To make experiments on the electric equilibrium between uranium and a metal in its neighbourhood, we connected an insulated horizontal metal disc to the insulated pair of quadrants of an electrometer. We placed the uranium opposite this disc, and connected it and the other pair of quadrants of the electrometer to sheaths. The surface of the uranium was parallel to that of the insulated metal disc, and at a distance of about 1 cm. from it. It was so arranged as to allow of its easy removal.

With a polished aluminium disc as the insulated metal, and with a similar piece of aluminium placed opposite it, in place of the uranium, no deviation from the metallic zero was found when the pairs of quadrants were insulated from one another. With the uranium opposite the insulated polished aluminium, a deviation of -84 sc. divs. corresponding to ·59 volt from the metallic zero was found in about half a minute. After that the electrometer reading remained steady at this point, which we may call the uranium rays-zero for the two metals separated by air which was traversed by uranium rays. If, instead of having the uranium opposite to the aluminium, with only air between them, the uranium was wrapped in a piece taken from the same aluminium sheet, and then placed opposite to the insulated polished aluminium disc, no deviation was produced. Thus in this case the rays-zero agreed with the metallic zero.

With polished copper as the insulated metal, and the uranium separated only by air from this copper, there was a deviation of about $+10$ sc. divs. With the uranium wrapped in thin sheet aluminium and placed in position opposite the insulated copper disc, a deviation from the metallic zero of $+43$ sc. divs. was produced in two minutes, and at the end of that time a steady state had not been reached.

With oxidised copper as the insulated metal, opposed to the

uranium with only air between them, a deviation from the metallic zero of about + 25 sc. divs. was produced.

When the uranium, instead of being placed at a distance of one centimetre from the insulated metal disc, was placed at a distance of two or three millimetres, the deviation from the metallic zero was the same.

These experiments show that two polished metallic surfaces connected to the sheath and the insulated electrode of an electrometer, when the air between them is influenced by the uranium rays, give a deflection from the metallic zero, the same in direction, and of about the same amount, as when the two metals are connected by a drop of water.

248. Continuation of Experiments on Electric Properties of Uranium. By Lord Kelvin, Dr J. Carruthers Beattie, and Dr M. Smulochowski de Smolan.

[From *Nature*, Vol. LVI. May 6, 1897, p. 20. Read before the Royal Society of Edinburgh, April 4, 1897.]

In a paper read before the Society on March 1, we had the honour to communicate some preliminary results on the electric properties of uranium. We propose now to give other results on the same subject, bearing on the conductance induced in air by uranium.

To measure the leakage in air at ordinary pressure at different voltages, we used in our first experiments the two-Leydens method described in a former paper. We found that the leakage was not proportional to the electro-motive force. It was not perceptibly increased when the uranium was heated, or when the sunlight fell on it.

We also observed the leakage in hydrogen, oxygen, and carbonic acid. The experimental arrangements necessary for this are described in a paper published by the Royal Society of Edinburgh. We found that the rate of leakage is greater in oxygen than in air. The ratio of the rates depends on the voltage chosen. The leakage in hydrogen is less than in air. In carbonic acid it is less for four volts per two mms. but greater for ninety volts per two mms. than it is in air; for the latter voltage the leakage in carbonic acid is greater even than the corresponding leakage for oxygen at ordinary pressure. We also made experiments with air, hydrogen, oxygen, and carbonic acid at different atmospheric pressures. We found that the leakage in air at pressures ranging from 760 mms. to 23 mms. was very nearly proportional to the atmospheric pressure. The rate of leakage for lower pressures was so slow as to make the results not very reliable. At pressures under 2 mms. no appreciable leakage with 4 or with 90 volts per 2 mms. was observed. With hydrogen, oxygen, and carbonic acid the rate of leakage at higher pressures was somewhat approximately proportional to the pressure, at lower ones to the square root of the pressure.

We found that at ordinary atmospheric pressure, sparking took place in air at 4800 volts, between a rough fragment of uranium and a metal tube around it, connected to the two electrodes of a vacuum-tube within which they were fixed. At 232 mms. pressure, the potential necessary to produce a spark fell to between 1500 and 2000 volts. At 127 mms. it had fallen to between 1100 and 1300 volts. At 54 mms. it was 700 volts; at 7 mms. 420 volts; at 2 mms. about 400 volts. At 1/1000 mm. the voltage necessary to produce sparking rose again to 2000 volts.

To measure the potential difference between two mutually insulated metals when the air between them is rendered conductive by the presence of uranium, we used two methods, which are described more particularly in the paper above referred to. The steady reading obtained when the quadrants of an electrometer were in metallic connection we shall call the metallic-zero. The deviation from the metallic-zero, when the quadrants were insulated to a steady point—the uranium-conductance-zero, as we shall call it—depended on the volta difference between the two opposed surfaces of metals, more or less tarnished as they generally were. This deviation took place gradually in about half a minute with one arrangement of apparatus, and in about four minutes with a second arrangement. On the other hand, if the insulated metal had a charge given to it of such an amount as to cause the electrometer reading to deviate from the metallic zero beyond the uranium-conductance-zero, the reading quickly fell to this conductance-zero, and there remained steady.

The following table gives the potential differences between the electrometer wires, when one of them is connected with uranium, and the other with a plate of one or other of the named metals opposed to it:—

Metal	Volt
Polished aluminium (1) immediately after being polished	−1·13
Polished aluminium (1) next day	−0·90
Polished aluminium (2)	−1·00
Amalgamated zinc	−0·80
Polished zinc	−0·71
Unpolished zinc	−0·55
Polished lead	−0·54
Tinfoil	−0·49
Unpolished aluminium (1)	−0·41
Polished copper	−0·17
Silver coin	+0·05
Unpolished copper	+0·07
Carbon	+0·20
Oxidised copper (*a*)	+0·42
Oxidised copper (*b*)	+0·90

It will be observed that the difference of potential observed depends very much on the state of polish of the metal concerned. With a third specimen of oxidised copper a potential difference of +0·35 of a volt was obtained. This specimen was afterwards connected to sheaths; a piece of polished aluminium was placed opposite it, and connected to the insulated terminal of the electrometer. The uranium disc, insulated on paraffin, was then placed between them, and the deviation observed was equivalent to a potential difference of −1·53 volts; that is, we obtained an effect equivalent to the sum of the effects we had when the metals were separately insulated in air opposite to uranium.

We observed also the effect of various screens on the rate of reaching the conductance-zero. For example, when a sheet of lead about 2 mms. in thickness was used as screen, no deviation from the metallic-zero was obtained. In other words, lead 2 mms. thick is not transparent to the uranium influence. Glass 3 mms. thick did not entirely stop the deviation; it reduced the deviation in the first minute, however, to $\frac{1}{6}$ of the amount obtained with no screen. A copper screen, 0·24 mm. in thickness, reduced the rate to $\frac{1}{3}$; two copper screens, total thickness 0·48 mm., reduced it to $\frac{1}{12}$; three copper screens, 0·72 mm., reduced it to $\frac{1}{40}$. A mica screen did not reduce the rate at all. A zinc screen, 0·235 mm. thick, reduced it to $\frac{1}{2}$. Two zinc screens, total thickness 0·47 mm., reduced it to $\frac{1}{7}$. Paraffin, 3 mms. thick, when placed between the two mutually insulated metals, stopped the deviation from the metallic to the conductance-zero.

The final difference of potential observed between the electrometer wires connected to two mutually insulated metals, when the air between them was made conductive by uranium, was found to be independent of the distance between the metals through distances ranging from less than $\frac{1}{2}$ cm. to 8 cms.

The difference of potential observed when two mutually insulated metals were brought into electric connection with one another by a drop of water, was in the same direction as the uranium conductance-zero between the two surfaces when dry, and was smaller in magnitude. On the other hand, when the uranium surface was covered with water to the depth of about a millimetre, and an air space left above the water, between the submerged uranium surface and the opposed insulated metal, so that we had uranium-water-air-metal, the rate of deviation from the metallic-zero was reduced so much as to be scarcely observable.

We found that the uranium-conductance-zero between zinc and uranium was the same in air, hydrogen, and oxygen. And that the final steady reading did not depend on the atmospheric pressure, though the rate at which this steady reading was reached did largely depend on the atmospheric pressure.

249. On Electrical Properties of Fumes Proceeding from Flames and Burning Charcoal. By Lord Kelvin and Magnus Maclean.

[From *Edinb. Roy. Soc.* Vol. XXI. [read April 5, 1897], pp. 313—322; *Nature*, Vol. LV. April 22, 1897, pp. 592—595.]

1. Many experimenters have investigated the electrical properties of flames and incandescent solids. The methods usually employed have been (1) to examine the electric conductivity of different parts of the flame*; (2) to measure the difference of potential between platinum wires in different positions in the same flame†; (3) to find the leakage of a charged conductor when placed near, or in view of, a flame or an incandescent solid‡; (4) to observe the leakage of a conductor, raised to a red or white heat, by an electric current, and electrically charged§; and (5) to observe the production of electrification or diselectrification by a glowing wire, through which a current is passing, in neighbouring insulated conductors separated from it by different gases||.

2. This short communication divides itself into three separate inquiries: (1) to test by one of our electric filters¶ the electric

* Account of experiments in Wiedemann's *Lehre von der Elektricität*, Vol. IV. B. Carl's *Rep.* XVII. pp. 269—294, 1881. J. J. Thomson, *Phil. Mag.* pp. 358, 441, 1890.

† Hankel, *Phil. Mag.* p. 542, December 1851; *Phil. Mag.* p. 9, January 1860. Elster and Geitel, *Wied. Ann.* Vol. XVI. 1882; also *Phil. Mag.* September 1882. Maclean and Goto, *Phil. Mag.* August 1890.

‡ Guthrie, *Phil. Mag.* p. 308, April 1873. Giese, *Wied. Ann.* Vol. XVII. 1882; Worthington, "On the Discharge of Electrification by Flames," *Brit. Assoc. Report*, 1889, pp. 225—227; Schuster, Lecture Royal Institution, February 22, 1895.

§ Guthrie, *Phil. Mag.* p. 237, October 1873.

|| Elster and Geitel, *Wied. Ann.* Vol. XXXVII. p. 315, 1889; Vol. XXXVIII. p. 27, 1889.

¶ Kelvin, Maclean, Galt, "Electrification and Diselectrification of Air," *Proceedings of the Royal Society, London*, Vol. LVII. February and March 1895; also *B. A. Report*, 1895; *supra*, p. 49.

quality of the fumes from different flames and burnings (this method has not, we believe, been tried before); (2) to observe the difference of potential maintained between two wires of the same metal connected with a copper plate and a zinc plate when the fumes from different flames and burnings at different distances from the plates passed between them and round them; and (3) to observe the leakage between two parallel metal plates with any difference of electric potential when the fumes from flames and burnings were allowed to pass between them.

3. To test the electrification of fumes from different flames and burnings, the arrangement shown diagrammatically in fig. 1 was used. The flame is kept burning at the mouth of a large vertical iron funnel A, closed at its upper end; and the heated air,

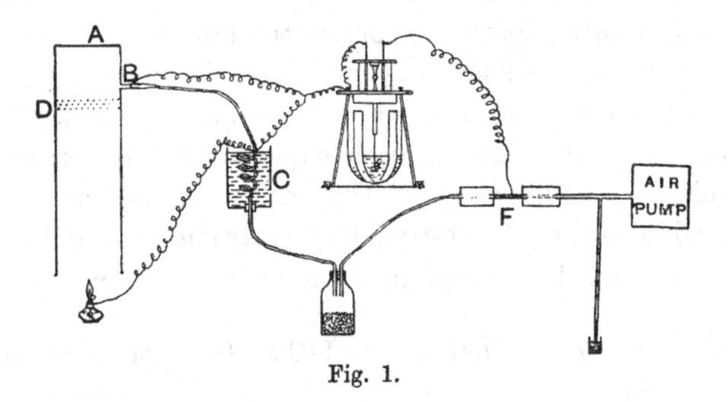

Fig. 1.

along with the products of combustion, is drawn off by an air-pump through a small aperture, B, near the upper end. Before reaching the pump the air has to pass through three circular pieces of brass wire gauze, D, 1 centimetre apart, which are fixed across the funnel about 5 centimetres below the exit tube B; and through a worm of block-tin pipe, 90 centimetres long, which is kept surrounded by cold water in a vessel C. The electrification was tested by a quadrant electrometer (sensitiveness of the electrometer 111 scale divisions per volt), and an electric filter F. The filter F was of block-tin tube, 5 centimetres long and 1 centimetre bore, and full of fine brass filings kept in position by a plug of cotton wool and a piece of brass wire gauze at each end. Between the filter and the air-pump is a T-shaped piece of glass tubing with lower end of the vertical tube dipping into a basin of mercury. This served as a pressure gauge to indicate the difference of air pressures on the two sides of the filter when the air-pump was worked. The flame, the iron funnel, the worm, and the case of the electrometer are all metallically connected.

4. The following flames and burnings were tried:—

(1) Candle.
(2) Paraffin lamp.
(3) Spirit flame.
(4) Portable electrometer matches.
(5) Coal-gas (Bunsen flame).
(6) Hydrogen flame.
(7) Glowing charcoal.
(8) Glowing coal.

5. The method of experimenting was to place the burning substance in position at the bottom of the funnel, to insulate the quadrant of the electrometer in connection with the electric filter, and to start working the air-pump at the rate of one stroke per three seconds. The time of each experiment was ten minutes (200 strokes of the air-pump). The results obtained are given in the following table. In testing the electrometer matches, four matches were stuck in holes in a metallic plate, and the plate

Sensitiveness of the Electrometer 110 *scale divisions per volt.*

	Number of Experiments	Mean Deflection in Scale Divisions of Electrometer	Potential in Volts
		negative	negative
1. One candle	2	90	0·81
2. One paraffin lamp—			
(*a*) without glass funnel . .	2	84	0·76
(*b*) with glass funnel . .	2	30	0·27
3. One spirit lamp . . .	4	109	0·99
4. Four portable electrometer matches	2	224	2·03
5. One Bunsen flame	4	30	0·27
6. One hydrogen flame . . .	At low pressure gave small negative; at higher pressures large positive. No electrification was found from the jet at any pressure when not burning		
7. Charcoal	Both gave negative electrification when there was a flame; and both gave positive electrification when they were glowing without flame		
8. Coals			

connected by a wire to the case of the electrometer. These matches, according to a suggestion made more than thirty years ago by Faraday, are made of white blotting-paper soaked in a solution of nitrate of lead, and rolled up with paste into little rods of about five millimetres diameter. The hydrogen was generated in an ordinary Woulffe's bottle from zinc and hydrochloric acid. The rise of the dilute hydrochloric acid in the long vertical tube through which the acid was admitted, indicated the pressure under the nozzle, above which the hydrogen was burning.

6. In the case of the charcoal and coal, the burning fuel was placed at the bottom of the iron funnel in a thin rectangular metallic vessel with small holes perforated in the bottom and in the sides. A wire from the case of the electrometer passed through one of these holes, and was thrust into the burning fuel. It was noticed that when the burning charcoal was first put in position below the funnel it always produced negative electrification, which ultimately changed to positive. Thus, in four experiments, the electrification, which was at first negative, became positive after 8, 10, 14, and 18 minutes respectively. On investigation it was found that as long as any flame* was visible in the burning char-

* In a paper on "Electrification of Air by Combustion," by Magnus Maclean and Makita Goto, communicated to the Philosophical Society of Glasgow on November 20, 1889, is a statement of results of many observations to find the potential to which the insulated quadrant of a quadrant electrometer is raised

Substances giving Flames or Burnings	Electrification of Insulated Fuel	Greatest observed Potential in Volts
Charcoal	Negative	3·0
Lucifer match, wood, and paper glowing	"	3·0
Hydrogen	"	0·6
Iron burning in vapour of sulphur	"	...
Copper " " "	"	...
Paraffin lamp	Positive	0·6
Alcohol lamp	"	0·3
Sulphur	"	2·0
Phosphorus exposed to air	"	1·5
Magnesium	"	...
Iron burning in oxygen	"	...
Lucifer match, wood, and paper burning with flame	"	...
Bisulphide of carbon	"	0·6
Sulphuric ether	"	0·9
Turpentine	"	0·5
Beeswax	"	0·7
Camphor	"	...

coal the electrification was negative; but as soon as all the flame disappeared, leaving only the red glow, the electrification became positive. To test this the heated charcoal was kept away from the funnel till all flame had disappeared. Then the vessel was put in position, and the deflections obtained in two experiments were—

51 scale divisions positive in 10 minutes.
100 " " " "

7. Next an experiment was made with the burning charcoal put in position while a flame was visible. The flame remained visible for 7 minutes, and in that time a negative electrification of 34 divisions was obtained. Then the deflection came back to the metallic zero in 1 minute, and in 10 minutes more a positive electrification of 87 divisions (0·78 volt) was obtained.

8. Glowing coals taken from the fire and put at once in the vessel in position, repeatedly gave negative electrification; but when they were kept away from the funnel till all flame had disappeared, the electrification obtained was slightly positive. Glowing coals remained glowing a very short time after all flame ceased, and the smallness of the observed effect is probably due to this cause.

9. A few experiments have also been tried to find to what positive potential the flame must be raised so as to overcome the negative electrification it gives to the air. Hitherto the only flame tried was a spirit flame. The positive electrode of a secondary cell was put into the flame of the lamp, and the negative electrode was joined to the iron funnel and to the case of the electrometer. The results obtained are not very regular, but we found that one storage cell was not sufficient to overpower

when in metallic connection with various kinds of flames and fires. It is there said: "The effect of an ordinary lucifer match is very interesting. While the match is burning with a flame the deflection indicates positive electrification; but after the flame ceases the electrification becomes negative, the effect now being that of glowing charcoal." The above table is quoted from that paper. In some cases the burnings lasted so short a time that quantitative determinations of the potential were not obtained. It is *conceivable* that all of the complementary opposite electricity separated from that which went to the electrometer in those experiments went to uninsulated solids in the neighbourhood. The experiments described in the text demonstrate that *some of it* was lodged in the air and fumes proceeding from the fire or flame.

the electrifying effects of the spirit flame. With one cell we got 45 divisions negative in 10 minutes, instead of 109 divisions with metallic connection; with two cells we got 10 divisions positive in 10 minutes; and with six cells we got 83 divisions positive in 4 minutes.

10. The filter, pump, and worm were now removed, and two plates—one of polished copper, and the other of polished zinc—were fixed 0·9 centimetre apart in a block of paraffin, as represented in fig. 2. The arrangement was such that either plate could be insulated, while the other was kept in metallic connection with the case of the electrometer. Observations were made to find the deflection from metallic zero with one plate insulated, and fumes from different flames and burnings at different distances from the plates passing up between them. This may be called the fumes-zero. When the top of the flame was within 5 or 6 centimetres from the plates, the results were very irregular. The results in the following table for spirit flame are in accord-

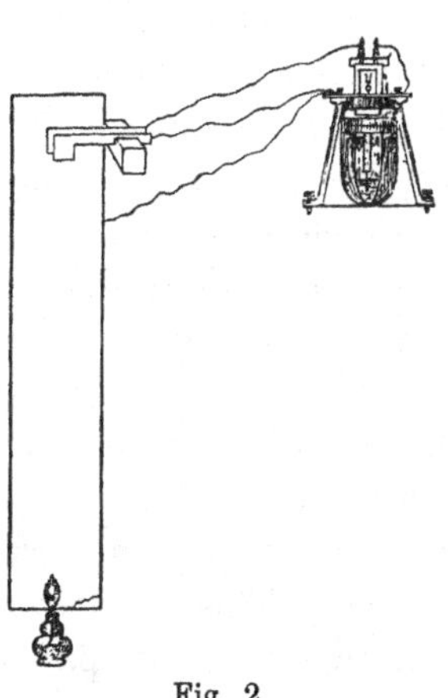

Fig. 2.

Sensitiveness of the Electrometer 136 *scale divisions per volt.*

Flame	Distance of top of Flame below the Plates in Centimetres	Metal connected to Insulated Terminal of Electrometer	Difference between Fumes Zero and Metallic Zero in Scale Divisions	Potential in Volts
Spirit lamp . . .	23	Copper	81 pos.	0·60
" . . .	"	Zinc	101 neg.	0·74
" . . .	11	Copper	53 pos.	0·39
" . . .	"	Zinc	76 neg.	0·56
Paraffin lamp without glass funnel . . .	7	Copper	141 pos.	1·04
" . . .	"	Zinc	138 neg.	1·01
" . . .	15	Copper	90 pos.	0·66
" . . .	"	Zinc	103 neg.	0·76
" . . .	23	Copper	108 pos.	0·79
" . . .	"	Zinc	112 neg.	0·82
" . . .	30	Copper	83 pos.	0·61
" . . .	"	Zinc	83 neg.	0·61

ance with what Maclean and Goto obtained from unguarded fumes from a spirit lamp 30 centimetres below the plates, as stated in their paper published in the *Philosophical Magazine* for August 1890. The effect is of the same kind as if the plates were connected by a drop of water*.

11. To observe the leakage between two parallel metal plates, the zinc plate was removed, and a polished copper plate, equal and similar to the other copper plate, was substituted for it. The distance between their parallel planes was 0·9 cm. The experiments were conducted as follows: One pair of quadrants of the electrometer, with one of the copper plates in metallic connection with it, was insulated. There was now no deviation from metallic zero. A small charge, positive or negative, was given to it, producing a deflection of about 450 scale divisions. This corresponds to over 9 volts, as the sensitiveness of the electrometer now used was 48·2 scale divisions per volt. In two or three minutes the ordinary leakage of the arrangement was observed. This did not amount to more than one division, or at most two divisions, per minute. Then the flame was lit, and readings were taken every half-minute. This was done with the variations in the funnel described in the last column of the following table, and illustrated by fig. 3. For comparison, the numbers in the following table show the leakage for two minutes after the reading was 300 scale divisions (6·2 volts) from metallic zero. This gives us the leakage at diminishing electric potentials during the time of observation. We intend to continue these experiments, and to arrange to find the leakage at different constant electric pressures.

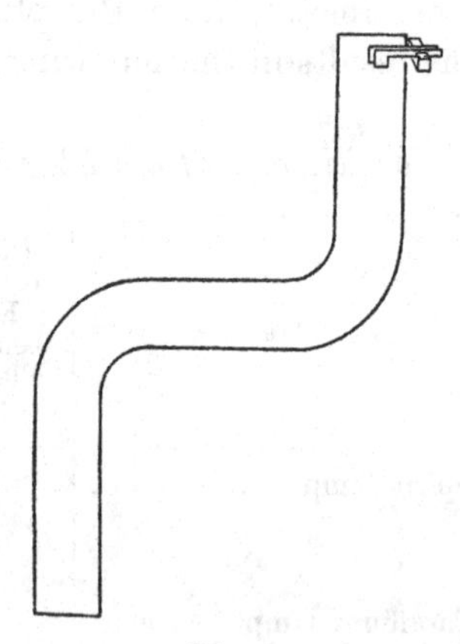

Fig. 3.

12. The marked difference in the leakage obtained when the horizontal tube was of small bore (3·8 cms.) and when it was of larger bore (15·3 cms.) may be contrasted as indicated in the last four results given for spirit flame. We also tried how long the fumes retained this conductive quality, but in every case we found

* Kelvin, *Electrostatics and Magnetism*, §§ 413, 414, pp. 332, 333.

that the leakage stopped in less than a quarter of a minute after the flame was extinguished, or removed from the bottom of the funnel. Closing the top and bottom of the funnel immediately after the flame was removed, we still found that the conductive quality of the air and fumes ceased within a quarter of a minute.

300 *Scale Divisions, equivalent to* 6·2 *volts, to begin with in each case.*

Flame, or Fire	Length of Funnel between Burning Substance and Copper Plates	Leakage in Two Minutes	Remarks
	Centimetres	Scale Divisions	
Spirit flame	66	292 pos.	Funnel of 15·3 cms. bore all vertical
”	”	287 neg.	” ” ”
”	112	253 pos.	” ” ”
”	”	254 neg.	” ” ”
”	343	22 pos.	Funnel 114 cms. vertical of 15·3 cms. bore; and 229 cms. horizontal of 3·8 cms. bore
”	”	20 neg.	
”	236	24 pos.	Same vertical, and 122 cms. horizontal of 3·8 cms. bore
”	”	20 neg.	
”	160	40 pos.	Same vertical, and 46 cms. horizontal of 3·8 cms. bore
”	”	46 neg.	
”	244	165 pos.	Same vertical, and 130 cms. horizontal of 15·3 cms. bore
”	”	187 neg.	
Charcoal	”	54 pos.	” ” ”
”	”	57 neg.	” ” ”

13. In connection with these last experiments, attention may be directed to an experiment described by Prof. Schuster, in which he uses an insulated metallic tube bent round at the upper end, to prove that "it is not only the flame itself which conducts, but also the gases rising from the flame*." He discovers electric conductance in products of combustion mixed with air quite out of sight from the flame.

* Prof. Schuster, on "Atmospheric Electricity," at Royal Institution, February 22, 1895.

250. CONTACT ELECTRICITY OF METALS*.

[From *Roy. Instit. Proc.* Vol. XV. 1897 [May 21, amplified Feb. 1898], pp. 521—554; *Phil. Mag.* Vol. XLVI. July 1898, pp. 82—120.]

1. WITHOUT preface two 95 years old experiments of Volta's were, one of them shown, and the other described. The apparatus used consists of: (*a*) a Volta-condenser of two varnished brass plates, of which the lower plate is insulated in connection with the gold leaves of a gold leaf electroscope, and the upper plate is connected by a flexible wire with the sole plate of the instrument; (*b*) two circular discs, one of copper and the other of zinc, each polished and unvarnished. I hold one in my right hand by a varnished glass stem attached to it, while in my left hand I hold the other, which is kept metallically connected with the sole plate of the electroscope by a thin flexible wire.

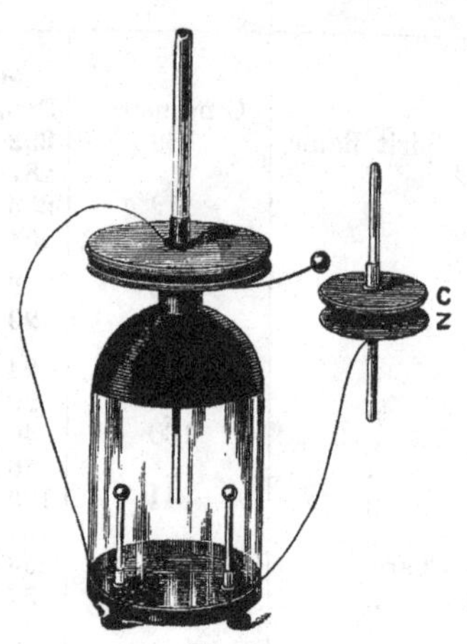

Fig. 1.

To commence the experiment I place one disc resting on the other, and lift the two till the upper touches a brass knob connected by a stiff metal wire with the lower plate of the Volta-condenser. I break this contact and then lift the upper plate of the condenser; you see no divergence of the gold leaves. This proves that no disturbing electric influence sufficient to show any perceptible effect on our gold leaf electroscope is present. Now I repeat what I did, with only this change—I hold the lower plate of the Volta-condenser with the upper plate resting on it two or

* [For the thermodynamic treatment of this subject cf. *supra*, Vol. v. pp. 29—35.]

three centimetres below the knob. I then with my right hand lift the upper plate of the Volta-condenser; you see a very slight divergence between the shadows of the gold leaves on the screen. I can just see it by looking direct at the leaves from a distance of about half a metre. Still holding the lower plate firmly in my left hand in the same position, and holding the upper plate by the top of its glass stem in my right, at first resting on the lower plate, I lift it and let it down very rapidly a hundred times, so as to produce one hundred cycles of operation—break contact between discs, make and break contact between upper disc and knob, make contact between discs. Lastly, I lift the upper plate of the condenser; you see now a great divergence of the gold leaves, many of you can see it direct on the leaves, while all of you can see it by their shadows on the screen. Now, keeping the upper plate of the condenser still unmoved, I bring a stick of rubbed sealing-wax into the neighbourhood of the electroscope; you see the divergence of the leaves is increased. I remove the sealing-wax and the divergence diminishes to what it was before. This proves that the gold leaves diverge in virtue of resinous electricity upon them, and therefore that the insulated plate of the condenser received resinous electricity from the copper disc. If now I interchange the two discs so that the upper is zinc and the lower copper, and repeat the experiment, you see that the rubbed sealing-wax diminishes the divergence as it is brought from a distance into the neighbourhood, and that a glass rod rubbed with silk increases the divergence. Hence we conclude that in the separation of two discs of copper and zinc the copper carries away resinous electricity and the zinc vitreous electricity.

2. Experiment 2. The same apparatus as in Experiment 1, except that the polished zinc and copper discs have their opposed faces varnished with shellac, and are provided with wires soldered to them for making metallic connection between them when the upper rests on the lower, as shown in Fig. 2. All operations are the same as in Experiment 1, but now with this addition—when the upper disc rests on the lower, make and break metallic contact by hand as shown in the diagram. The results are the same as those of Experiment 1, except that the quantity of electrification given to the gold leaves by a single cycle of operations is generally greater than in Experiment 1, for this reason: In Experiment 1

at the instant of breaking contact between the zinc and copper there is generally some degree of inclination between the two discs, while at the corresponding instant of Experiment 2 they are parallel and only separated by the insulating coats of varnish. If great care is taken to keep the discs as nearly as possible parallel at the instant of separation, the effect of a single separation may be made greater in Experiment 1 than in Experiment 2 (see § 3 below).

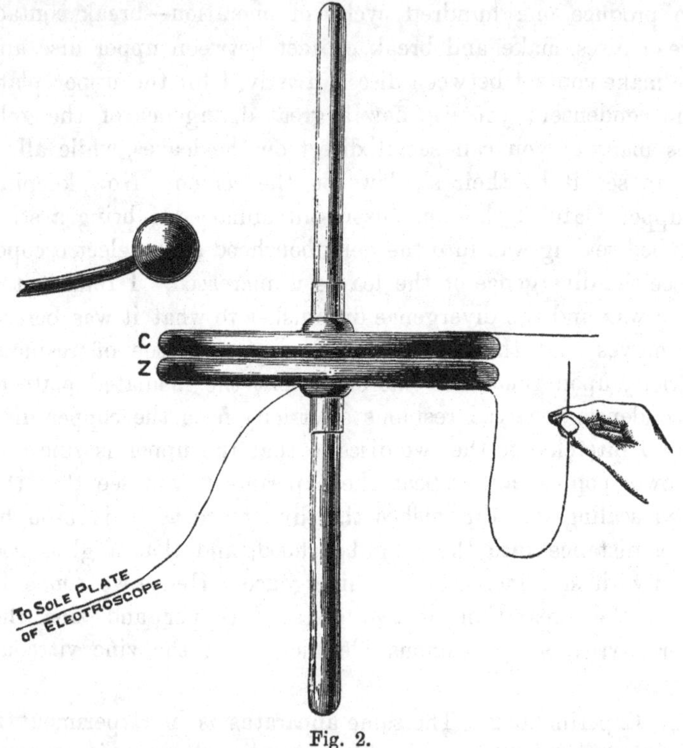

Fig. 2.

3. An instructive variation of Experiment 1 may be made by giving a large inclination, 5°, or 10°, or 20°, of the upper plate to the lower, while still in contact and at the instant of separation. By operating thus the experiment may be made to fail so nearly completely that no divergence of the leaves will be observed even after one hundred cycles.

4. These two experiments, with the variation described in § 3, put it beyond all doubt that Volta's electromotive force of contact between two dissimilar metals is a true discovery. It

seems to have been made by him about the year 1801; at all events he exhibited his experiments, proving it in that year to a Commission of the French Institute (Academy of Sciences). It is quite marvellous that the fundamental experiment (§ 1 above), simple, easy and sure as it is*, is not generally shown in courses of lectures on electricity to students, and has not been even mentioned or referred to in any English text-book later than 1845, or at all events not in any one of a large number in which I have looked for it, except in the *Elementary Treatise on Electricity and Magnetism,* founded on Joubert's *Traite Élémentaire d'Électricité,* by Foster and Atkinson, 1896 (p. 136). The only other places in which I have seen it described in the English language are Roget's article in the *Encyclopædia Metropolitana* referred to above; Tait's *Recent Advances in Physical Science,* 1876; and Professor Oliver Lodge's most valuable, interesting and useful account of all that had been done for knowledge of contact electricity from its discovery by Volta till 1884, in his Report to the British Association of that year, "On the Seat of the Electromotive Forces in the Voltaic Cell."

5. The reason for this unmerited neglect of a great discovery regarding properties of matter is that it was overshadowed by an earlier and greater discovery of its author, by which he was led to the invention of the voltaic pile and crown of cups, or voltaic battery, or, as it is sometimes called, the galvanic battery. Knowing, as we now know, both Volta's discoveries, we may describe the earlier most shortly by saying that the simple experiment (§ 1 above), demonstrating the later discovery, is liable to fail if a drop of water is placed on the lower of the two polished plates. It fails if (see Fig. 4 below) the last connection between the zinc and copper, when the upper disc is lifted, is by water. It would not fail (see Fig. 6 below) nor be sensibly altered from what is found with the dry polished metals, if the upper disc is slightly tilted in the lifting, so as to break the water arc before the separation between the metals, and secure that the last connection is contact of dry metals. To show this to you more readily than by a Volta condenser with gold leaf electroscope, I shall now use instead my quadrant electrometer without condenser.

* Fully and clearly described in Roget's article on "Galvanism," in the *Encyclopædia Metropolitana,* Vol. IV. edition 1845, p. 210.

(1) Holding the copper disc connected with the metal case of the electrometer in one hand, with my other hand I hold by a glass handle the zinc disc, which you see is connected by a fine wire with the insulated quadrants of the electrometer. I first place the zinc resting on the copper, both being polished and dry. You now see the spot of light at the point marked *O* on the scale, which I call the metallic zero. I now lift the zinc disc two or three millimetres from resting on the copper, and you see the spot of light travelling largely to the right, which proves that vitreous electricity has passed from the zinc disc through the connecting wire to the insulated quadrants of the electrometer. I lower the zinc disc down to rest again on the copper disc; you see the spot of light again comes back to the metallic zero.

(2) I now raise the zinc disc, and with a little piece of wet wood (or a quill pen) place a little mound of water on the copper disc, as shown in Fig. 3. I bring down the zinc disc to touch the

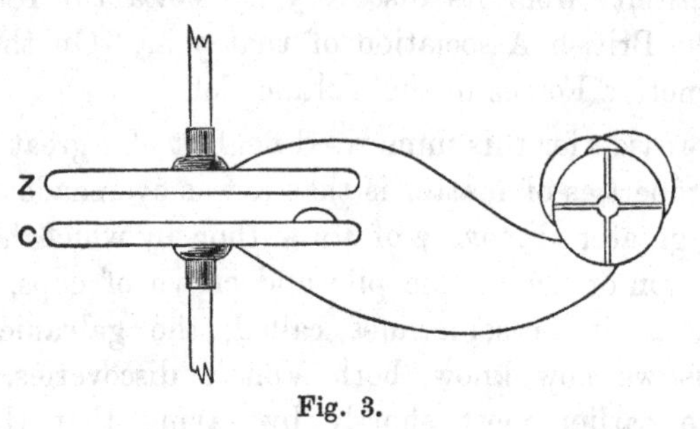

Fig. 3.

top of the little mound of water, keeping it parallel to the copper disc so that there is no metallic contact between them (Fig. 4);

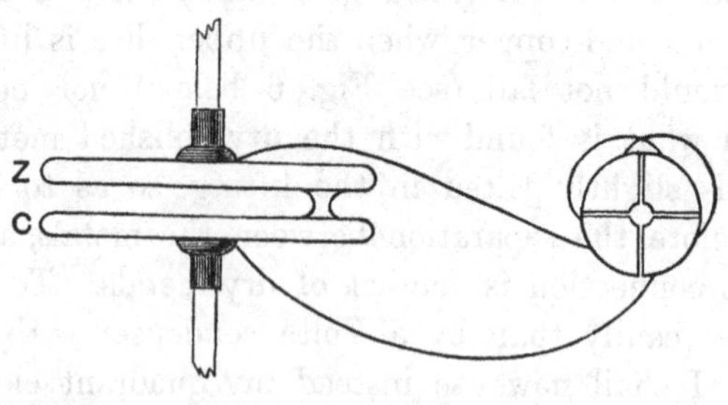

Fig. 4.

you see that the spot of light moves to the left and settles at a point marked *E* (which I call the electrolytic zero of our circumstances), a few scale divisions to the left of the metallic zero. This motion and settlement is the simplest modern exhibition of Volta's greatest discovery.

(3) Now that the spot of light has settled, I lift the zinc disc a millimetre till the water column is broken, and then two or three centimetres farther (Fig. 5); the spot of light does not

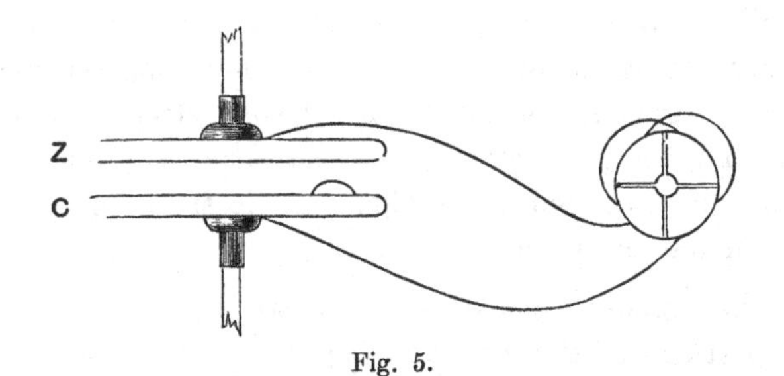

Fig. 5.

move, it remains at *E*. I lower the zinc disc again; still no motion of the spot of light, not even when the zinc again touches the little mound of water.

(4) Now I tilt the zinc disc slightly till it makes a dry metallic contact with the copper, as shown in Fig. 6; while the

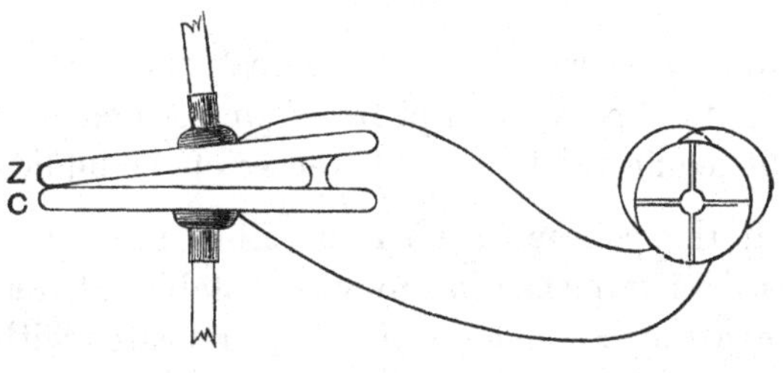

Fig. 6.

water arc still remains unbroken. You see the spot of light, at the instant of metallic contact, suddenly leaves *E* and moves to the right, and settles quickly at the metallic zero after a few vibrations through diminishing range.

(5) Lastly, I break the metallic contact, and hold the zinc disc again parallel to the copper (Fig. 4) with the water connection still remaining unbroken between them; the spot of light shows no sudden motion; it creeps to the left till, in half a minute or three-quarters of a minute, it reaches its previous

steady position on the left. This is the now well-known phenomenon (never known to Volta) of the recovery of a voltaic cell from electrolytic polarisation after a metallic short-circuit.

6. The succession of experiments described in § 5, interpreted according to elementary electrostatic law, proves the following conclusions:—

(1) When the dry and polished discs of zinc and copper are metallically connected and held parallel, their opposed faces are oppositely electrified, the zinc with vitreous electricity, and the copper with resinous electricity, in quantities varying inversely as the distance between them when this is small in comparison with the diameter of each.

(2) The opposed polished faces are non-electrified when polished portions of the zinc and copper surfaces are connected by water, and when there is no metallic connection between them. Or, if not absolutely free from electrification, they may be found slightly electrified, zinc resinously or vitreously, and copper vitreously or resinously, according to differences in respect to cleanness, polish, or scratching or burnishing, as explained in § 16 below; and according to polarisational or other difference in the wetted portions of the surfaces.

If instead of pure water we take a weak solution of common salt, or carbonate of soda, or sulphate of zinc or ammonia, we find results but little affected by the differences of the liquids.

7. But if the polished surface of either the copper or the zinc is oxidised, or tarnished in any way, notably different results are found when the experiments of § 5 are repeated with the disc or discs thus altered.

For example, hold the copper disc, with its polished side up, over a slab of hot iron, or a spirit lamp, or a Bunsen burner, till you see a perceptible change of colour, due to oxidation of the previously polished face. Then allow the copper to cool, and repolish a small area near one edge; place a little mound of water upon this area, and operate as in § 5 (2), (3). The water connection between polished zinc and polished copper brings the spot of light to the same electrolytic zero E as before. But now when we lift the zinc disc and break the water connection, the spot of light moves to the right, instead of remaining steady as it does when both the dry opposed surfaces are polished. If next we tarnish the zinc disc by heat, as we did for the copper disc, and repeat the experiment with wholly polished copper, and with the zinc disc oxidised where dry, and polished only where wet by the water connection, we find still the same electrolytic zero E; but now the spot of light moves to the left when we lift the zinc disc and break the water connection.

8. The experiments of § 7, interpreted in connection with those of § 5, prove that there are dry contact voltaic actions between metallic copper and oxide of copper in contact with it, and between metallic zinc and oxide of zinc in contact with it; according to which, dry oxide of copper is resinous to copper in contact with it, and dry oxide of zinc is resinous to zinc in contact with it, just as copper is resinous to zinc in contact with it. We may verify this conclusion by another interesting experiment. Taking, for instance, the oxidised copper plate, with a little area polished for contacts; put a little mound of copper, instead of the mound of water, on this area for contact with the upper plate; and for the upper plate take polished copper instead of polished zinc. If we operate now as in § 7, the spot of light settles at the metallic zero O when the metallic contact is made, instead of at the electrolytic zero E, as it did when we had water connection between zinc and copper. But now, just as in § 7, the spot of light moves to the right when the contact is broken and the upper plate lifted, which proves that vitreous electricity flows into the electrometer from the upper plate, when its distance from the lower plate is increased after breaking the metallic contact. We conclude that when the two plates were parallel, and very near one another, and when there was metallic connection between them, vitreous and resinous electricities were induced upon the opposed surfaces of metallic copper and oxidised copper respectively. This statement, which we know from § 7 to be also true for zinc compared with oxidised zinc, is probably also true for every oxidisable metal compared with any one of its possible oxides. It is true, as we shall see later (appended paper of 1880–1; also Erskine Murray's paper referred to in § 15), even for platinum in its ordinary condition in our atmosphere of 21 per cent. oxygen and 79 per cent. nitrogen, voltaically tested in comparison with platinum which has been recently kept for several minutes or several hours in an atmosphere of pure oxygen,

or even in an atmosphere of 95 per cent. oxygen and 5 per cent. nitrogen.

9. Hitherto we have had no means of measuring the amount of the Volta-contact electric force between dry metals, except observation of the degrees of deflection of the gold leaves of an electroscope, or of the spot of light of the quadrant electrometer consequent upon operations performed upon different pairs of metals, with dimensions and distances of motion exactly the same, and comparison of these deflections with the steady deflection from the metallic zero given by polished zinc and copper connected conductively with one another by water, and connected metallically with the two electrodes of an electroscope or electrometer. Kohlrausch, in 1851*, devised an apparatus for carrying out this kind of investigation systematically, and with a good approach to accuracy, by aid of a Dellman's electrometer and a Daniell's cell, as more definite and constant than a zinc-water-copper cell. This method of Kohlrausch's for measuring the Volta electromotive forces between dry metals, "has been employed with modifications by Hankel, by Gerland, by Clifton, by Ayrton and Perry, by von Zahn, and by most other experimenters on the subject†." About thirty-seven years ago, in repetitions of Volta's fundamental experiment proving contact electricity by electroscopic phenomena resulting from change of distance between parallel plates of zinc and copper, I found a null method for measuring electromotive forces due to metallic contact between dissimilar metals, in terms of the electromotive force of a Daniell's cell, which is represented diagrammatically in Fig. 7, and in perspective in Fig. 8 [omitted]. The two discs are protected against disturbing influences by a metal sheath. The lower disc is permanently insulated in a fixed position, and is kept connected with the insulated pair of quadrants of a quadrant electrometer. The upper disc is supported by a metal stem passing through a collar in the top of the sheath, so that it is kept always parallel to the lower disc and metallically connected to the sheath, while it can be lifted a few centimetres at pleasure from an adjustable lowest position in which its lower face is

* *Poggendorff Annalen*, Vols. LXXV. p. 88; LXXXII. pp. 1 and 45; and LXXXVIII. p. 465, 1851 and 1853.

† Prof. O. J. Lodge, "On the Seat of the Electromotive Forces in the Voltaic Cell," *Brit. Assoc. Report*, 1884, pp. 464—529.

about half a millimetre or a millimetre above the upper face of the lower disc. A portion of the wire connecting the lower plate to the insulated quadrants of the electrometer is of polished platinum, and contact between this and a platinum-tipped wire connected to the slider of a potential divider is made and broken at pleasure. For certainty of obtaining good results it is necessary that these contacts should be between clean and dry polished metals, because if the last connection on breaking contact is through semi-moist dust, or oxide, or "dirt" (defined by Lord Palmerston to be matter in a wrong place), or if it is anything other than metallic, vitiating disturbance is produced.

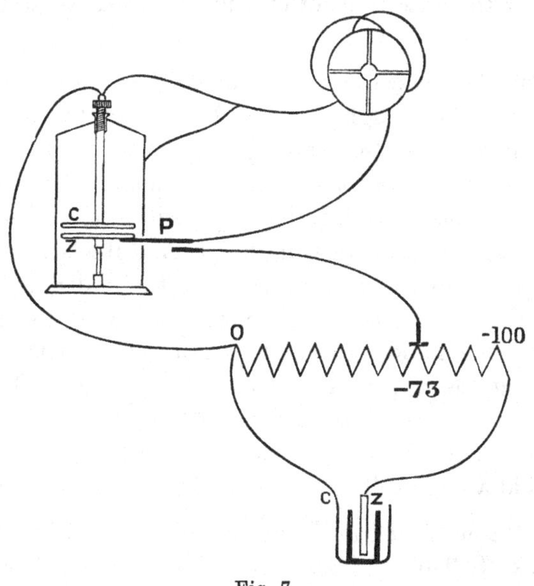

Fig. 7.

10. To make an experiment, first test the insulation with the upper plate held up in its highest position, and after that with it let down to its lowest position, in each case proceeding thus: Holding by hand the wire connected to the slider, run the slider to zero, make contact at P, observe on the screen the position of the spot of light from the electrometer mirror for the metallic zero, and then run the slider slowly to the top of its scale and break contact; the spot of light should remain steady, or at all events should not lose more than a very small percentage of its distance from metallic zero, in half a minute. Repeat the

test with the cell reversed. If the test is satisfactory with the upper plate high, the insulation of the insulated quadrants in the electrometer and of the lower disc in the Volta-condenser is proved good. If after that the test is not satisfactory with the upper disc at its lowest, we infer that there are vitiating shreds between the two plates, and we must do what we can to remove them; or, if necessary, we must alter the screw-stop at the top so as to increase the shortest distance between the plates sufficiently to prevent bridges of shred or dust between them, and so to give good insulation. The smaller we make the shortest distance with perfect enough insulation, the more sensitive is the apparatus for the measurement of contact electricity performed as follows.

11. Run the slider to zero; make and keep made the contact at P till the spot of light settles at the metallic zero; break contact at P, and lift the upper plate slightly. (If you lift it too far, the spot of light may fly out of range.) If the spot of light moves in the direction showing positive electricity on the insulated quadrants (as it does if the lower plate is zinc and the upper copper), connect the cell to make the slider negative (as shown in Fig. 7). Repeat the experiment with the slider at different points on the scale, until you find that, with contact P broken, lifting the upper plate causes no motion of the spot of light. If the compensating action with the slider at the top of the range is insufficient, add a second cell; if it is still insufficient, add a third cell; if still insufficient, add a fourth*.

12. By this method I made an extended series of experiments in the years 1859–61, as stated in a short paper communicated to Section A of the British Association at its Swansea meeting in August 1880, which with additions published in *Nature* for April 14, 1881, is appended to the present article.

* The only case hitherto tested by any experimenter, so far as known to me, in which more than two Daniell cells would be required for the compensation, is bright metallic sodium, guarded against oxide by glass, in Mr Erskine Murray's experiments (§ 18 below), showing volta-difference of 3·56 volts from his standard gold plate. For direct test this would require four Daniell cells on the potential divider. The greatest volta-difference of potentials observed by Pellat was 1·08 volts, for which a Daniell's cell would rather more than suffice. About 1862 I found considerably more than the electromotive force of a single Daniell's element required to compensate the Volta electromotive force between polished zinc and copper oxidised by heat to a dark purple or slate colour.

13. Quite independently*, Mr H. Pellat found the same method, and made admirable use of it in a series of experiments described in theses presented to the Faculty of Sciences in Paris in 1881†, of which the results, accurate to a degree of minuteness unknown in previously published researches on the electrical effects of dry contacts between metals, constitute in many respects the most important and most interesting extension of our knowledge of contact electricity since the times of Volta and Pfaff. One of his results (I shall have to speak of others later) was that Pfaff was right in 1829‡ when he described experiments in which he found no difference in the Volta-contact-electromotive force between zinc and copper, whether tested in dry or damp air, oxygen, nitrogen, hydrogen, carburetted hydrogen, or carbonic acid, so long as no visible chemical action occurred; and that De la Rive was not right when he "asserted that there was no Volta effect in the slightly rarefied air then known as vacuum§." Pfaff experimented with varnished plates; Pellat arrived at the same conclusion with polished unvarnished plates of zinc and copper. He found slight variations of the Volta electromotive force due to the nature of the gas surrounding the plates, and to differences of its pressure, of which he says: "Ces variations sont très faibles, par rapport à la différence de potentiel totale....Ces variations dans la différence de potentiel sont toujours en retard sur les changements de pression. Elles ne paraissent donc pas dépendre directement de celle-ci, mais bien des modifications qui en résultent dans la nature de la surface métallique, modifications qui mettent un certain temps à se produire." The smallest pressures for which Pellat made his experiments were from 3 to 4 or 5 cms. of mercury.

14. The same method was used by Mr J. T. Bottomley in an investigation by which he demonstrated with minute accuracy the equality of the Volta-contact-difference measured in a glass

* *Ann. de Chimie et de Physique*, Vol. XXIV. 1881, p. 20, footnote.

† "Thèses présentées à la Faculté des Sciences de Paris, pour obtenir le Grade de Docteur-ès-Sciences Physiques," par M. H. Pellat, Professeur de Physique au Lycée Louis le Grand, No. 461, juin 22, 1881. See also *Journal de Physique*, 1881, XVI. p. 68, and May 1880, "Différence de potentiel des couches électriques qui recouvrent deux métaux en contact."

‡ *Ann. de Chim.* 2 series, Vol. XLI. p. 236.

§ Lodge, *Brit. Assoc. Report*, 1884, pp. 477–8.

tube exhausted to less than $\frac{1}{523}$ mm. of mercury* ($2\frac{1}{2}$ millionths of an atmosphere), and immediately after in the same tube filled with air to ordinary atmospheric pressure; and again exhausted and filled with hydrogen to atmospheric pressure three times in succession; and again exhausted and filled to atmospheric pressure with oxygen. In some cases the electrical test was repeated several times, while the gas was entering slowly. The actual apparatus which he used is before you, and in it I think you will see with interest the little Volta-condenser, with plates of zinc and copper a little larger than a shilling, the upper hung on a spiral wire by a long hook carrying also a small globe of soft iron. Thus you see by aid of an external magnet I can lift and lower the upper plate without moving the vacuum tube which, during the experiments, was kept in connection with a Sprengel pump and phosphoric acid drying tubes. Mr Bottomley sums up thus: "The result of my investigation, so far as it has gone, is that the Volta contact effect, so long as the plates are clean, is exactly the same in common air, in a high vacuum, in hydrogen at small and full pressure, and in oxygen. My apparatus, and the method of working during these experiments, was so sensitive that I should certainly have detected a variation of 1 per cent. in the value of the Volta contact effect, if such a variation had presented itself†."

15. With the same method further researches have been carried on by Mr Erskine Murray, and important and interesting results obtained, within the last four years, in the Physical Laboratories of the Universities of Glasgow and Cambridge. He promises a paper for early communication to the Royal Society, and, from a partial copy of it which he has already given me, I am able to tell you of some of his results. Taking generally as standard a gilt brass disc which he found among the apparatus remaining from my experiments of 1859–61, he measured Volta-differences from it in terms of the modern standard *one volt*. These differences are what we may call the Volta-potentials of the different metallic surfaces, or surfaces of metallic oxides, iodides, &c., or metallic surfaces altered by cohesion to them of gases or vapours, or residues of liquids which had been used for

* A very high exhaustion had been maintained for two days, and finally perfected by two and a half hours' working at the pump immediately before the electric testing experiment.

† *Brit. Assoc. Report*, 1885, pp. 901—3.

washing them; if for simplicity we agree to call the Volta-potential of the gold, zero. As a rule he began each experiment by polishing the metal plate to be tested on clean glass paper or emery cloth, and then measured its difference of potential from the standard gold plate. After that the plate was subjected to some particular treatment, such as filing or burnishing; or polishing on leather or paper; or washing with water, or alcohol, or turpentine, and leaving it wet or drying it; or heating it in air, or exposing it to steam or oxygen, or fumes of iodine or sulphuretted hydrogen; or simply leaving it for some time under the influence of the atmosphere. The plate as altered by any of these processes was then measured for potential against the standard gold. Very interesting and instructive results were found; only of one can I speak at present. Burnishing by rubbing it firmly with a rounded steel tool, or by rubbing two plates of the same metal together, increased the potential in every case; that is to say made the metallic surface more positive if it was positive to begin with; or made it less negative or changed it from negative to positive, if it was negative to begin with. Thus:—

Zinc immediately after being scratched sharply by polishing on clean glass paper was found	+ ·70 volt.
After being burnished with hard steel burnisher it was found	+ ·94 volt.
After being left to itself for 2 hours it was found . .	+ ·92 volt.
After further burnishing	+1·00 volt.
After still further burnishing	+1·02 volt.
It was then scratched by polishing on glass paper, and its surface potential returned to its original value of . .	+ ·70 volt.

16. This seems to me a most important result. It cannot be due to the removal of oxygen, or oxide, or of any other substance from the zinc. It demonstrates that change of arrangement of the molecules at the free surface, such as is produced by crushing them together, as it were, by the burnisher, affects the electric action between the outer surface of the zinc and the opposed parallel gold plate. It shows that the potential* in zinc (uniform

* There has been much of wordy warfare regarding potential in a metal, but none of the combatants has ever told what he means by the expression. In fact the only definition of electric potential hitherto given has been for vacuum, or air, or other fluid insulator. Conceivable molecular theories of electricity within a

throughout the homogeneous interior) increases from the interior through the thin surface layer of a portion of its surface affected by the crushing of the burnisher, more by ·32 volt than through any thin surface-layer of portions of its surface left as polished and scratched by glass paper. The difference of potentials of copper and zinc across an interface of contact between them is only about 2½ times the difference of potential thus proved to be produced between the homogeneous interior of the zinc and its free surface, by the burnishing. Pellat had found that polished metallic surfaces, seemingly clean and free from visible contamination of any kind, became more positive by rubbing them forcibly with emery paper, zinc showing the greatest effect, which was ·23 volt. Murray's burnished surface of zinc actually *fell* ·32 volt when scratched by polishing on glass paper.

17. With two copper plates (*a*), (*b*) polished on emery (*a*) − ·11 volt.
and each compared with standard gold, Murray found. (*b*) − ·06 volt.
They were then burnished by rubbing them forcibly (*a*) − ·02 volt.
together, and again tested separately; he found . (*b*) − ·02 volt.

Rises of Volta-potential of about the same amount were produced by burnishing with a steel burnisher copper plates which

solid or liquid conductor might admit the term potential at a point in the interior; but the function so called would vary excessively in intermolecular space, and must have a definite value for every point, whether of intermolecular space, or within the volume of a molecule, or within the volume of an atom, if the atom occupies space. It would also vary intensely from point to point in the ether or air outside the metal at distances from the frontier small or moderate in comparison with the distance from molecule to molecule in the metal.

But when, setting aside our mental microscopic binocular which shows us atoms and molecules, we deal with the mathematical theory of equilibrium and motion of electricity through metals with outer surfaces bounded by ether or air or other insulating fluids or solids, we find it convenient to use a mathematical function of position called potential in the interior of each metal. This function must, for the case of equilibrium, fulfil the condition that it is of uniform value through each homogeneous portion of metal. Its value must, as a rule, change gradually (or abruptly) with every gradual (or abrupt) change of quality of substance occupying space.

To illustrate the difficulty and complexity of expression with which I have struggled, and to justify if possible my ungainly resulting sentence in the text, consider the case of a crystal of pure metal: suppose, for example, an octahedron with truncated corners, all natural faces and facets. In all probability Volta-differences of potential would be found between the octahedronal and truncational faces. We might arbitrarily define the uniform interior potential as the potential of the air either near an octahedronal face or near a truncational face; or, still arbitrarily, we might define it as some convenient mean or average related to measurements of Volta-differences of potential between the different faces.

had been polished and scratched in various ways. Such experiments as those of Murray with burnishing ought to be repeated with hammering or crushing by a Bramah's press. Indeed Pellat* suggested that metals treated bodily "par le laminage ou le martelage" (rolling or hammering) might probably show Volta-electric properties of the same kind as, but more permanent than, those which he had found to be produced by violent scratching with emery paper.

18. It is interesting to remark that Murray's most highly burnished zinc differed from his emery-polished copper (a) by 1·13 volts. This is considerably greater, I believe, than the highest hitherto recorded Volta-difference between pure metallic surfaces of zinc and copper.

By far the greatest Volta-difference between two metallic surfaces hitherto measured is, I believe, 3·56 volts, which Murray, in another part of his work, found as the Volta-difference between bright sodium protected by glass and his standard gold. He had previously found a copper surface after exposure to iodine vapour to be − ·34 relatively to his standard gold. The difference between this iodised surface and the bright metallic surface of sodium was therefore 3·90 volts: which is the highest dry Volta electromotive force hitherto known.

19. Seebeck's great discovery of thermo-electricity (1821) was a very important illustration and extension of the twenty years' earlier discovery of the contact electricity of dry metals by Volta. It proved independently of all disturbing conditions that the difference of potentials between two metals in contact varies with the temperature of the junction. Thus, for instance, in the

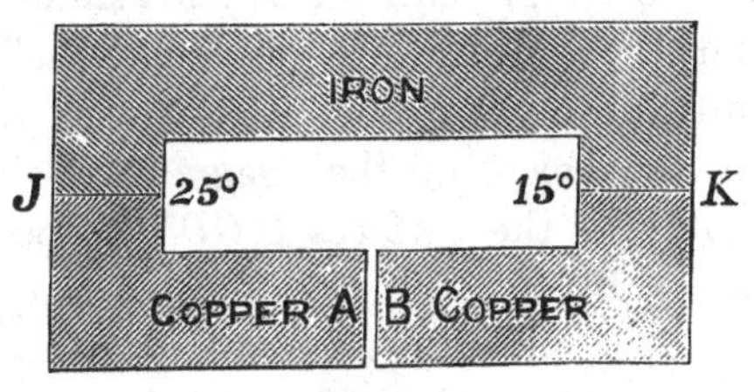

Fig. 9.

copper-iron arrangement represented in Fig. 9, with its hot junction at 25° and its cold at 15°, the electromotive force tends

* *Ann. de Chimie et de Physique*, 1881, Vol. xxiv. footnote on p. 83.

to produce current from copper to iron through hot, and its amount is ·00148 volt: that is to say, if the circuit is broken at AB the two opposed faces A, B, at equal temperatures, present a difference of electric potential of ·00148 volt, with B positive relatively to A. This is not too small a difference to be tested directly by the Volta-static method, worked by two exactly similar metal discs connected to A and B, when they are at their shortest distance from one another, and then disconnected from A and B, and separated and tested by connection with a delicate quadrant electrometer. But the test would be difficult, because of the difficulty of preparing the opposed surfaces of two equal and similar discs, so as to make them equal in their surface-Volta-potentials within one one-thousandth of a volt, or even to make their difference of potentials constant during the time of experiment within one one-thousandth of a volt. There would, however, be no interest in making the experiment in this way, because by the electromagnetic method we can with ease exhibit and measure with great accuracy the difference of potentials between A and B, by keeping them exactly at one temperature and connecting them by wires of any kind with brass or other terminals of a galvanometer of high enough resistance not to sensibly diminish the difference of potentials between A and B, provided all the connections between metals of different quality except J and K are kept at one and the same temperature (or pairs of them, properly chosen, kept at equal temperatures).

20. Suppose, now, instead of breaking a circuit of two metals at a place in one of the metals, as AB in copper in Fig. 9, we break it at one of the junctions between the two metals, as at $C'C$, $I'I$, Fig. 10. CD represents a movable slab of copper which (for § 22 below) may be pushed in so as to be wholly opposite to $I'I$, or at pleasure drawn out to any position, still resting on the copper below it as shown in the diagram. Calling zero the uniform potential over the surfaces $C'CD$, the potential at $I'I$ would be about +·16 volt (according to Murray's results for emery-polished copper and iron surfaces) if the temperature at J and throughout the system is uniform at about 15° C. Keeping now the temperature of $C'C$, $I'I$ exactly at 15°, let the temperature of J be raised to 25°. The difference of potentials between $C'C$ and $I'I$ would be increased to ·16148 volt, supposing ·16000 to have been exactly the difference of potentials when the temperature

of J was 15°. This difference of differences of potentials would be just perceptible on the most delicate quadrant electrometer connected as indicated in the diagram. Lastly, raise the temperature of $C'C$ and $I'I$ to exactly 25°, J being still kept at this temperature: the spot of light of the electrometer will return exactly to its metallic zero. But, would the Volta-difference of potentials between the surfaces $C'C$, $I'I$ remain unchanged, or would it return exactly to its previous value of ·16000, or would it come to some other value? We cannot answer this question without experiment. The proper method, of course, would be to use the metal-sheathed Volta-condenser and compensation (§ 9 above), and with it measure the Volta-differences between copper and iron at different temperatures,

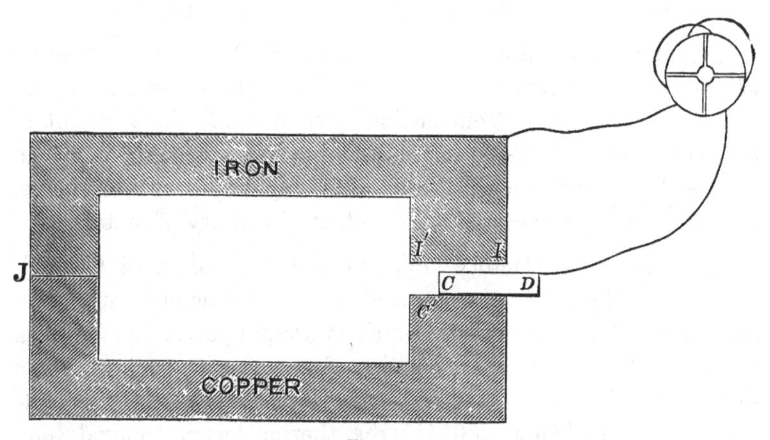

Fig. 10.

the same for the two metals in each case. The sheath and everything in it should, in each experiment, be kept at one and the same constant temperature. But it would probably be very difficult to get a decisive answer, because of the uncertainties and time-lags of changes in the Volta-potential of metallic surfaces with change of temperature, which, if we may judge from Pellat's and Murray's experiments on effects of temperature when the two metals are unequally heated, would probably also be found when the temperatures of the two metals, kept exactly equal, are raised or lowered at the same time.

21. The thermoelectric difference between bismuth and antimony is about ten times that between copper and iron for

temperature differences of ten or twenty degrees on the two sides of 20° C., and their Volta-contact difference is exceedingly small (according to Pellat, just one one-hundredth of a volt when both their surfaces are strongly scratched by rubbing with emery). It would be very interesting, and probably instructive, to find how much their Volta-contact difference varies with temperature by the method at present suggested. The great variations of Volta-surface potentials, found by Pellat and Murray, when one of the two metals is heated, may have been due to difference of temperatures between the two opposed plates with air between them; and it is possible that no such large variation, or that large variation only due to changes of cohering gases, may be found when the two metals are kept at equal temperatures, and these temperatures are varied as in the experiment I am now suggesting.

22. Peltier's admirable discovery (1834) of cold produced where an electric current crosses from bismuth to antimony, and heat where it crosses from antimony to bismuth, in a circuit of the two metals, with a current maintained through it by an independent electromotive force, is highly important in theory, or in attempts for theory, of the contact electricity of metals.

From an unsatisfactory* hypothetical application of Carnot's principle to the thermodynamics of thermoelectric currents I long ago inferred† that probably electricity crossing a contact between copper and iron in the direction from copper to iron would produce cold, and in the contrary direction heat when the temperature is below 280° C. (the thermoelectric neutral temperature of copper and iron)‡, and I verified this conclusion by experiment§. Hence we see, looking to Fig. 10, if the movable copper plate *CD* is allowed to move inwards (in the position shown in the diagram it is pulled inwards by the Volta-electrifications of the opposed surfaces of iron and copper), cold will be produced at the junction *J*, all the metal being at one temperature

* *Mathematical and Physical Papers*, Vol. I. Art. XLVIII. § 106, reprinted from *Transactions of the Royal Society of Edinburgh*, May 1854.

† *Ibid.* § 116 (19).

‡ In a thermoelectric circuit of copper and iron the current is from copper to iron through hot when both junctions are below 280° C. It is from iron to copper through hot when both junctions are above 280° C.

§ "Experimental Researches in Thermoelectricity," *Proc. R. S.* May 1854; republished as Art. LI. in *Mathematical and Physical Papers*, Vol. I. (see pp. 464—465).

to begin with; and if we draw out the copper plate *CD*, heat will be produced at *J*. The thermodynamics of this action*, because it does not involve unequal temperatures in different parts of the metals concerned, is a proper subject for unqualified application of Carnot's law, and has nothing of the unsatisfactoriness of the thermodynamics of thermoelectric currents, which essentially involves dissipation of energy by conduction of heat through metals at different temperatures in different parts. At present we cannot enter further into thermodynamics than to remark that when the plate *CD* is drawn out, the heat produced at *J* is not the thermal equivalent of the work done by the drawing out of the copper plate, but in all probability is very much less than the thermal equivalent. Probably by far the greater part of the work spent in drawing out the plate against the electric attraction goes to storing up electrostatic energy, and but a small part of it is spent on heat produced at *J*; or on excess (positive or negative) of this Peltier heat above quasi-Peltier (positive or negative) absorptions of heat in the surface layers of the opposed surfaces when experiencing changes of electrification.

23. Returning to Fig. 9; suppose, by electrodes connected to *A*, *B* and an independent electromotive force, a current is kept flowing from copper to iron through one junction, and from iron to copper through the other; the Peltier heat produced where the current passes from iron to copper is manifestly not the thermal equivalent of the work done. In fact, if the two junctions be at equal temperatures the amounts of Peltier heat produced and absorbed at the two junctions will be equal, and the work done by the independent electromotive force will be spent solely in the frictional generation of heat.

24. Many recent writers†, overlooking the obvious principles

* [March, 1898.] It has been given in a communication to the Royal Society of Edinburgh entitled "The Thermodynamics of Volta-contact Electricity"; Feb. 21, 1898. [*Supra*, Vol. v. p. 29.]

† Perhaps following Clerk Maxwell, or perhaps independently. At all events we find the following in his splendid book of 1873: "Hence $J\Pi$ represents the electromotive contact force at the junction acting in the positive direction....... Hence the assumption that the potential of a metal is to be measured by that of the air in contact with it must be erroneous, and the greater part of Volta's electromotive force must be sought for, not at the junction of the two metals, but at one or both of the surfaces which separate the metals from the air or other medium which forms the third element of the circuit."—*Treatise on Electricity and Magnetism*, Vol. I. § 249.

of §§ 22, 23, have assumed that the Peltier evolution of heat is the thermal equivalent of electromotive force at the junction. And in consequence much confusion, in respect to Volta's contact electricity and its relation to thermoelectric currents, has largely clouded the views of teachers and students. We find over and over again the statement that thermoelectric electromotive force is very much smaller than the Volta-contact electromotive force of dry metals. The truth is, Volta-electromotive force is found between metals all of one temperature, and is reckoned in volts, or fractions of a volt, without reference to temperature. If it varies with temperature, its *variations* may be stated in fractions of a volt per degree. On the other hand, thermoelectric electromotive force depends essentially on difference of temperature, and is essentially to be reckoned *per degree*; as for example, in fraction of a volt per degree.

25. Volta's second fundamental discovery, that is, his discovery (§ 5 above) that vitreous and resinous electricity flow away from zinc and copper to insulated metals connected with them (for example, the two electrodes of an insulated electrometer) when the two metals are separated after having been in metallic contact, makes it quite certain that there must be electric force in the air or ether in the neighbourhood of two opposed surfaces of different metals metallically connected. This conclusion I verified about thirty-six years ago by experiments described in a letter to Joule, of January 21, 1862, which he communicated to the Literary and Philosophical Society of Manchester, published in the Proceedings of the Society and in *Electrostatics and Magnetism* (§ 400) under the title of "A New Proof of Contact-electricity."

26. Volta's second fundamental discovery also makes it certain that movable pieces of two metals, metallically connected, attract one another, except in the particular case when their free surfaces are Volta-electrically neutral to one another. This force, properly viewed, is a resultant of chemical affinity between thin surface layers of the two metals. And the work done by it, when they are allowed to approach through any distance towards contact between any parts of the surfaces, is the dynamical equivalent of the portion of their heat of combination due to the approach towards complete chemical combination constituted by the diminution of distance between the two bodies. To fix the ideas,

let the metals be two plane parallel plates of zinc and copper, with distance between them small in comparison with their diameters, and let us calculate the amount of the attractive force between them at any distance. Let V be the difference of potentials of the air or ether very near the two metallic frontiers, but at distances from these frontiers amounting at least to several times the distance from molecule to nearest molecule in either metal (see footnote on § 16 above). The electric force in air or ether between these surfaces will be V/D, if D denotes the distance between them. Hence (our molecular microscopic binocular set aside) if ρ is the electric density of either of the opposed surfaces, A the area of either of the two, and P the attraction between them, we have

$$\frac{V}{D}=4\pi\rho, \qquad P=\frac{1}{2}\rho\frac{V}{D}A.$$

Hence $$P=V^2A/8\pi D^2.$$

Hence the work done by electric attraction in letting them come from any greater distance asunder D' to any smaller distance D is

$$\frac{V^2A}{8\pi}\left(\frac{1}{D}-\frac{1}{D'}\right), \text{ or approximately, } \frac{V^2A}{8\pi D},$$

if D is very small in comparison with D'.

27. For clean sand-papered copper and zinc* we may take V as $\frac{3}{4}$ of a volt c.g.s. electromagnetic, or $\frac{1}{400}$ c.g.s. electrostatic.

Let now A be 1 sq. cm. and D be ·001 of a centimetre. We find P equal to ·249 dyne, and the work done by attraction to this distance from any much greater distance is ·000249. This is sufficient to heat $5{\cdot}9\times10^{-12}$ grammes of water by 1°.

The table on the next page shows corresponding calculated results for various distances ranging from 1/100 of a centimetre to $1/10^{10}$ of a centimetre.

Columns 5 and 6 are introduced to illustrate the relation between the electric attraction we are considering and chemical

* Pellat's measured values range from ·63 to ·92, according to the physical condition left by less or more violent scrubbing with emery paper. The mean of these numbers is ·77. Murray's range was still wider, from ·63 volt to 1·13, the smallest being for copper burnished, opposed to zinc scratched and polished with glass paper; and the largest, copper polished merely with emery paper, opposed to zinc polished and burnished.

METALLICALLY CONNECTED PARALLEL DISCS OF ZINC AND COPPER, EACH OF 1 SQUARE CENTIMETRE AREA, ATTRACTING ONE ANOTHER.

1 Distance between plates D	2 Force of attraction in dynes* P	3 Work in ergs* W	4 Equivalent of W in heat-units (gramme-water-1° Cent.) H	5 (See § 27) Heat-units per gramme of brass disc of thickness D and area 1 sq. cm. $H \div 8D$	6 (See § 27) Rise of temperature produced by giving H to copper and zinc discs of thickness $\frac{1}{2}D$, or to brass disc of thickness D and area 1 sq. cm. if specific heat constant at ·093 $H \div (8 \times D \times \cdot 093)$
10^{-2} of centimetre	$10^{-4} \times 25$ of dyne	$10^{-4} \times \cdot 25$ of erg	$10^{-12} \times \cdot 59$ of heat-unit		
10^{-3} „	$10^{-2} \times 25$ „	$10^{-3} \times \cdot 25$ „	$10^{-11} \times \cdot 59$ „		
10^{-4} „	25 dynes	$10^{-2} \times \cdot 25$ „	$10^{-10} \times \cdot 59$ „		
10^{-5} „	$10^{2} \times 25$ „	$10^{-1} \times \cdot 25$ „	$10^{-9} \times \cdot 59$ „		
10^{-6} „	$10^{4} \times 25$ „	·25 „	$10^{-8} \times \cdot 59$ „	·00074	·0079°
10^{-7} „	$10^{6} \times 25$ „	$10 \times \cdot 25$ ergs	$10^{-7} \times \cdot 59$ „	·074	·79°
10^{-8} „	$10^{8} \times 25$ „	$10^{2} \times \cdot 25$ „	$10^{-6} \times \cdot 59$ „	7·4	79°
10^{-9} „	$10^{10} \times 25$ „	$10^{3} \times \cdot 25$ „	$10^{-5} \times \cdot 59$ „	740	7,900°
10^{-10} „	$10^{12} \times 25$ „	$10^{4} \times \cdot 25$ „	$10^{-4} \times \cdot 59$ „	74,000	790,000°

* The dyne is ·981 of a milligramme heaviness in the latitude of Greenwich. For approximate estimate it may be taken as 2 per cent. less than 1 milligramme heaviness in any latitude. The erg is the work done by a force of 1 dyne acting through the space of 1 centimetre.

affinity as manifested by heat of combination. The "brass" referred to is an alloy of equal parts of zinc and copper, assumed to be of specific gravity 8 and specific heat ·093.

28. It would be exceedingly difficult, if indeed possible at all, to show by direct experiment, at any distance whatever, the force of attraction between the discs; as we see from the table at a distance of 1/100 of a centimetre it amounts to only 1/400 of a milligramme-heaviness; and to only $2\frac{1}{2}$ grammes-heaviness at the distance 10^{-5} of a centimetre, which is about $\frac{1}{6}$ of the wave-length of ordinary yellow light. At the distances 10^{-7}, 10^{-8}, 10^{-9} of a centimetre the calculated forces of attraction are 25 kilogrammes, $2\frac{1}{2}$ tons*, and 250 tons. This last force is 2 or 3 times the breaking weight per square centimetre of the strongest steel (pianoforte wire), 6 times that of copper, 15 times that of zinc. We are, therefore, quite sure that the increase of attraction according to the inverse square of the distance is not continued to such small distances as 10^{-9} of a centimetre; and at distances less than this, the electric attraction merges into molecular force between the two metals.

29. Consider, now, a large number of discs of zinc and copper, each of 1 square centimetre area, and thickness D, and polished on both sides. On one side of each disc attach three very small columns, of length D, of glass or other insulating material, and place one disc on top of the insulators of another, zinc and copper alternately, so as to make a dry insulated pile of the metal discs, separated by air spaces each equal to the thickness D. If in the building of this pile each disc is kept metallically connected with the one over which it is placed, while it is being brought into its position, work will be done upon it by electric attraction to the amount shown in column 3, and the total work of electric attraction during the building of the pile will be the amount shown in column 3, multiplied by one less than the number of discs.

But if each disc, after being metallically connected with the one on which it is to be placed, till it comes within some con-

* The metrical ton is about 2 per cent. less than (·984 of) the British ton in general use through the British empire for a good many years before 1890, but destined, let us hope, to be rarely if ever used after the 19th century, when the French metrical system becomes generally adopted through the whole world.

siderable distance—say $300D$, for example, from the disc over which it is to rest—is then disconnected and kept insulated while carried to its position in the pile, no work will be done on it by electric attraction. And if now, lastly, metallic connection is made between all the discs of the pile, currents pass from each copper to each zinc disc, and heat is generated to an amount equal to that shown in column 4, multiplied by one less than the number of discs; and if this heat is allowed to become uniformly diffused through the metals, they rise in temperature to the extent shown in column 6.

All these statements assume that the electric attraction increases according to the inverse square of the distance between opposed faces of zinc and copper. We have already (§ 28) seen that this assumption cannot be extended to such small distances as 10^{-9} of a centimetre. We have now further proof of this conclusion beyond the possibility of doubt, because the large numbers in columns 5 and 6 for 10^{-9} are enormously greater than any rational estimate we can conceive for the heat of combination of equal parts of zinc and copper per gramme of the brass formed. (See § 32 below.)

30. When, on a Friday evening in February 1883—fourteen years ago—quoting from an article which had been published in *Nature** in 1870, I first brought these views before the Royal Institution, we had no knowledge of the amount of heat of combination of zinc and copper, nor indeed of any other two metals. It appeared probable to us, from Volta's discovery of contact electricity between dry metals, that there must be some heat of combination; but I could then only express keenly-felt discontent with our ignorance of its amount. Now, however, after twenty-seven years' endurance, I am happily relieved since yesterday by Professor Roberts Austen, who most kindly undertook to help me in my preparations for this evening, with an investigation on the heat of combination of copper and zinc, by which he has found that the melting together of 30 per cent. of zinc with 70 per cent. of copper generates about 36 heat-units (gramme-water-Cent.) per gramme of the brass formed. I am sure you will all join with me in hearty thanks to him, both for this result and for his further great kindness in letting us now

* *Nature*, Vol. I. p. 551, "On the Size of Atoms." [*Supra*, Vol. V. p. 289.]

see a very beautiful experiment, demonstrating a large amount of heat of combination between aluminium and copper, in illustration of his mode of experimenting with zinc and copper, which could not be so conveniently put before you, because of the dense white fumes inevitable when zinc is melted in the open air.

[Experiment: A piece of solid aluminium dropped into melted copper: large rise of temperature proved by thermoelectric test. Result seen by all in large deflection of spot of light reflected from mirror of galvanometer.]

31. Another method of investigating the heat of combination of metals, which I have long had in my mind, is to compare the heat evolved by the solution of an alloy in an acid with the sum of the heats of combination of its two constituents in mixed powders. The former quantity must be less than the latter by exactly the amount of the heat of combination. This investigation was undertaken a month ago by Mr Galt, in the Physical Laboratory of the University of Glasgow, and he has already obtained promising results; but many experimental difficulties, as was to be expected, have presented themselves, and must be overcome before trustworthy results can be obtained.

Added Feb. 1898. By dissolving a gramme of a powdered alloy, and again a gramme of mixed powders of the two metals in the same proportion, in dilute nitric acid, Mr Galt has now obtained approximate determinations of heats of combination for four different alloys, as shown in the following table:—

No.	Alloy	Heat of combination †
I.* ...	48 per cent. zinc 52 „ copper	... 77
II. ...	30 per cent. zinc 70 „ copper	... 34·6
III.* ...	76·7 per cent. silver 23·3 „ copper	... 18
IV. ...	51·6 per cent. silver 48·4 „ copper	... 7

The composition stated for the alloy in each case is the result of chemical analysis. No. I. was intended to be equal parts of

* The chemical combining proportions are—
(i) 50·8 zinc with 49·2 copper,
and (ii) 77·4 silver „ 22·6 „

† Per gramme of alloy in gramme-water-Cent. thermal units.

zinc and copper (as being approximately the chemically combining proportions); but the alloy, which resulted from melting together equal parts, was found to have 4 per cent. more copper than zinc, there having no doubt been considerable loss of the melted zinc by evaporation. No. III. turned out on analysis to be, as intended, very nearly in the chemically combining proportions of silver and copper. No. IV. was intended to be equal parts of silver and copper, but analysis showed the deviation from equality stated in the table. The proportions of No. II. were chosen for the sake of comparison with Professor Roberts Austen's result (§ 30), and the agreement (34·6 and 36) is much closer than could have been expected, considering the great difference of the two methods and the great difficulties in the way of obtaining exact results which each method presents.

From a chemical point of view it is interesting to see, from Mr Galt's results, how much more, both in the case of copper and zinc, and copper and silver, the heat of combination is, when the proportions are approximately the chemically combining proportions. than when they differ from these proportions to the extents found in Alloys II. and IV. Mr Galt intends, in continuance of his investigation, to determine as accurately as he can the heats of combination of many different alloys of zinc and copper and of silver and copper, and so to find whether or not it is greatest when the proportions are exactly the chemically "combining proportions." He hopes also to make similar experiments with bismuth and antimony, using *aqua regia* as solvent.

32. *Feb.* 1898. Looking now to column 5 of the table of § 27, we see from Professor Roberts Austen's result, 36 thermal units, for the heat of combination of 30 per cent. copper with 70 per cent. zinc, and from Galt's 77 thermal units for equal parts of copper and zinc, that the law of electric action on which the calculations of the tables are founded is utterly disproved for discs of metal of one one-thousand-millionth of a centimetre thickness, with air or ether spaces between them of the same thickness, but is not disproved for thicknesses of one one-hundred millionth of a centimetre.

Consider now our ideal insulated pile (§ 29) of discs 10^{-8} of a centimetre thick, with air or ether spaces of the same thickness between them. Suddenly establish metallic connection between all the discs. The consequent electric currents will generate 7·4 thermal units, and heat the discs by 79° C. Take again the insulated column with thicknesses and distances of 10^{-8} of a centimetre; remove the ideal glass separators and diminish the distance to 10^{-9} of a centimetre (the thicknesses of discs being still 10^{-8} of a centimetre). Now, with these smaller distances between two opposed areas, make metallic contact throughout the column by bending the corners (the discs for convenience being now supposed square); 74 thermal units will be immediately generated, and the discs will rise 790° C. in temperature and we have a column of hot brass—perhaps solid, perhaps liquid. This last statement assumes that the law of electric action, on which the table is founded, holds for discs 10^{-8} of a centimetre thick, with ether or air spaces between them of 10^{-9} of a centimetre. In reality it is probable that the law of electric action for discs 10^{-8} of a centimetre thick, begins to merge into more complicated results of intermolecular forces, before the distance is as small as 10^{-8} of a centimetre.

Resuming our mental molecular microscopic binocular (§ 16, footnote), we cannot avoid seeing molecular structures beginning to be perceptible at distances of the hundred-millionth of a centimetre, and we may consider it as highly probable that the distance from any point in a molecule of copper or zinc to the nearest corresponding point of another molecule is less than one one-hundred-millionth, and greater than one one-thousand-millionth of a centimetre.

33. In all that precedes I have, by frequent repetition of the phrase "air or ether," carefully kept in view the truth that the *dry* Volta contact-electricity of metals is, in the main, independent of the character of the insulating medium occupying space around and between the metals concerned in each experiment, and depends essentially on the chemical and physical conditions of molecules of matter in the thin surface stratum between the interior homogeneous metal and the external space, occupied by ether and dry or moist atmospheric air or any gas or vapour which does not violently attack the metal: or by ether with vapours only of mercury and glass and platinum and steel and vaseline (caulking the glass-stopcocks), as in Bottomley's experiments (§ 14 above).

This truth has always seemed to me convincingly demonstrated by Volta's own experiments, and I have never felt that that conviction needed further foundation; though of course I have not considered quite needless or uninstructive, Pfaff's and my own and Pellat's repetitions and verifications, in different gases at different pressures, and Bottomley's extension of the demonstration to vacuum of 2½ millionths of an atmosphere. I am now much interested to see by Professor Oliver Lodge's report, already referred to (§ 4 above), that in the Bakerian Lecture to the Royal Society in 1806*, Sir Humphry Davy, who had had contemporaneous knowledge of Volta's first and second discoveries, expressed himself thus clearly as to the validity of the second: "Before the experiments of M. Volta on the electricity excited by mere contact of metals were published, I had to a certain extent adopted this opinion," an opinion of Fabroni's; "but the new fact immediately proved that another power must necessarily be concerned, for it was not possible to refer the electricity exhibited by the opposition of metallic surfaces to any chemical alterations, particularly as the effect is more distinct in a dry atmosphere, in which even the most oxidisable metals do not change, than in a moist one, in which many metals undergo oxidation."

34. It is curious to find, thirty or forty years later, De la Rive explaining away Volta's second discovery by moisture in the atmosphere! Fifty-one years ago, when I first learned Volta's second discovery, by buying, in Paris, apparatus by which it has ever since been shown in the ordinary lectures of my class in the University of Glasgow, I was warned that De la Rive had found it wrong, and had proved it to be due to oxidation of the zinc by moisture from the air. I soon tested the value of this warning by the experiments of § 5 above, and a considerable variety of equivalent experiments, in one of which (real or ideal, I cannot remember which), a varnished zinc disc, scratched in places and moistened, sometimes on the scratched parts and sometimes where the varnish was complete, was tested in the usual manner by separating from contact with an unvarnished or varnished copper disc, with or without metallic connection when the discs were at their nearest.

* *Phil. Trans.* 1807.

§§ 35—40 *are added in Feb.* 1898.

35. Within the last eighteen or twenty years there has been a tendency among some writers to fall back upon De la Rive's old hypothesis, of which there are signs in expressions quoted by Prof. Oliver Lodge in his great and valuable report of 1884, and in some statements also of Prof Lodge's own views.

In what is virtually a continuation of this report in the *Philosophical Magazine* a year later*, we find the following with reference to writings of Helmholtz and myself on the contact-electricity of metals: "Both these contact theories, in explaining the Volta effect, ignore the existence of the oxidising medium surrounding the metals. My view explains the whole effect as the result of this oxygen bath, and of the chemical strain by it set up." With views seemingly unchanged, he returned to the subject at the end of 1897 with the following statement in the printed syllabus of his "Six Lectures adapted to a Juvenile Auditory, on the Principles of the Electric Telegraph" (Royal Institution, Dec. 28, 1897, Jan. 8, 1898).

"Chemical method of producing a current—Voltaic cell—Two differently oxidisable metals immersed in an oxidising liquid and connected by a wire can maintain an electric current, through the liquid and through the wire, so long as the circuit is closed. [The same two metals immersed in a potentially oxidising gas and connected by a wire, can maintain an electric force or voltaic difference of potential in the space between them.]

"N.B.—No one need try too hard to understand sentences in brackets."

And lastly, after some correspondence which passed between us in December, I have to-day (Feb. 14) received from him a "slightly amplified statement made in order to concentrate the differences," which he kindly gives me for publication as a supplement to the shorter statement from the syllabus.

Amplification, February, 1898.

"There is a true contact-force at a zinc-copper junction†, which on a simple and natural hypothesis (equivalent to taking an

* Prof. O. Lodge "On the Seat of the Electromotive Force in a Voltaic Cell," *Phil. Mag.* Oct. 1885, p. 383.

† See footnote on § 16 above. K. Feb. 14, 1898.

integration-constant as zero) can be measured thermoelectrically* and is about $\frac{1}{3}$ millivolt at 10° C.

"A voltaic force, more than a thousand times larger*, exists at the junction of the metals with the medium surrounding them; and in an ordinary case is calculable as the difference of oxidation-energies of zinc and copper; but it has nothing to do with the heat of formation of brass.

References:

"*Phil. Mag.* [5].

Vol. XIX. pp. 360 and 363, brass and atoms, pp. 487 and 494, summary.

Vol. XXI. pp. 270 and 275, thermoelectric argument.

Vol. XXII. p. 71, Ostwald experiment.

August 1878, Brown experiment."

36. With respect to the first of the two paragraphs of this last statement and the first two lines of the second, the wrongness of the view there set forth is pointed out in § 24 above. With respect to the last clause of the second paragraph and the statement quoted from the syllabus, I would ask any reader to answer these questions:—

(i) What would be the efficacy of the supposed oxygen bath in the experiments of § 2 above with varnished plates of zinc and copper? or in Erskine Murray's experiment, described in his paper communicated last August to the Royal Society, in which metallic surfaces, scraped under melted paraffin so as to remove condensed oxygen or nitrogen from them, and leave fresh metallic surfaces in contact with a hydro-carbon, are subjected to the Voltaic experiment? or in Pfaff's and my own and Pellat's experiments with different gases, at ordinary and at low pressures, substituted for air? or in Bottomley's high vacuum and hydrogen and oxygen experiments (§ 14 above)?

(ii) What would be the result of Volta's primary experiment, shown at the commencement of my lecture (§ 1 above), if it had been performed in some locality of the universe a thousand kilometres away from any place where there is oxygen? The insulators may be supposed to be made of rock-salt or solid paraffin, so that there may be no oxygen in any part of the apparatus. This I say because I understand that some anti-

* See § 24 above. K. Feb. 14, 1898.

Voltaists have explained Bottomley's experiments by the presence of vapour of silica from the glass, supplying the supposedly needful oxygen!

37. The anti-Voltaists seem to have a superstitious veneration for oxygen. Oxygen is entitled to respect because it constitutes 50 per cent. of all the chemical elements in the earth's crust; but this gives it no title for credit as coefficient with zinc and copper in the dry Volta experiment, when there is none of it there. Oxygen has more affinity for zinc than for copper; so has chlorine and so has iodine. It is partially true that different metals—gold, silver, platinum, copper, iron, nickel, bismuth, antimony, tin, lead, zinc, aluminium, sodium—are for dry Volta contact-electricity in the order of their affinities for oxygen; but it is probably quite as nearly true that they are in the order of their affinities for sulphur, or for oxysulphion (SO_4) or for phosphorus or for chlorine or for bromine. It may or may not be true that metals can be unambiguously arranged in order of their affinities for any of these named substances; it is certainly true that they cannot be *definitely and surely* arranged in respect to their dry Volta contact-electricity. Murray's burnishing, performed on a metal which has been treated with Pellat's washing with alcohol and subsequent scratching and polishing with emery, alters the Volta quality of its surface far more than enough to change it from below to above several metals polished only by emery; and, in fact, Pellat had discovered large differences due to molecular condition without chemical difference, before Murray extended this fundamental discovery by finding the effect of burnishing.

38. Returning to Prof. Lodge's supposed oxygen bath (§ 35); if it exists between the zinc and copper plates, it diminishes or annuls or reverses the phenomenon, to explain which he invokes its presence (see § 5 above).

39. Many years ago I found that ice, or hot glass, pressed on opposite sides by polished zinc and copper, produced deviations from the metallic zero of the quadrants of an electrometer metallically connected with them in the same direction as if there had been water in place of the ice or hot glass. From this I inferred that ice and hot glass, both of which had been previously known to have notable electric conductivity, acted as electrolytic conductors.

Experiments made by Maclean and Goto in the Physical Laboratory of the University of Glasgow in 1890*, proved that polished zinc and polished copper, with fumes passing up between them from the flame of a spirit-lamp 30 centimetres below, gave, when metallically connected to the quadrants of an electrometer, deviations from the metallic zero in the same direction, and of nearly the same amount, as if cold water had been in place of the flame. This proved that flame acted as an electrolytic conductor. They also found that hot air from a large red-hot soldering bolt, put in the place of the spirit lamp, had no such effect; nor had breathing upon the plates, nor the vapour of hot water, any effect of the kind. In fact hot air, and either cloudy or clear steam, act as very excellent insulators; but there is some wonderful agency in fumes from a flame, remaining even in cooled fumes, in virtue of which the electric effect on zinc and copper is nearly the same as if continuous water, instead of fumes, were between the plates and in contact with both†.

A similar conclusion in respect to air traversed by ultra-violet light was proved by Righi‡, Hallwachs§, Elster and Geitel||, Branly¶. The same was proved for ordinary atmospheric air, with Röntgen rays traversing it between plates of zinc and copper, by Mr Erskine Murray, in an experiment suggested by Prof. J. J. Thomson, and carried out in the Cavendish Laboratory of the University of Cambridge**.

40. The substitution for ordinary air between zinc and copper, of ice or hot glass, or of air or gas modified by flame or by ultra-violet rays, or by Röntgen rays, or by uranium (§§ 41, 42 below), gives us, no doubt, what would to some degree fulfil Prof. Lodge's idea of a "potentially-oxidising" gas, and each one of the six fails wholly or partially to "maintain electric force or voltaic difference of potential in the space between them." In fact, Prof. Lodge's bracketed sentence, so far as it can be understood, would be nearer the truth if in it "cannot" were substituted for "can." I hope no reader will consider this sentence too short or sharp. I am quite sure that Prof. Lodge will approve of its tone, because in

* *Phil. Mag.* Aug. 1890.

† Kelvin and Maclean, *R.S.E.* 1897.

‡ *Rend. R. Acc. dei Lincei*, 1888, 1889.

§ *Wiedemann's Annalen*, Vol. XXXIV. 1888.

|| *Ibid.* Vols. XXXVIII. XLI. 1888.

¶ *Comptes Rendus*, 1888, 1890.

** *Proc. R. S.* March 1896.

his letter to me of the 14th, he says, "In case of divergence of view it is best to have both aspects stated as crisply and distinctly as possible, so as to emphasise the difference." I wish I could also feel sure that he will agree with it, but I am afraid I cannot, because in the same letter he says, "I am still unrepentant."

Continuation of Lecture of May 21, 1897.

41. In conclusion, I bring before you one of the most wonderful discoveries of the century now approaching its conclusion, made by the third of three great men, Antoine Becquerel, Edmond Becquerel, Henri Becquerel—father, son and grandson—who by their inventive genius and persevering labour have worthily contributed to the total of the scientific work of their time; a total which has rendered the nineteenth century more memorable than any one of all the twenty-three centuries of scientific history which preceded it, excepting the seventeenth century of the Christian era.

You see this little box which I hold in my right hand, just as I received it three months ago from my friend Professor Moissan, who will be here this day week to show you his isolation of fluorine. It induces electric conductivity in the air all round it. If I were to show you an experiment proving this, you might say it is witchcraft. But here is the witch. You see, when I open the box, a piece of uranium of about the size of a watch. This production of electric conductance in air is only one of many marvels of the "uranium rays" discovered a year ago by Henri Becquerel, of no other of which can I now speak to you, except that the wood and paper of this box, and my hand, are to some degree transparent for them.

I now take the uranium out of its box and lay it on this horizontal copper plate, fixed to the insulated electrode of the electrometer. I fix a zinc plate, supported by a metal stem which is in metallic connection with the sheath of the electrometer, horizontally over the copper plate at a distance of about one centimetre from the top of the uranium. Look at the spot of light; it has already settled to very nearly the position which you remember it took when we had a water-arc between the copper and zinc plates, connected as now, copper to insulated quadrants and zinc to the sheath. I now lift the uranium, insulating it from the copper plate by three very small pieces of solid paraffin, so as to touch neither plate, or, again, to touch the zinc but not the copper. This change makes but little difference to the spot of light. I tilt the uranium now to touch the zinc above and the copper below; the spot of light comes to the metallic zero as nearly as you can see. I leave it to itself now, resting on its paraffin supports and not touching the zinc, and the spot of light goes back to where it was; showing about three-quarters of a volt positive.

42. I now take this copper wire, which is metallically connected with the zinc plate and the sheath of the electrometer, and bring it to touch the under side of the copper shelf on which the uranium is supported by its paraffin insulators. Instantly the spot of light moves towards the metallic zero, and after a few vibrations settles there. I break the contact; instantly the spot of light begins to return to its previous position, where it settles again in less than half a minute. You see, therefore, that if I re-make and keep made the metallic contact between the zinc and copper plates, a current is continuously maintained through the connecting wire, by which heat is generated and radiated away, or carried away by the air; as long as the contact is kept made. What is the source of the energy thus produced? If we take away the uranium, and send cool fumes from a spirit-lamp, or shed Röntgen rays or ultra-violet light, between the zinc and copper, the results of breaking and making contact would be just what you see with uranium. So would they be—you have already, in fact, seen them (§ 5)—without either Röntgen rays or ultra-violet light, but with the copper and zinc a little closer together and with a drop of water between them: and so would they be with dry ice, or with hot glass, between and touched by the zinc and copper. In each of these six cases we have a source of energy; the well-known electro-chemical energy given by the oxidation of zinc in the last mentioned three cases; and the energy drawn upon by the cooled fumes, or by the Röntgen rays or ultra-violet light, acting in some hitherto unexplained manner, in the three other cases. We may conjecture evaporations of metals; we have but little confidence in the probability of the idea. Or does it depend on metallic carbides mixed among the metallic uranium? I venture on no hypothesis. Mr Becquerel has given irrefragable proof of the truth of his discovery of

radiation from uranium of something which we must admit to be of the same species as light, and which may be compared with phosphorescence. When the energy drawn upon by this light is known, then, no doubt, the *quasi* electrolytic phenomena, induced by uranium in air*, which you have seen, will be explained by the same dynamical and chemical principles as those of the previously known electrolytic action of cooled fumes from a spirit-lamp, and of air traversed by Röntgen rays or ultra-violet light.

CONTACT-ELECTRICITY AND ELECTROLYSIS ACCORDING TO FATHER BOSCOVICH.

[From *Nature*, Vol. LVI. May 27, 1897, pp. 84, 85.]

YESTERDAY evening, in the Royal Institution, I spoke of an ideal one-fluid electricity subject to attractions of solid substance, to account for contact-electricity of metals; and I said that before the end of our meeting I might speak of it further and might have to reverse the conventional language I was using as to positive and negative, and call resinous electricity positive, and vitreous negative. My allotted hour was woefully overpast, and half an hour more gone, before I could return to the subject; and I felt bound to stop. What I wished to say may be said in the columns of *Nature* in fewer words than I could have found, to make it intelligible, last night.

Varley's fundamental discovery of the kathode torrent, splendidly confirmed and extended by Crookes, seems to me to necessitate the conclusion that resinous electricity, not vitreous, is *the electric fluid*, if we are to have a one-fluid theory of electricity. Mathematical reasons, to which I can only refer without explanation at present, prove that if resinous electricity is a continuous homogeneous liquid, it must, in order to produce

* Experiments made in the Physical Laboratory of the University of Glasgow [§ 33 of Kelvin, Beattie and Smolan, *Proc. R. S. E.*; also *Nature*, March 11, 1897, and *Phil. Mag.* March 1898] show this electrolytic conductivity to be produced by uranium to nearly the same amount in common air, oxygen and carbonic acid; and to about one-third of the same amount in hydrogen, at ordinary atmospheric pressure; but only to about $\frac{1}{100}$ of this amount in each of these four gases at pressures of 2 or 3 millimetres. There seems every reason to believe that it would be non-existent in high vacuum, such as that reached by Bottomley in his Volta-contact experiments (§ 14 above).

the phenomena of contact-electricity which you have seen this evening, be endowed with a cohesional quality such as that shown by water on a red-hot metal, or mercury on any solid other than a metal amalgamated by it. It is just conceivable. though it does not at present seem to me very probable, that this idea may deserve careful consideration. I leave it, however, for the present, and prefer to consider an atomic theory of electricity foreseen as worthy of thought by Faraday and Clerk Maxwell, very definitely proposed by Helmholtz in his last lecture to the Royal Institution, and largely accepted by present-day theoretical workers and teachers. Indeed, Faraday's law of electro-chemical equivalence seems to necessitate something atomic in electricity, and to justify the very modern name *electron*. The older, and at present even more popular, name *ion* given sixty years ago by Faraday, suggests a convenient modification of it, *electrion*, to denote an atom of resinous electricity. And now, adopting the essentials of Aepinus' theory, and dealing with it according to the doctrine of Father Boscovich, each atom of ponderable matter is an electron of vitreous electricity; which, with a neutralising electrion of resinous electricity close to it, produces a resulting force on every distant electron and electrion which varies inversely as the cube of the distance, and is in the direction determined according to the well-known requisite application of the parallelogram of forces.

In a solid metal the ponderable atoms must exert such other mutual forces, compounded with the electric forces, that the assemblage in equilibrium shall have the crystalline configuration, and the elasticity-moduluses, of the metal. The electrions must be perfectly mobile among the ponderable atoms, subject only to the condition that the electric attraction ceases to increase according to the inverse square of the distance and becomes zero (or, perhaps, strong repulsion) when the distance is diminished below some definite limit. For simplicity we may arbitrarily assume the following conditions:

(1) Each electrion is a point-atom of resinous electricity and repels every other electrion with a force varying inversely as the square of the distance between them.

(2) Each electrion is attracted by each ponderable atom with a force which varies inversely as the square of its distance from

the centre of the ponderable atom when the distance exceeds a certain limit r and is zero when the distance is less than r.

(3) The shortest distance between two centres of ponderable atoms need not be limited to be $> 2r$: it may be whatever we find convenient for the structure and properties to be realised. It will be $> 2r$ in an insulating solid and $< 2r$ in a conductor.

Two pieces of metal, M, M', each constituted as I have now explained, will behave in respect to contact-electricity just as two pieces of metal behave in a perfect vacuum. For example, if $r > r'$, M will behave to M' as zinc behaves to copper.

To illustrate electrolysis, consider an ideal case of a detached compound zinc-copper atom, composed of two single atoms with their centres at C, C'; and two electrions e, e' which must, for equilibrium, be in the positions shown in the diagram [omitted], if r, r' be of such magnitudes as the radii of the circles showing the shortest distances to which C and C' attract electrions. Let now electrified bodies at great distances (such as the vitreously and resinously electrified plates indicated in the diagram) act in the manner indicated by the dotted arrows relatively to the ponderable atoms, and the full arrows relatively to the electrions. The ponderable atom C will be drawn away to the right by the electric force on itself: and the ponderable atom C' will be dragged away to the left by the two electrions overcoming the rightward force which itself experiences in virtue of the electric field. Lastly, to take a real case, the electrolysis of copper-sulphate, let C' be the centre of an atom of copper in combination with oxysulphion (SO_4), not shown in the diagram; with, in all, six electrions. The copper atom C' will be drawn away to the right, with no electrion attached to it: and the oxysulphion will be pushed and dragged to the left by the excess of leftward electric forces on the six electrions above rightward electric forces on the five ponderable atoms.

251. NOTE TO "THE ELECTRIFICATION OF AIR BY URANIUM AND ITS COMPOUNDS. BY J. CARRUTHERS BEATTIE."

[From *Edinb. Roy. Soc. Proc.* Vol. XXI. [read June 7, 1897], pp. 471, 472; *Phil. Mag.* Vol. XLIV. July 1897, pp. 107, 108.]

IN some of our previous experiments with high voltages we found sparks to pass between uranium and other metals* apparently according to the laws of disruptive discharge subject to but little modification by the special quasi-conductivity induced in air by the "uranium rays." On the other hand, all our experiments with voltages, less than 500 or 600 volts per cm. of line of force in the air at ordinary atmospheric pressure seem to be not sensibly influenced by disruptive charges or by brushes; and the quasi-conductivity of air induced by uranium was the dominant factor. This is undoubtedly the case in the experiments now described by Dr Beattie, and I assume it to be so in what follows, except when I give express warning of possible liability to disruptive discharges.

The effective conductivity induced in the air by the uranium influence is of course greatest in the immediate neighbourhood of the uranium, but there is something of it throughout the enclosure. Hence it may be expected that electricity of the same kind as that of the uranium will be deposited in the air close around it, and electricity of the opposite kind in the air near the enclosing metal surface. According to our former experiments* the quantity flowing from either the uranium or from the surrounding metal per sq. cm. of its surface increases but little with increased voltage when this exceeds 5 or 10 volts per cm. Now, when the greatest diameter of the uranium is small in comparison with distances to the outer metal surface, the voltage per cm. is much greater along the lines of force near

* Kelvin, Beattie, and Smolan, *Proc. R. S. E.* 1897.

the uranium surface than near their outer ends on the surrounding metal. Hence the rate of discharge of electricity into the air from the uranium will cease to increase sensibly with the difference of potential between the uranium and the surrounding metal, while the rate of discharge of the opposite electricity from the large surrounding metal surface is still notably increasing. Hence if the dimensions and shapes of the uranium and of the surrounding metallic surface are such that for small voltages, such as 10 or 20 volts of difference between the uranium and the surrounding metal, the electricity lodged in the air by discharge from the uranium preponderates over that discharged from the surrounding metal, the excess must come to a maximum and diminish, possibly even down to zero, with greater differences of potential: and at potential differences still greater the electricity lodged in the air from the outer metal may preponderate, and the electricity in the air drawn off and given to the filter be of opposite sign to that of the uranium which was found with the lower voltages: *provided the configurations are such, and the voltages are so moderate that, disruptive discharge does not intervene to any practically disturbing extent.*

252. ELECTRIFICATION OF AIR, OF VAPOUR OF WATER, AND OF OTHER GASES. By Lord KELVIN, MAGNUS MACLEAN, and ALEXANDER GALT.

[From *Roy. Soc. Proc.* Vol. LXI. [read June 17, 1897], pp. 483—485; *Phil. Trans.* (*A.*), Vol. CXCI. 1898, pp. 187—228.]

1. IN this paper we describe a long series of experiments on the electrification of air and other gases, with which we have been occupied from May, 1894, up to the present time (June, 1897). Some results of our earlier experiments, and of preliminary efforts to find convenient methods of investigation, have from time to time been communicated to the Royal Society, the British Association, and the Glasgow Philosophical Society*.

2. The method for testing the electrification of air, which we used in our earliest experiments, was an application of the water-dropper†, long well-known in the ordinary observation of atmospheric electricity. Its use by Maclean and Goto‡, in 1890, led to an interesting discovery that air in an enclosed vessel, previously non-electrified, becomes electrified by a jet of water falling through it. An investigation of properties of matter concerned in this effect, related as it is to the "development of electricity in the breaking up of a liquid into drops," which had

* "On the Electrification of Air," *Proc. Roy. Soc.* May 31, 1894, and *Phil. Mag.* August, 1894. "Preliminary Experiments to find if Subtraction of Water from Air Electrifies It," *Brit. Assoc. Report*, 1894, p. 554. "Electrification of Air and other Gases by Bubbling through Water and other Liquids," *Proc. Roy. Soc.* February, 1895. "On the Diselectrification of Air," *Proc. Roy. Soc.* March, 1895. "On the Electrification of Air," *Proc. Glasgow Phil. Soc.* March, 1895. "On the Electrification and Diselectrification of Air and other Gases," *Brit. Assoc. Report*, 1895, p. 630.

† Kelvin and Maclean, *Proc. Roy. Soc.* 1894, and Kelvin, Maclean, and Galt, *Proc. Roy. Soc.* February, 1895. "Electrostatics and Magnetism," § 262 (from *Proc. Lit. and Phil. Soc. of Manchester*, October 18, 1859).

‡ "Electrification of Air by Water-Jet." By Magnus Maclean and Makita Goto, *Phil. Mag.* August, 1890.

been discovered by Holmgren* as early as 1873, and to the later investigations and discoveries described by Lenard†, in his paper on the "Electricity of Waterfalls," forms the subject of §§ 25—37 of the present communication.

3. The electrification of air by drops of water, breaking from a jet in it, or falling through it, or striking on the ground, or on water, or on metal below it, produces absolutely no practical disturbance of the electric potential measured by the water-dropper in its use for the observation of open-air atmospheric electricity, but constitutes a serious objection to its application for investigating atmospheric electricity within doors, unless in a very large room or hall, and renders it altogether unsuitable for the experimental investigations with which we are now concerned.

4. We were, therefore, early led to abandon it; and, for testing the electrification of air, we have used three different methods, one or other of which we have found convenient in different cases.

Method (1). Observation of electrification of the substance receiving the electricity equal and opposite to that taken by air in any case of electrification of air.

Method (2). Observation of the electricity of a hollow metal vessel into which electrified air is introduced, or from which electrified air is removed.

Method (3). Observation of the electricity taken out of air by the electric filter (§ 9).

5. Method (1) was used in the experiments described in our communication to the Royal Society of February, 1895, from which we concluded that air, and several other gases tried, became electrified by blowing them in bubbles through water, and through solutions of various salts, acids, and alkalies in water. We verified this conclusion, for the case of common air and pure water, by collecting into a large reservoir over water, air which had been

* "Sur le développement de l'électricité à l'occasion de la dissolution en gouttes des liquides," *Kongl. Svensk. Vet. Ak. Handl.*, 1 c. Vol. 11, No. 8, pp. 14—43 (pour l'an 1873).

† "Ueber die Electricität der Wasserfälle." By P. Lenard, *Annalen der Physik und Chemie*, 1892.

bubbled through pure water in a U-tube. We tested the electrification of the air thus collected by a water-dropper taking the same potential as the air at the centre of the reservoir. We thus proved that the electrification of the air was negative, as was to be expected from the positive electrification which we had found on insulated vessels containing water through which air had been bubbled.

6. Method (2) was used in the first experiments described in the present paper (§§ 16—24), which were undertaken for the purpose of determining approximately in absolute measure the total quantity of electricity in a given mass of electrified air, and particularly for finding the greatest electrification which we could communicate to a large quantity of air by needle points supplied with electricity from an electric machine. The result thus found in § 23 below, $3{\cdot}7 \times 10^{-4}$ C.G.S. electrostatic, is the greatest electric density (quantity of electricity per cubic centim.) which we have been able to communicate to air by electrified needle points. But, by an electrified hydrogen flame a density of 22×10^{-4} C.G.S. electrostatic unit was obtained (§ 65).

7. In all the experiments described in our paper after § 24, method (3) was used; but, probably, we must return to method (2) if, in future, we undertake further experiments to find the greatest electric density which we can measure in air or other gases.

8. Lenard's important discovery of very strong electric effects produced by drops of water falling on a hard surface, gave us a very convenient method for obtaining a steady and strong negative electrification of air, which we used in §§ 30—32 for preliminary efforts for finding a good and convenient form of electric filter to be used in further investigations on the electrification and diselectrification of air.

9. In testing the efficiencies of the electric filters used in method (3), we at first used the filter described in our paper on "Diselectrification of Air," *Proc. Roy. Soc.*, vol. 57, [*supra*, p. 33] and which consisted of twelve discs of brass wire cloth, fixed in a short metal pipe, supported in a paraffin tunnel. This filter was joined to the insulated quadrants of a quadrant electrometer, and electrified air was sucked through it (§ 25) till a convenient deflection was obtained. Then the filter to be tested was connected to the sheath of the electrometer, and so placed that the

electrified air passed through the tested filter before it passed through the filter attached to the electrometer (§ 68). In this way, by drawing equal quantities of electrified air, as nearly as may be equally electrified, through the different tested filters, a comparison of their relative diselectrifying powers was obtained. For example, d being the deflection obtained when no tested filter was used, and d_1, d_2 ... d_n when the tested filters were successively used, then the relative diselectrifying powers of the filters would be $\frac{d-d_1}{d}$, $\frac{d-d_2}{d}$, ... $\frac{d-d_n}{d}$, if the primary electrifications were equal.

10. In other sets of experiments (§§ 32, 69) we successively joined each separate tested filter to the insulated quadrants of the electrometer, and sucked approximately equal quantities of electrified air through them. The diselectrifying efficiencies of the filters were now approximately in simple proportion to the final readings on the electrometer.

11. But none of these methods gave us a means of determining the absolute diselectrifying power of any filter without realizing an equality of primary electrification of the air in different experiments. We therefore, a long time later, used two insulated filters and two electrometers, as described in § 55 and fig. 7. Let the filters be called AB and $A'B'$ and their diselectrifying powers n and n'. In a first experiment the electrified air was sucked through AB and $A'B'$ in immediate succession, and in a second experiment the electrified air was sucked through $A'B'$ and AB.

12. On the assumption that the two filters took out the same proportions (respectively n or n') of the electricity of the electrified air entering them in the two experiments, we get the following equations:

In the first experiment

Let Q = total quantity of electricity in air entering,

q_1 = total quantity of electricity taken out by filter AB,

q_2 = total quantity of electricity taken out by filter $A'B'$.

Then $$q_1 = nQ \quad \text{...(1)},$$

$$q_2 = n'(Q - q_1) = n'Q(1-n) \quad \text{..............(2)}.$$

In the second experiment

Let Q' = total quantity of electricity in air entering,

q_1' = total quantity of electricity taken out by $A'B'$,

q_2' = total quantity of electricity taken out by AB.

Then $$q_1' = n'Q' \quad\text{..................................(3)},$$

$$q_2' = n(Q' - q_1') = nQ'(1 - n') \quad\text{..............(4)}.$$

From these four equations we find

$$n = \frac{q_1q_1' - q_2q_2'}{q_1'(q_1 + q_2)}; \quad n' = \frac{q_1q_1' - q_2q_2'}{q_1(q_1' + q_2')}.$$

13. By taking a movable plate of a small air-condenser charged to a known potential, and applying it to the insulated terminal of the quadrant electrometers used as described in § 18, we could calculate q_1, q_2, q_1', and q_2', and hence find Q and Q', the absolute density of the electrified air or gases in C.G.S. electrostatic units. It was thus that we found 11×10^{-4} and 22×10^{-4} mentioned in §§ 64, 65.

14. Up till the middle of December of 1895 the most efficient filter we tried had a diselectrifying power of about 0·8. During the Christmas holidays of 1895, we succeeded in obtaining a filter of fine brass filings, as described in § 62, which abstracted so much of the electricity from the electrified air passing through it, that what was left was not sufficient to show on a similar filter, $A'B'$, attached to an electrometer, E' (see fig. 7). It was not necessary now to have two experiments, first with electrified air in one direction through two filters to be tested, and then with electrified air in the reverse direction through them. It was sufficient to take a filter, the diselectrifying power of which, determined as above (§ 12), is found to be very nearly unity, and attach it to electrometer E'. Then join the tested filter to another electrometer, E, and allow the electrified air to pass this filter first, and thence through the almost perfectly diselectrifying filter.

With the same notation as in § 12, we get

$$Q = \frac{n'q_1 + q_2}{n'}, \text{ and } n = \frac{n'q_1}{n'q_1 + q_2} = \frac{1}{1 + \dfrac{1}{n'}\dfrac{q_2}{q_1}}.$$

If $n' =$ unity, as it is for a filter of fine brass filings (§ 62),

$$Q = q_1 + q_2, \text{ and } n = q_1/(q_1 + q_2).$$

15. With such a filter as this it is possible to determine the quality of the natural electricity of the atmosphere, and it may be desirable that it should be used for that purpose in meteorological observatories.

Greatest Electrification which we could communicate to a large quantity of Air by one or more Electrified Needle Points (§§ 16—24).

16. The first apparatus used is shown in fig. 1. It consisted of a metal can, D, 48 centims. high and 21 centims. in diameter,

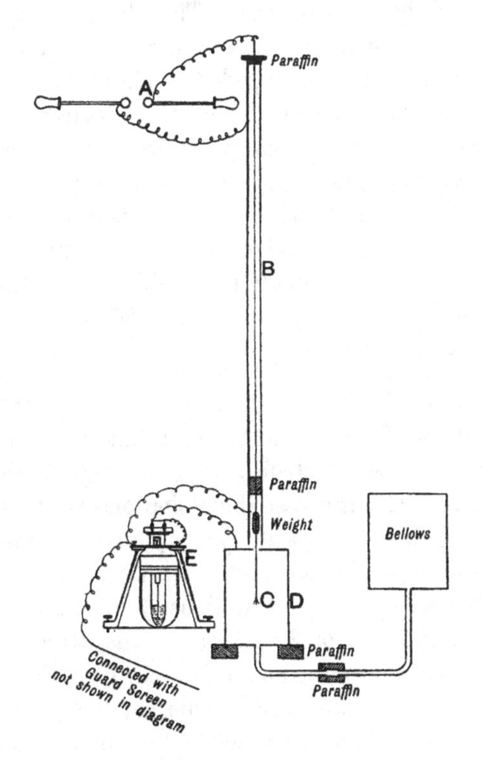

Fig. 1.

supported by paraffin blocks, and connected to one pair of quadrants of a quadrant electrometer. It had a hole at the top to admit the electrifying wire, which was 531 centims. long, hanging vertically within a metallic guard-tube, *B*. This guard-tube was always metallically connected to the other pair of quadrants of the electrometer, and to its sheath and to a metallic screen surrounding it, which is not shown in the diagram. This prevented any external influences from sensibly affecting the electrometer, such as the working of the electric machine, *A*, which stood on a shelf five metres above it.

17. The experiment is conducted as follows:—One terminal of the electric machine is connected with the guard-tube and the other with the electrifying wire, which is tipped with needle points or tinsel, and which is let down to place the point or points nearly in the centre of the can. The can is temporarily connected to the sheath of the electrometer. The electric machine is then worked for some minutes, so as to electrify the air in the can. As soon as the machine is stopped, the electrifying wire is lifted clear out of the can. The can and the quadrants in metallic connection with it are disconnected from the sheath of the electrometer, and the electrified air is very rapidly drawn away from the can by a blowpipe bellows, arranged to suck. This releases the opposite kind of electricity from the inside of the can, and allows it to place itself in equilibrium on the outside of the can and on the insulated quadrants of the electrometer in metallic connection with it.

18. We tried different lengths of time of electrification and different numbers of needles and tinsel, but we found that one needle and four minutes' electrification gave as great electrification as we could get. The greatest deflection observed was 936 scale divisions on a half millim. scale put up in the usual way, with lamp at a distance of about a metre from the electrometer. To find from this reading the electric density of the air in the can, we took a metallic disc of 2 centims. radius, attached to a long varnished glass rod and placed at a distance of 1·45 centims. from another and larger metallic disc. This small air condenser was charged from the electric light conductors in the laboratory to a difference of potential amounting to 100 volts, or $\frac{1}{3}$ of an electrostatic unit. The insulated disc thus charged was removed and laid upon the roof of the large insulated can. This addition to the metal in connection with it does not sensibly influence its

electrostatic capacity. The deflection obtained was 122 scale divisions. The capacity of the condenser is approximately

$$\frac{\pi \times 2^2}{4\pi \times 1{\cdot}45} = \frac{1}{1{\cdot}45} \text{ C.G.S. electrostatic unit.}$$

The quantity of electricity with which it was charged was therefore $\frac{1}{1{\cdot}45} \times \frac{1}{3} = \frac{1}{4{\cdot}35}$ C.G.S. electrostatic unit.

Hence, the quantity on the can and connected metal to give 936 scale divisions was $\frac{1}{4{\cdot}35} \times \frac{936}{122} = 1{\cdot}7637$ C.G.S. electrostatic units.

The capacity of the can was 16,632 cub. centims., which gives, for the quantity of electricity per cub. centim.,

$$\frac{1{\cdot}7637}{16632} = 1{\cdot}06 \times 10^{-4} \text{ of the C.G.S. electrostatic unit.}$$

19. This is about four times the electric density which we roughly estimated as about the greatest given to the air in the inside of a large vat, electrified by a needle point and then left to itself, and tested by the potential of a water-dropper with its nozzle in the centre of the vat, in experiments made more than three years ago and described in a communication to the Royal Society of date May, 1894*.

20. To enable us to remove the electrified air quickly from the can, the following modification was adopted:—The can was suspended vertically by three stout silk threads (*S*, *S*, *S*, fig. 2) which had been previously soaked in melted paraffin; it was quite open at the bottom but closed at the top, with the exception of a central aperture for the piston-rod of a piston, *P*. The piston was of wood encased in lead, and was free to move up and down in the can by the movement of the paraffined silk cord, *C*, over the pulleys, *F*, *F*. The can and the piston and piston-rod were connected metallically by spiral springs of fine wire. The can was surrounded by a metallic guard-screen, *G*, kept in connection with the sheath of the quadrant electrometer, and with the sheath of a vertical electrostatic voltmeter, and with one terminal, *B*, of an electric machine. The other terminal of the machine was

* "On the Electrification of Air," by Lord Kelvin and Magnus Maclean[, *supra* p. 6].

connected to an insulated needle point inside the can and to the insulated terminal of the voltmeter.

21. By working the machine the needle was kept charged positively or negatively at 12,000 volts for four minutes. The air inside the can became charged similarly by the brush discharge from the needle point. As soon as possible after stopping the machine, the needle was removed and *A* and *B* were joined. The wire, *W*, from the can was disconnected from the guard-screen, *G*, and then attached to the electrometer terminal, *O*, after which this terminal was insulated and the downward movement of the piston begun. Usually about thirteen seconds were required to make these changes.

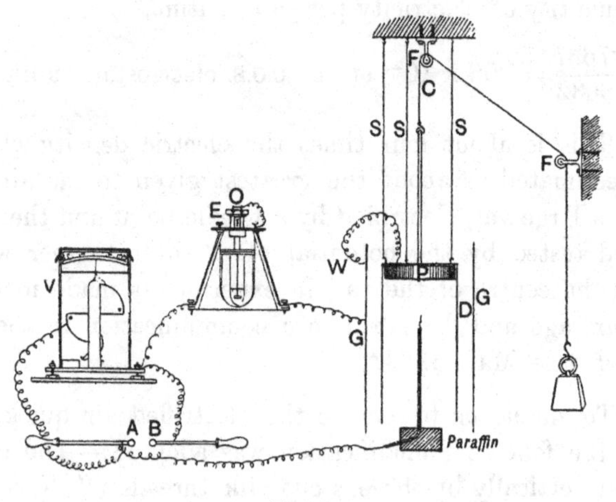

Fig. 2.

22. When the electrified air inside the can was expelled by dropping the piston to the bottom, the reading of the electrometer went off the scale, and a shorter drop had to be used to get a convenient deflection.

23. A drop of 11·5 centims., by which 3979 cub. centims. of air were expelled, gave a deflection of 1060 scale divisions. The quantity of electricity on the can and connecting wire and insulated pair of quadrants of the electrometer which gives this deflection, was (by the method of § 18) found to be 1·47 C.G.S. electrostatic. This, therefore, was the quantity of electricity of the opposite kind in the 3979 cub. centims. of air expelled from

the open bottom of the can; and the electric density of this air was therefore $3{\cdot}7 \times 10^{-4}$ C.G.S. electrostatic per cubic centimetre.

24. In preliminary experiments before electrifying the air inside the can by needle point, it was found that the dropping of the piston produced no deflection on the electrometer.

Electrification of Air by Water Jet (§§ 25—37).

25. In previous experiments* we found air to be negatively electrified by water falling in drops through it, and to pursue the investigation further the arrangement shown in fig. 3 was put up.

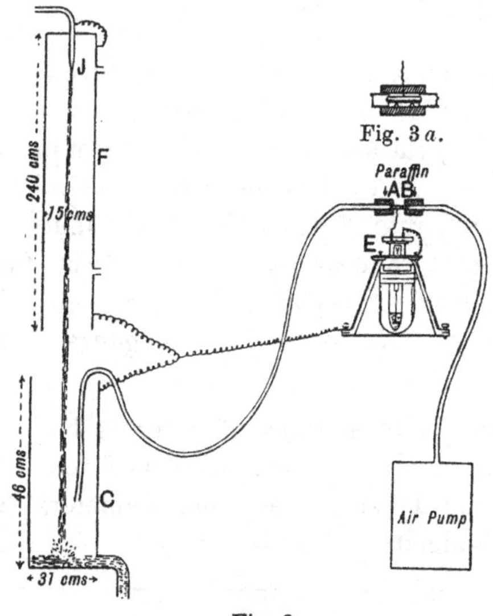

Fig. 3.

Loch Katrine water, under full pressure, issues from a jet, *J*, fixed in the lid of a funnel, *F*, and falls down the funnel centrally into a metal can, *C*, below. By means of an air-pump some of the air is withdrawn from the can or funnel through metallic tubing, as shown in the diagram, to a metallic filter, *AB*, containing 50 fine brass wire gauzes. In each experiment 200 strokes of the pump are taken.

* "Electrification of Air by Water Jet," by Maclean and Goto, *Phil. Mag.* August, 1890. "Electrification of Air," by Lord Kelvin and Magnus Maclean, *Proc. Roy. Soc.* May, 1894[, *supra* p. 6].

26. The greatest effect, 4·5 volts negative, was obtained, as was to be expected from Lenard's discovery, when the air was withdrawn from a point well down in the can, the can being close to the funnel, and the falling water rattling on the bottom of the can. Decreasing effects were observed (1) when the air was withdrawn from the can at increasing distances from the bottom; (2) when several inches of water were kept in the can*; (3) when air was drawn from the funnel, and the distance between the can and the funnel was gradually increased.

27. A filter of 100 wire gauzes gave at the rate of 26 volts, and a filter of 2 gauzes, with a loose plug of cotton wool between them, gave 6·3 volts, in the same time and under the same conditions as the filter of 50 wire gauzes gave (§ 26) 4·5 volts.

28. A *sloping metallic plate* was next fitted to the bottom of the funnel in such a way that the falling water on striking the plate passed out by the aperture between the funnel and the lower edge of the plate. In each experiment 120 strokes of the pump were taken at the rate of 40 strokes per minute. Drawing the air from the aperture near the bottom of the funnel gave, in 4 experiments, results averaging about 8 volts. These results were given by a filter containing 2 brass gauzes with cotton wool between them.

29. Two simple brass tubes of different bores with no wire gauze or cotton wool were now tried as filters. Each was 10 centims. long and 1 centim. external diameter. The following results were obtained:

Internal diameter of brass filter	Air drawn from side aperture of funnel: Near the top		Near the bottom	
	Aperture near the bottom:		Aperture near the top:	
	Shut	Open	Shut	Open
centims.	volts neg.	volts neg.	volts neg.	volts neg.
0·3	1·5	1·0	3·2	3·0
0·18	3·0	2·8	4·5	4·3

* See a Paper by Lenard on "Electricity of Waterfalls," *Wied. Ann.* 1892, Vol. 46, pp. 584—636.

Using the brass filter of 0·18 centim. bore, drawing from the aperture near the bottom, and varying the water pressure, we found mean results as follows:

Full pressure	3·4 volts.
Diminished pressure	2·6 „
Very low pressure (200 drops per minute) .	0·17 „

30. In the long metallic tube between the funnel and the testing filter we placed, in successive experiments, an increasing number of brass gauzes and plugs of wool. The air had to pass through these and the long length of tubing before reaching the testing filter at the electrometer. The testing filter consisted of a block-tin tube with two wire gauzes and one plug of cotton wool. Twenty experiments were made on air drawn from the aperture of the funnel near its lower end, and with jet from full pressure of water. The following are mean results, for 120 strokes of the pump:

No gauze between the funnel and the testing filter	2 gauzes and 1 wool plug	3 gauzes and 2 wool plugs	4 gauzes and 3 wool plugs	5 gauzes and 4 wool plugs	6 gauzes and 5 wool plugs	7 gauzes and 6 wool plugs
volts 7	volts 2·4	volts 1·25	volt 0·4	volt 0·3	volt 0·1	volt 0·04

These results show that a large proportion of the electricity was taken, by the 7 gauzes and 6 plugs, from the air before it reached the testing filter.

31. Extracting the air with no water falling gave no perceptible electrification.

32. With the water again at full pressure, and falling on the sloping plate fixed into the bottom of the funnel, the negatively electrified air was drawn from the bottom aperture through different filters at different speeds. Two experiments at each speed were usually made, and whenever possible the deflection for 120 strokes of the pump was noted. In some cases, however, the reading exceeded 8 volts, and went off the scale with much fewer strokes; but to preserve uniformity the tabular results given below are all calculated for 120 strokes.

Results 2 and 4 are in accordance with Results 2 and 3 of § 68.

Filter used	Duration of each stroke of the pump in seconds					
	0·5	1	2	3	4	5
	volts neg.	volts neg.	volts neg.	volts neg.	volts neg.	volts neg.
1. Platinum tube 2·4 cms. long, 0·1 cm. bore .	...	4·9	9·9	14·5	18·7	
2. Brass tube 4·1 cms. long, 0·2 cm. bore .	7·8	12·5	21·8	25·8	26	25
3. Brass tube 4·1 cms. long, 0·34 cm. bore .	4	7·2	12	13·2	12·7	
4. Solid brass cylinder with rounded ends, 8 cms. long and 1·8 cm. diameter, insulated within a paraffin tunnel (fig. 3*a*)	20·1	34·1	52·7	59	62·1	

33. A metallic water-dropper was now fixed into the lid of the funnel, and the metallic plate at the bottom removed. A strong solution of common salt was placed in the dropper, and

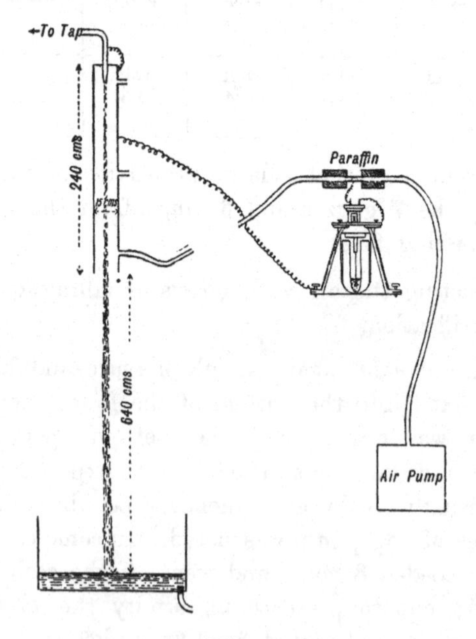

Fig. 4.

allowed to fall down the centre of the funnel into a basin below. On drawing the air from the funnel by the side aperture near its lower end, and testing it by the brass tubular filter, 0·18 centim. diameter (4 of § 32), a mean of four experiments showed 2·5 volts positive.

34. Arrangements were now made to test the effect of falling water upon air only, uninfluenced by impact of drops on any hard solid. The metallic plate at the bottom of the funnel was removed, and the funnel, 240 centims. long, was placed vertically in a position giving a clear fall of 640 centims. from the lower end of the funnel to a water-trough below (see fig. 4). A new aperture was made in the funnel near the centre. The filter used had 12 wire gauzes and 11 plugs of cotton wool.

The following results were obtained from 120 strokes of the pump:

Water pressure	Air drawn from aperture in funnel near the—			Remarks
	Top	Middle	Bottom	
	volts neg.	volts neg.	volts neg.	
Full . . .	0·5	0·7	0·5	
Reduced . .	0·16	0·23	0·33	
Full . . .	21·0	37·0	21·0	Sloping plate inserted in the funnel at the bottom
Full . .	6·7	7·4	4·8	Plate removed. Funnel tilted so that the water struck against the side

35. The lower half of the funnel was now removed and the upper half used. The water now fell clear through the whole length of the funnel, and the extracted air gave, by 120 strokes of the pump as usual, $\frac{1}{4}$ to $\frac{1}{2}$ volt negative. No electrification in the extracted air could be detected if no water was falling.

36. Putting the water-dropper (§ 33) in the top of the shortened funnel and allowing a strong solution of salt water to fall down from the dropper, $\frac{1}{4}$ volt *positive* was got from the extracted air. Placing pure water in the dropper and testing again, $\frac{1}{4}$ volt *negative* was found.

Electrification of Air by an Insulated Water-Dropper at Different Potentials (§ 37).

37. The water-dropper was now insulated and connected with the positive terminal of 1 or more, up to 12, cells of a secondary battery, the negative terminal of which was connected with the funnel, and *vice versa.* On letting water fall from the dropper, and testing the electrification of the air in the funnel by drawing it through a testing filter, the results were not sensibly affected by substituting metallic connections for the connection of the battery terminals with the dropper and funnel. Hence, the large positive and negative electrifications thus given to the drops as they fell from the nozzle did not sensibly diminish or increase the negative electrification which they produced in the air through which they fell.

Effect of Heat on Electrified and Non-Electrified Air (§§ 38, 39).

38. The apparatus shown in fig. 5 was designed and used for the purpose of trying to diselectrify air by heat. Air is admitted

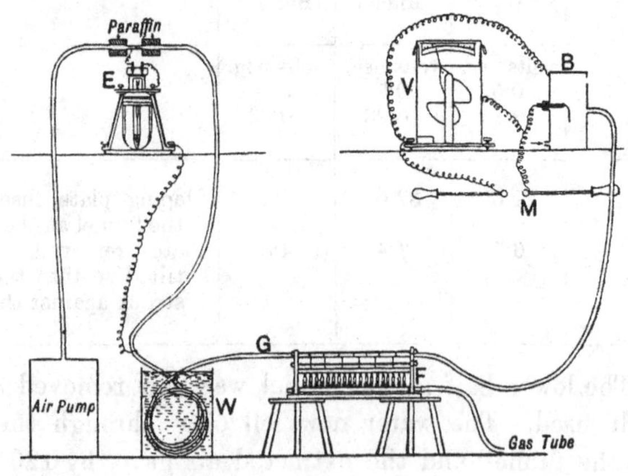

Fig. 5.

into a tin plate biscuit canister, *B*, near the bottom. Two metallic tubes are fixed into it at the middle opposite each other. One of these two is plugged with paraffin through which passes a wire, ending in a needle point inside, and connected outside with the insulated terminal of an electric machine, *M*. By means of an air-pump air is drawn into the canister, where it is electrified by the needle. It passes thence through a few metres of indiarubber pipe, to a 2-metre length of glass combustion tubing, *G*, 2 centims. internal diameter, heated to a high temperature in a gas furnace, *F*. The hot air passes on through a length of 3¾ metres of block-tin piping coiled in a large vessel of cold water, *W*. The air thus cooled passes through two paraffin tunnels between which is the insulated filter consisting of block-tin pipe with two wire gauzes and a plug of cotton wool. There were altogether 10½ metres of tubing between the canister and the filter. The air in the canister is kept electrified by an electrified needle point during an experiment.

39. Beginning with the glass tube cold, the air gave an electrification at the rate of 14 volts positive for 200 strokes of the pump. On gradually increasing the temperature of the tube the electrification correspondingly diminished to less than 3 volts. In cooling, the electrification, now become negative by an accidental reversal in the inductive machine used, increased to 4·5 volts negative. On another occasion 5 volts negative were got with the tube cold, decreasing to 2 volts as the temperature was raised, increasing again to 5 volts as the tube cooled. Occasionally irregular results were noted, especially with positively charged air.

Non-Electrified Air Passed over Hot Copper and Hot Charcoal (§ 40).

40. Passing air through the apparatus without first electrifying it, but keeping the glass tube at a high temperature, we found no deflection on the electrometer. But on repeating this experiment with copper foil in the tube an electrification of 9 volts *positive* for 200 strokes was observed. Replacing the copper foil by charcoal, with temperature high enough to keep the charcoal visibly burning, we found a *negative* electrification of 7 volts for 160 strokes.

Electrification Produced by Shaking Air and Other Gases with Water and with Solutions of Different Substances (§§ 41—46).

41. An ordinary Winchester glass bottle, *A*, fig. 6, of capacity 2500 cub. centims., has two broad strips of tin foil cemented on its outer surface on opposite sides, from top to bottom. To the foil at the shoulder is attached a metallic disc, *D*, of 5 centims.

diameter, and having a small hole to allow a connecting wire from the quadrant electrometer to be quickly hooked or removed. 500 cub. centims. of Loch Katrine water having been put into the bottle, its mouth is stopped by hand, and the bottle vigorously shaken for 5 seconds, thoroughly and violently mixing the enclosed 2000 cub. centims. of air and 500 of water. It is now immediately placed on a block of paraffin, *P*, and the disc, *D* is connected to the electrometer. A bent metallic tube, *T*, supported by an insulating paraffin stopper, is placed in the bottle, and a foot length of rubber tube connects it to one end of a paraffin tunnel,

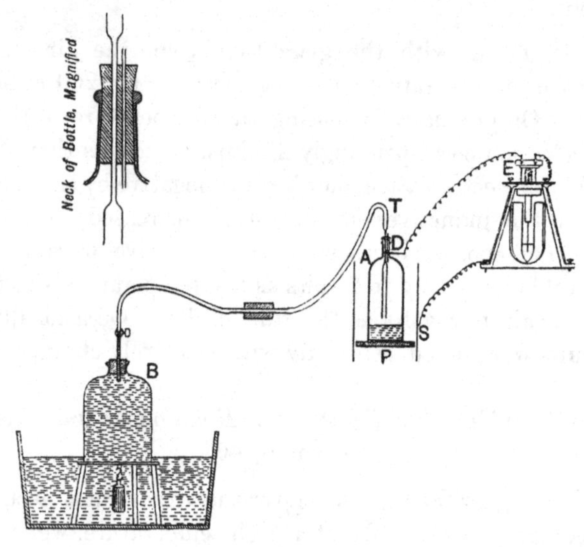

Fig. 6.

from the other end of which a rubber tube, 2 metres long, passes to an aspirator consisting of a large bell jar, *B*, of capacity 8500 cub. centims., filled with water, and resting on supports near the water surface. The metallic guard-screen, *S*, which surrounds the bottle, is always connected to the sheath of the electrometer. The outer surface of the bottle is always wet. Less than half a minute is required after shaking the bottle to make the necessary arrangements and connections. The electrometer terminal connected with the bottle is now insulated, and then, by opening the stop-cock of the aspirator, air is drawn rapidly out of the insulated bottle, its place being taken by air flowing in through a small vertical slit in the paraffin stopper. The electrometer shows positive electricity, which proves the withdrawn air to have been negatively electrified.

42. Very many experiments were made to test the effect of shaking up the air in the bottle with solutions of different substances, the solutions being varied from saturated (100 per cent.) down to practically pure water. For this purpose three acids (sulphuric, hydrochloric, and acetic), three alkalies (sodium hydrate, lime, and ammonia), and three salts (sodium chloride, sodium carbonate, and zinc sulphate), were used.

Curves 1, 2, 3.

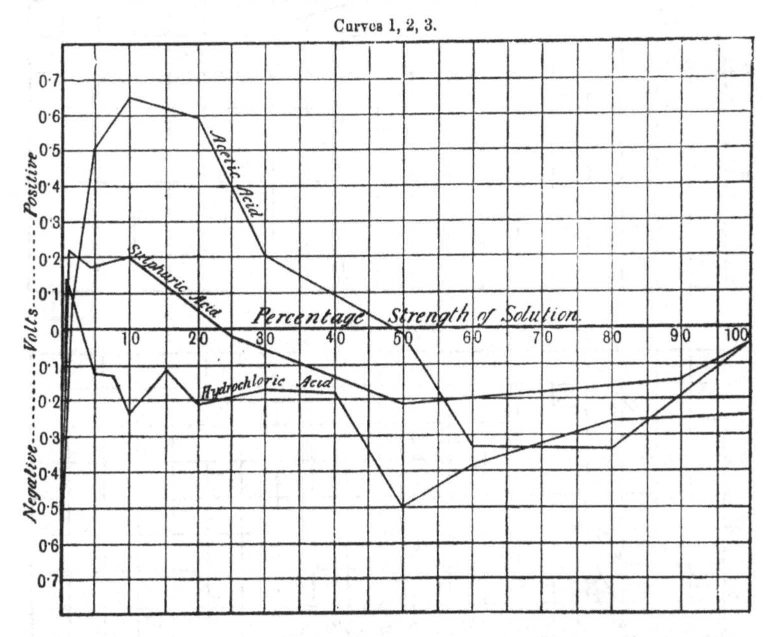

43. Curves 1 to 9 show the results obtained. Generally, each test was repeated a large number of times, and the curves are drawn for mean values. It will be noticed that, with the exception of lime water, all the acids, alkalies, and salts, when added to the water in the bottle in very minute quantities, showed a rapid diminution in the negative electrification of the air produced by shaking it up with the liquid. In some cases a single drop of a saturated solution of the substance added to the water and shaken up with the air was almost enough to entirely neutralize the negative electrification of the air which is obtained by shaking up with pure water. On gradually increasing the strength of any of the solutions named, the zero line is crossed

Curves 4, 5, 6.

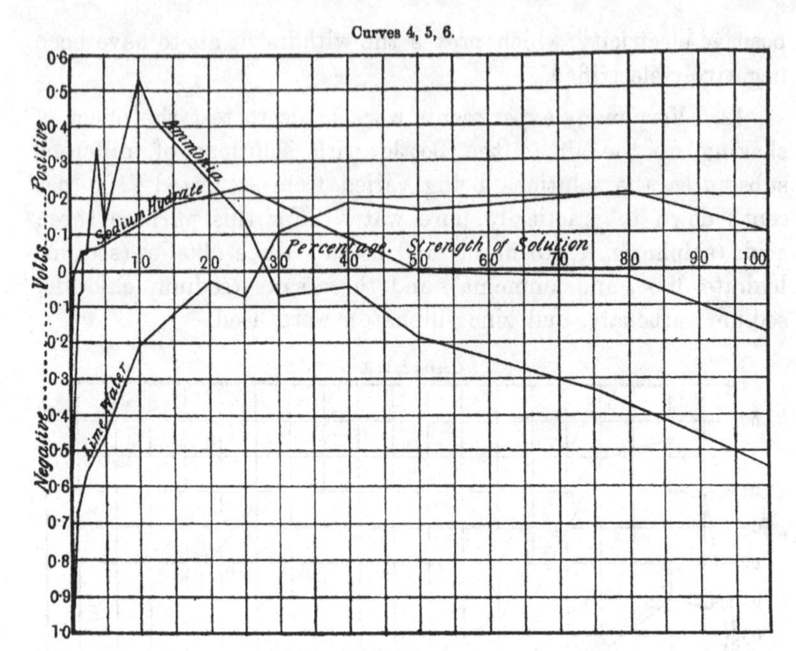

Curves 7, 8, 9.

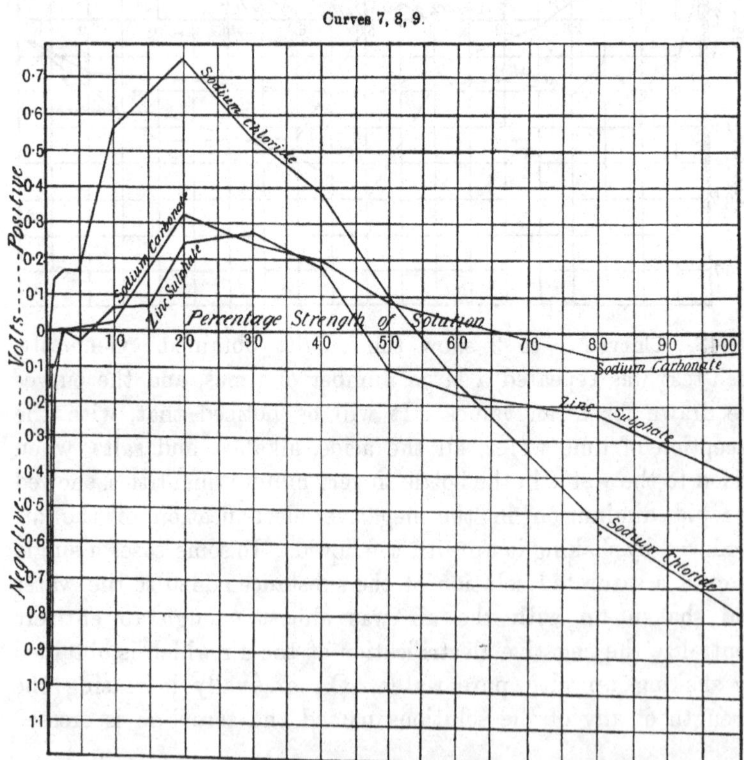

and the electrification of the air now becomes positive, increasing to a maximum and then diminishing till the zero line is again reached, the air becoming negative again. The air continues to receive negative electrification, on shaking it with the solution, for all further increase of strength of the solution up to saturation. All the acids, all the alkalies except lime water, and all the salts tested, show two zero points.

44. Experiments were now tried in which different liquids (500 cub. centims.) and gases (2000 cub. centims.) were shaken up in the bottle, and the electrification of the gas in each case was tested. The following is a tabular summary of the results. In all the experiments, except No. 1, the gas was got into the bottle by displacement of water. The bottle was first filled with water and inverted in a large trough of water and 2000 cub. centims. of the gas admitted.

	Substances shaken up together in the bottle		Deflection on electro-meter. 1 volt=69 divisions	Quantity of electricity found in the gas per cub. centim., C.G.S. electrostatic unit
	Liquid	Gas		
1	Water	Coal gas from gas mains . .	37 pos.	$0{\cdot}17 \times 10^{-4}$ neg.
2	"	Coal gas from pressure cylinder	89 "	$0{\cdot}42 \times 10^{-4}$ "
3	"	Oxygen from cylinder . . .	52 "	$0{\cdot}24 \times 10^{-4}$ "
4	"	Carbonic acid gas from cylinder	50 "	$0{\cdot}23 \times 10^{-4}$ "
5	"	Carbonic acid gas from marble and hydrochloric acid, the gas being passed direct from generator to bottle . . .	34 "	$0{\cdot}16 \times 10^{-4}$ "
6	"	Carbonic acid gas, as in 5, but allowed to pass into a gas-holder first	30 "	$0{\cdot}14 \times 10^{-4}$ "
7	"	Hydrogen from zinc and dilute sulphuric acid, direct from generator	33 "	$0{\cdot}15 \times 10^{-4}$ "
8	"	Hydrogen, as in 7, but allowed to pass into gasholder first .	20 "	$0{\cdot}09 \times 10^{-4}$ "
9	"	Nitrogen from air, the oxygen being removed by burning phosphorus	71 "	$0{\cdot}33 \times 10^{-4}$ "
10	Water impregnated with carbonic acid from a gazogen or siphon . . .	Carbonic acid gas, as in 4 . .	40 "	$0{\cdot}19 \times 10^{-4}$ "
11	Strong solution of common salt	Carbonic acid gas, as in 4 . .	68 neg.	$0{\cdot}32 \times 10^{-4}$ pos.
12	"	Coal gas from mains	71 "	$0{\cdot}33 \times 10^{-4}$ "
13	"	Oxygen, as in 3	43 "	$0{\cdot}20 \times 10^{-4}$ "

45. Some experiments were made to ascertain how long the air, on being shaken up with Loch Katrine water, retains *its* negative electrification.

Time between stopping the shaking of the bottle and beginning to draw out the air	Deflection of the electrometer. 1 volt = 69 divisions
	Divisions
25 secs.	52 pos.
27 „	50 „
1 min.	26 „
2 mins.	28 „
5 „	17 „
15 „	7 „
15 „	10 „
30 „	1 „

46. A strong solution of ammonia (500 cub. centims.) was placed in the bottle. Without shaking the bottle, the mixed air and ammonia evaporating from the surface of the liquid were aspirated. They were found to be negatively electrified to a fraction of a volt.

Electrification of Different Gases by Electrified Needle Points and Flames. Absolute Efficiencies of Filters (§§ 47—66).

47. The arrangement shown in fig. 7 was put up to test the electrification of different gases by needle points and by flames.

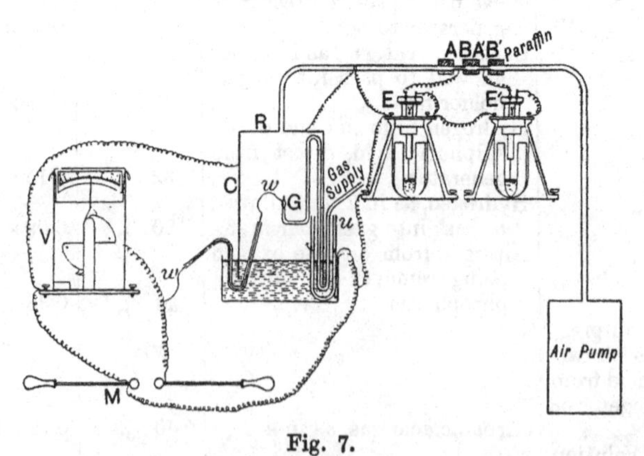

Fig. 7.

At first we only used one electrometer, *E*, and one filter, *AB*. The filter used was a block-tin pipe 4 centims. long, 1 centim.

bore, with two brass gauzes and one plug of cotton wool between them. A large glass cylinder, *C*, with a removable metal roof, *R*, has strips of tinfoil pasted on its inside and outside. These strips are kept in metallic connection with *R*, with one terminal, *M*, of an electric machine, and, with the sheath of the vertical electrostatic voltmeter, *V*, and with the sheath of the quadrant electrometer, *E*. A pump was used for drawing the electrified gas from the cylinder, *C*, through the electric filter, *AB*. By this means, calculating the effective volumes of the two cylinders of the pump, we knew the volume of electrified gas that was drawn through the electric filter in each experiment. Placing the cylinder, *C*, over water, as shown in the diagram, we found that 10 strokes of the pump raised the water inside to a height of 8·1 centims. The cylinder was 38 centims. high and 81 centims. in circumference. Hence the volume of gas drawn through the filter was 422·8 cub. centims. per stroke of the pump. This agrees with the measured effective volume of the two cylinders of the pump.

48. An electrifying wire, *ww*, was put inside a glass tube full of paraffin and bent as shown. A gas burner, *G*, was fixed in a brass tube which was connected by sealing wax to a bent glass tube varnished with shellac leading in from the gas supply.

49. The method of experimenting was as follows: The gas was lit at the burner, *G*, and the machine started. The machine was worked for 4 minutes at potentials from 3000 to 10,000 volts in different experiments. At the instant of stopping the machine the gas supply was stopped, and an indiarubber cork put into the tube, *u*, which was used to keep the gas in the cylinder at atmospheric pressure during the electrification. Ten strokes of the pump were taken in every experiment, immediately or after the lapse of certain intervals of time from stopping the machine.

The results of a large number of experiments performed in May, 1895, are briefly summarised in the following table:—

Electrification of Air by Electrified Coal-Gas Flame.

Interval between stopping electric machine and starting pump	Number of experiments	Average deflection by 10 strokes of the pump (47 divs. per volt)	Remarks
minutes		divs.	
0	3	+210	Potential of electrification ±3000 volts
0	3	−234	
1	2	+108	
1	2	−110	
2	2	+ 76	
2	2	− 94	
3	3	+100	
3	3	− 82	
5	2	+110	
5	2	−107	
30	1	+ 40	
30	1	− 36	
60	1	+ 53	
0	3	+360	Potential of electrification ±4000 volts
0	2	−395	
1	1	− 90	
5	1	− 52	
10	1	− 45	
15	1	− 36	

50. We now electrified the air in the glass cylinder by six needle points without flame. The method of experimenting was in other respects exactly the same as before. A very wonderful result was noticed, namely, that short times of electrification at 5000 volts gave greater results as indicated by electrometer and filter than longer times of electrification.

Time of electrification	Number of experiments	Average deflection for 10 strokes of pump
30 secs.	6	30·6
1 min.	6	22·1
5 „	6	0

51. A conjecture occurred to us that this very surprising result might possibly be due to the formation of lines of conductance from the electrifying needle to the inside of the containing cylinder, so that all the electricity from the needle might be

passing to the glass and tinfoil strips without electrifying the intervening air, and that if means were provided for continually breaking these conjectural lines of conductance the results would be different. Accordingly a hexagonal ring of sheet tin an inch broad was constructed to serve as stirrer. Its outer diameter was slightly less than the diameter of the cylinder. Two rods which passed through two air-tight holes in the metallic top of the glass cylinder were fixed at the ends of a diameter of this stirrer. Several experiments were tried (1) with the stirrer not moved but resting near the surface of the water at the bottom of the cylinder, (2) with the stirrer kept moving up and down during the electrification, which was by needle points at 5000 volts in each experiment. The results obtained disprove our conjecture. They are summarised as follows:

Time of electrification	Number of experiments	Average deflection per 10 strokes of the pump	Remarks
30 secs.	5	16·0 neg.	No stirring
30 "	6	16·3 "	Stirring
1 min.	6	11·6 "	No stirring
1 "	6	13·3 pos.	"
1 "	5	9·3 neg.	Stirring
1 "	6	10·4 pos.	"
5 "	No deflection, stirring or no stirring!		

Electrification of Carbonic Acid Gas from Pressure Cylinder.

52. One of the carbonic acid cylinders of the Scottish and Irish Oxygen Company was taken and laid on its side near the glass cylinder, *C* (fig. 7). Carbonic acid gas from it was let in by a tube into the glass cylinder, and electrified by six needle points. To start with, the stop-cock of the carbonic acid gas cylinder was kept shut, and ten strokes of the pump caused the water in the cylinder to rise. The stop-cock was then opened and carbonic acid was allowed into the cylinder, the average time of letting it in being one minute. This was performed five or six times, and then the experiments were done exactly as already described in § 49, except that carbonic acid gas was let in instead of air. The machine, in the two sets of experiments summarised below, kept the needle points at a potential of 8000 volts positive.

Electrification of Carbonic Acid Gas by Electrified Needle Points.

Time of electrification	Interval between stopping machine and starting pump	Deflection for 10 strokes
min.	min.	
1	0	463 pos.
1	2	303 ,,
1	5	158 ,,
1	10	97 ,,
1	30	43 ,,
0	...	4 pos.*
1	0	414 ,,
5	0	106 ,,
2	0	309 ,,
10	0	2 ,,
1	0	554 ,, in 9 strokes
0	...	128 neg.†
1	0	554 pos. in 7 strokes
1	1	338 pos.
1	2	254 ,,
1	5	160 ,,
1	10	89 ,,
1	30	37 ,,
1	60	6 ,,

53. The nearly perfect annulment of the electrification by 10 minutes of the electrified needle point in carbonic acid, and by 5 minutes (§§ 50, 51) in common air, is very wonderful. So far as we can see at present, its explanation seems to be a conductive quality, such as that first discovered, we believe, by Schuster, produced throughout the carbonic acid, and throughout the air, by continued electric disruptive action.

54. It was noticed in some of the experiments with the carbonic acid cylinder that, when it was lying on its side on the table beside the glass vessel, *C*, and when the gas issued in a jerky manner, the electrification found was invariably negative, even when the needle points were positive. This was, of course, due

* This first experiment was made with the carbonic acid gas which had been in the glass cylinder overnight.

† For explanation of this negative electricity in the gas with no electrification of the needle points, see § 54.

to the boiling and freezing, and ultimate evaporation from the liquid carbonic acid in the cylinder. Various experiments were tried, with the cylinder (1) lying on its side, (2) standing, straight up: in both cases without any electrification of the needle points. Very *slight positive electrification* was found with the cylinder straight up; and *very large negative electrification* when the cylinder was lying on its side. In the latter case, the more rapid the rush of the carbonic acid gas from its cylinder, and the shorter the time it was left in the glass vessel before observations were taken, the greater was the electrification observed.

Testing Efficiency of different Filters by using two at a time in Series and two Electrometers (§§ 55—60, with reference to §§ 11, 12 above).

55. For the purpose of testing the efficiencies of various filters, two electrometers were fitted up side by side, as shown in fig. 7, each with a filter of block-tin pipe, 4 centims. long and 1 centim. bore, containing six wire gauzes and five plugs of cotton wool. These filters were put in series, with a paraffin tunnel between them to insulate the one from the other. Thus the electrified air passed the second receiver immediately after it passed the first. Call the one filter AB and the other filter $A'B'$; then, in the first experiment, the electrified gas passed in the direction $ABA'B'$, and in the second experiment it passed in the direction $B'A'BA$. Each filter is kept metallically connected always to the same electrometer. The sensitiveness of electrometer, E, with the filter, AB, was 52·3 divisions per volt, and 100 divisions per 0·154 electrostatic unit of electricity. The sensitiveness of electrometer, E', with the filter, $A'B'$, was 158·3 divisions per volt, and 100 divisions per 0·071 electrostatic unit of electricity.

56. Air was electrified negatively by six needle points in the glass cylinder and the pump worked at the same time with the U-tube open, so that the water did not rise inside. 100 strokes of the pump gave a deflection of 160 divisions on electrometer, E, and 32 divisions on electrometer, E'. This gives

$$q_1 = 0{\cdot}246, \quad q_2 = 0{\cdot}023 \text{ electrostatic unit.}$$

Reversing the direction of the current of air through the filters, 100 strokes of the pump gave a deflection of 156 divisions on electrometer, E', and 123 divisions on electrometer, E. This gives

$$q_1' = 0·111, \quad q_2' = 0·189 \text{ electrostatic unit.}$$

Hence (§ 12)

$$n = 0·77 \text{ (filter } AB\text{), and } n' = 0·31 \text{ (filter } A'B'\text{).}$$

Similar numbers were got when the air was electrified positively.

For carbonic acid gas, electrified positively and negatively, the same filters gave

$$n = 0·82 \text{ (filter } AB\text{)}, \quad n' = 0·42 \text{ (filter } A'B'\text{).}$$

57. We now used the filter ($n = 0·77$) for determining electric density of electrified air or gas. A known volume of the electrified gas (§ 47) was sucked through the filter in connection with an electrometer whose constant was determined as in § 18 (0·154 C.G.S. electrostatic quantity per 100 divisions deflection). We thus found—

(1) For air electrified, positively or negatively, by six needles at a potential of 5000 volts, an electric density of $0·92 \times 10^{-4}$ C.G.S. per cub. centim.

(2) For air electrified, positively or negatively, by electrified gas flame at a potential of 5000 volts, an electric density of $1·98 \times 10^{-4}$ C.G.S.

(3) For carbonic acid gas, electrified negatively by gas out of a cylinder lying on its side (§ 54), or positively by six needle points at a potential of 5000 volts, an electric density of

$$2·4 \times 10^{-4} \text{ C.G.S.}$$

58. We now set about to definitely determine the relative efficiencies of various forms of filters. A standard filter of block-tin pipe, 4 centims. long and 1 centim. bore, with 6 brass gauzes and 5 plugs of cotton wool was used, and it was permanently kept in metallic connection with electrometer, E'. The filter to be tested was joined to electrometer, E. Air electrified positively or negatively was sucked through in one direction, passing through the tested filter first, and then through the standard filter, the diselectrifying power of which was 0·77 for electrified air. Hence it is possible to determine the diselectrifying power of the tested filter by § 14. Thus, from the numbers in the following section,

we get for the diselectrifying power, for positive electricity, of the 7 millim. brass tube—the least effective of those mentioned—$n = 1 \div (1 + 4/0{\cdot}77) = 0{\cdot}16$; and for the block-tin pipe, 90 centims. long, and coiled into a spiral $n = 1 \div (1 + 0{\cdot}63/0{\cdot}77) = 0{\cdot}55$.

59. Let q_2 = quantity of electricity taken out by the standard filter of block-tin pipe with six brass gauzes and five plugs of cotton wool; and

q_1 = the quantity of electricity taken out by the tested filter from the air before passing the standard filter. A long series of experiments with no wire gauze or cotton wool in the tested filters is summarised in the following tables. The potential of the machine was in each experiment 10,000 volts.

Bore of brass tube in millims. (length = 4 centims.)	$\frac{q_2}{q_1}$ for positive	$\frac{q_2'}{q_1'}$ for negative
2·0	3·0	2·75
3·4	2·1	3·3
4·2	3·5	3·1
6·0	3·9	3·1
7·0	4·0	5·5

Thus the 3·4 millims. filter is the most effective and the 7·0 millims. filter is the least effective of these five filters of equal length. The following table shows results for different lengths and different bores:—

Length of tube	Bore of tube in millims.	$\frac{q_2}{q_1}$ positive	$\frac{q_2'}{q_1'}$ negative
1. Brass, 9·9 centims.	2·0	2·96	3·05
2. " 10·0 "	4·5	2·85	4·99
3. " 9·9 "	8·0	4·0	6·45
4. Block tin, 2·0 "	6·0	3·1	1·6
5. " 10·0 "	6·0	3·26	7·5
6. " 90·0 " (coiled in spiral)	6·0	0·63	1·15
7.* " 4·0 "	10·0	1·37	1·31
8. " 4·0 "	10·0	2·96	2·8

* Tube No. 7 had twelve wire gauzes inside it, and, as the table shows, its filtering efficiency was more than that of the equal and similar tube No. 8, which was clear inside.

60. A glass tube filter, 4 centims. long and 3·3 millims. in bore, covered outside with strips of tin-foil along its length, was similarly compared with the standard filter. When newly put up, and as long as the glass was dry, it took out very little electricity from the air; but as the experiment proceeded, and the glass became less dry by taking up moisture on its inner surface, the quantity of electricity taken out by the glass tube became greater and greater. Thus in a first experiment $q_2/q_1 = 28{\cdot}5$; but after working the pump for one hour $q_2/q_1 = 3{\cdot}8$.

61. Up to this time (December, 1895), we had not been able to find a filter which could take all the electricity from the air, and we now proceeded to search for a filter which would be able to practically do so. The first filters tried with this object were tubes filled with very small pieces of fine copper wire, and closed at each end by a plug of cotton wool and a disc of brass gauze. The diameter of the wire was 0·00296 centim. The containing tube was in one case block-tin pipe 10 centims. long and 1 centim. diameter: and in another it was a glass tube of the same length and diameter, coated both inside and outside with longitudinal strips of tin-foil. The diselectrifying power of each was calculated from the observations by the formulæ in § 12. That of the block tin was thus found to be 0·93 for negative electricity and 0·9 for positive, and that of the glass nearly the same, 0·9 for negative and 0·84 for positive. The block-tin filter contained 18·927 grms. of wire.

62. The next filter tried was a block-tin pipe, 10 centims. in length and 1 centim. in diameter, containing 8·33 grms. of clean, very fine brass filings, enclosed at each end by a plug of cotton-wool and a piece of brass gauze. These brass filings, which were got from a brass-finishing workshop, were poured into a glass tube about 4 feet long and about three-quarters full of water. The filings and water were well shaken up, and the tube was then allowed to stand for several hours, so as to give the filings time to settle. After washing the filings three times in this manner, the top portion was taken off and dried before a fire, and used for filling the filter. It was found that when the electrified air passed through this filter of brass filings, before it passed through the copper-wire filter attached to electrometer, E, no deflection was obtained on the latter. This showed that the brass-filings filter

deprived the air of practically all its electricity. We tried also a filter with sawdust instead of brass filings, but its efficiency was comparatively low.

Filter of Brass Filings used to find Effects of Spirit Flame, Coal Gas Flame, and Hydrogen Flame in Electrifying Air (§§ 63—66).

63. Having found the brass-filings filter thoroughly satisfactory, we used it to investigate the effect of various kinds of flames in electrifying air. First of all, we electrified the air of the laboratory by means of an insulated spirit flame, joined to the insulated positive terminal of a Voss electric machine. The machine was worked for 40 minutes, and then, 2 minutes after it was stopped, the electrification of the air in the vicinity was tested by drawing some of it through a tube leading to the brass-filings filter, joined to electrometer, *E*. The pump was worked at the rate of 1 stroke per 4 seconds, and after 200 strokes the electrometer read 174 divisions, or about 3·3 volts positive. After the lapse of half-an-hour, the air of the laboratory was found to be still strongly charged with positive electricity.

64. Removing the water vessel from below the glass cylinder, *C*, in fig. 7, and substituting for it a metal plate kept in metallic connection with the sheath of the electrometer and the disinsulated terminal of the machine, we kept a coal-gas flame burning within the glass cylinder, while the machine and pump were worked and observations taken by electrometer, *E*, and filter, *AB*. With the machine at 10,000 volts, we found that, after two or three strokes of the pump, the deflection was about 500 scale divisions, positive or negative, according as the machine was positive or negative. This gives (§ 13) for the electric density, per cub. centim., 11×10^{-4} C.G.S. electrostatic, which is much greater than any of our previous results (§§ 23, 57).

65. To burn hydrogen, the burner, *G* of fig. 7, was made of rolled platinum foil with a fine nozzle, which was kept in metallic connection with the insulated terminal of the electric machine. The hydrogen gas was generated from zinc and dilute sulphuric acid in a Woulff bottle. The electrification which we obtained in this way, with the machine at 10,000 volts, was very large, the greatest deflection being 500 divisions in one stroke of the pump.

This indicates an electric density in the air of the glass cylinder, of 22×10^{-4} C.G.S. electrostatic, which is about six times as great as that obtained by electrified needle points (§ 23). This electric density was got for both positive and negative electrification.

66. We next tried the effect of the insulated hydrogen flame alone, without working the electric machine, and we found that when the height of the liquid rising in the long, open glass tube of the Woulff bottle was not more than about 10 centims. above the level of the liquid in the bottle, there was a small negative electrification. When the liquid rose to a greater height than 10 centims. in the tube (indicating that the gas was issuing at a greater pressure to feed the flame), the electrification was positive. On one occasion, the positive electrification produced by the flame was $0{\cdot}84 \times 10^{-4}$ C.G.S. electrostatic unit per cub centim. of the air which carried the electricity to the filter. This was the greatest effect obtained from the flame without electrification by the machine, and the height of the liquid in the tube of the Woulff bottle was 14·5 centims.

The hydrogen gas, when not burning, gave no electrification at any pressure up to 26 centims. of water.

Platinum Tube Heated either by a Gas Flame or by an Electric Current (§ 67).

67. Through the kindness of Mr E. Matthey, we have been able to experiment with a platinum tube 96 centims. long and 1 millim. bore. It was put in between the glass cylinder, C, and the filter, AB, in the apparatus of fig. 7. The other filter, $A'B'$, was not used in these experiments. The platinum tube was heated either by a gas flame or an electric current. When the tube was cold, and non-electrified air drawn through it, we found no sign of electrification by our filter and electrometer. But when the tube was made red or white hot, either by gas burner applied externally or by an electric current through the metal of the tube, the previously non-electrified air drawn through it was found to be electrified strongly positive. To get complete command of the temperature, we passed a measured electric current through 20 centims. of the platinum tube. On increasing the current till the tube began to be at a scarcely visible dull red heat we found but little electrification of the air. When the tube was

a little warmer, so as to be quite visibly red hot, large electrification became manifest. Thus 60 strokes of the air-pump gave 45 scale divisions on the electrometer (0·86 of a volt) when the tube was dull red, and 395 scale divisions (7·5 volts) when it was a bright red (produced by a current of 36 amperes). With stronger currents raising the tube to white-hot temperature the electrification seemed to be considerably less. The following summary may be taken as a specimen of several experiments. It is a copy of our notes of an experiment made on 20th July, 1895:—

Time of starting			Deflection after 60 strokes. 52·3 divs. per volt	Time of 60 strokes		Current in amperes	Remarks
hrs.	mins.	secs.	Divisions	mins.	secs.		
11	6	0	1·0 pos.	5	15	0	Air left overnight in glass cylinder drawn through platinum tube
11	17	0	3·5 neg.	4	50	19·7	Tube hot, but not visibly red
11	27	0	1·0 "	5	30	26·4	Tube beginning to be dull red
11	34	15	395 pos.	5	18	33·5	Tube very bright red
11	42	30	45 "	5	40	28·6	One end dull red. End next ingress of air always duller
11	50	15	0	5	10	26·4	
11	57	30	9 pos.	5	30	27·5	Tube red before using pump
12	5	30	37·5 "	5	30	28·6	One end perceptibly red
12	14	10	190 "	5	47	30·5	One-third next ingress moderately red; other two-thirds (14 centims.) dull red
12	22	0	174 "	5	58	34·5	Nearly whole tube (20 centims.) bright red
12	29	30	86 "	5	51	39·0	White hot
12	46	15	58 "	5	1	46·6	White hot
12	58	30	0	3	57	0	

68. *The Diselectrifying Power of Various Filters* was farther tested as follows:—

Air, electrified in a metal vessel by a needle point kept electrified by a Voss electric machine, was drawn through 340 centims. of block-tin pipe (0·91 centim. bore), to one or other of the experimental filters which was connected to sheath (fig. 8). After passing through A, the air was drawn through an insulated testing filter, *B*, connected to the insulated terminal of the electrometer. *B* was a standard filter of block-tin pipe, 5 centims. long, 0·66 centim. bore, and filled with fine brass filings.

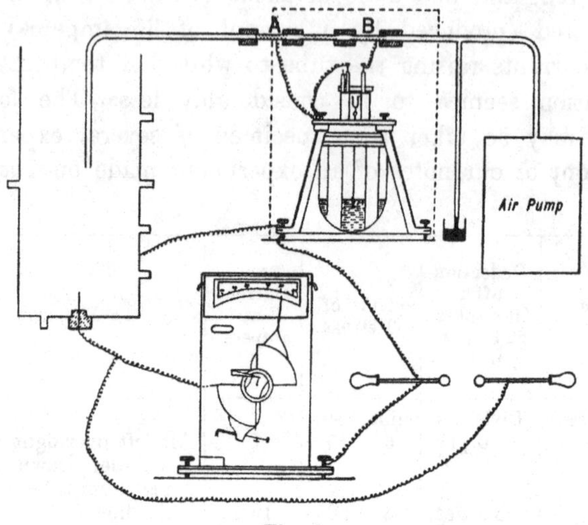

Fig. 8.

Tests were occasionally made with *A* removed, to ascertain the diselectrifying power of the standard filter (*B*) alone; then *A* being inserted, the effect on *B* was again noticed.

Filter *A*	Deflection for 6 strokes of pump		Percentage extracted by *A*
	A removed	*A* in position	
	divisions	divisions	
(1) Standard filter, similar to *B* .	220	35	84
(2) Brass tube, 5 centims. long, and 0·18 centim. bore . .	192	158	18
(3) Solid brass cylinder, rounded ends, 8·1 centims. long, and 1·8 centim. bore	222	132	40
(4) Block-tin tube, 5 centims. long, and 0·91 centim. bore . .	160	158	1·25

69. The standard filter, *B*, was now removed, and the various tubes used in the last experiments as tested filters (*A*'s) were now tried separately as testing filters (that is, insulated filters, *B*), connected to the insulated terminal of the electrometer. The

following are the particulars and deflections noted for 6 strokes of the pump in 1 minute:—

Filter used		Divisions of deflection	
		Positive	Negative
Block tin	5 centims. long and 0·91 centim. bore	none	none
	10 " " 0·91 "	"	"
	10 " " 0·66 "	4	4
Brass	4·1 " " 0·18 "	5	none
	10 " " 0·18 "	5	5
	10 " " 0·50 "	5	11
	17·5 " " 0·90 "	6	8
Solid brass (rounded ends) 8·1 centims. long and 1·8 centims. diameter		45	
Standard filter, block tin, 5 centims. long and 0·66 centim. bore, filled with brass filings		90	100

70. We then tried as filters tubes of different materials, but all of the same length (10 centims.), and bore (0·91 centim.). Glass, brass, block tin, copper, and zinc were used: the glass tube was covered externally with tin-foil, and also a little way inside at each end. As usual, the air in the can was charged from the insulated needle point at 4000 volts positive, and 3200 volts negative (§ 74), and drawn off through 340 centims. of block-tin pipe of 0·91 centim. bore, extending from the centre of the can to the insulated filter, which was either glass, brass, or block tin. Before testing the copper and zinc tubes, the can was brought nearer to the electrometer, so that the length of the block-tin pipe conveying the electrified air to the filter was reduced from 340 centims. to 100 centims. The mean of a large number of tests gave the following deflections for 6 strokes of the pump in 1 minute, the mean result in every case being very similar to the individual results.

Filter used	Deflection per 6 strokes of pump	
	Positive	Negative
Glass (covered outside with tin-foil)	1·1 mean of 3 experiments	7·0 mean of 3 experiments
Brass	1·4 " 2 "	3·8 " 2 "
Block tin	3·3 " 2 "	3·1 " 2 "
Copper	6·3 " 4 "	8·0 " 3 "
Zinc	8·4 " 6 "	2·3 " 6 "

71. The zinc filter was the only one which showed a distinctly greater deflection for positive than for negative electricity, a result which is opposite to one obtained by Rutherford*, experimenting with air which had been electrified from an electrified body under the influence of Röntgen rays. The previous experiments having shown that, on drawing electrified air over the insulated and non-electrified solid brass cylinder with rounded ends, the cylinder extracted from the air a large proportion of its charge, we now arranged to charge the cylinder and draw non-electrified air over it to the standard insulated filter connected with the electrometer, to see if the air would take up from the cylinder a part of its charge. The arrangements were as shown in fig. 9. The cylinder

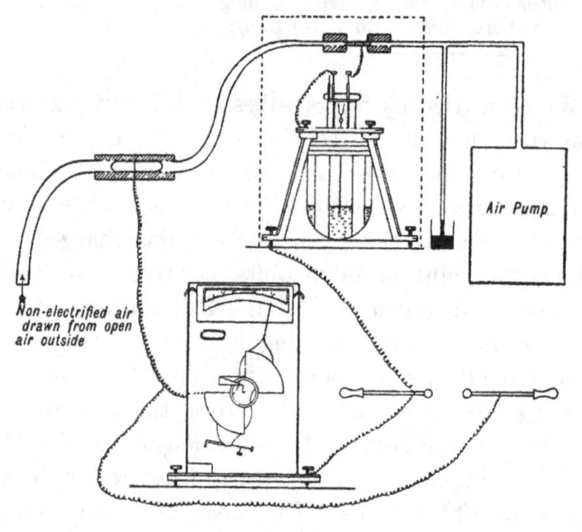

Fig. 9.

was charged positively and negatively at potentials varying from 1000 volts to 15,000 volts, *but no trace of electrification of the air after passing over the brass cylinder could be detected.*

The Effects of the Uranium "rays," discovered by Becquerel, and of a Candle Flame, on Electrified Air.

72. The air in the can (fig. 8) was charged in the usual way by a needle point at 3200 volts negative for 1 minute, and the electric machine was then stopped; the electrified air was drawn

* See *Phil. Mag.*, April, 1897, p. 246.

from a point half-way down the can to the insulated standard filter connected to the electrometer, and the deflection noted. The experiment was repeated with a uranium plate suspended three-quarters of the way down in the can by wires metallically connected with the can. The following results were noted:—

Without uranium, 50 strokes of the pump were required before the electrometer ceased to give a deflection, the total deflection being 271 divisions.

The uranium was now placed in position in the can, and the air was then charged for one minute. It was kept in position till all the electrified air was drawn off to the filter, the total deflection being 61 divisions. When the uranium was inserted after the electric machine was stopped, and before the air in the can was drawn to the filter, little more than 10 strokes were required before the electrometer ceased to give a deflection, and the deflection was now 121 divisions.

Using a very small lighted candle instead of the uranium plate, we found the following results:—

Without the candle, 195 divisions negative for 60 strokes were noted.

With the candle inserted when about to withdraw the electrified air to the filter, the deflection was, for 40 strokes, 81 divisions.

Best Charging Potential for Air and for Carbonic Acid Gas, in a Cylinder of Metal, 48 *centims. long and* 21 *centims. diameter. Greatest Electric Density Obtained, and Loss of the Electrification.*

73. Numerous experiments were made to find the charging potential which would give the greatest electric density to air drawn off from a metallic can, *A*, fig. 10. An insulated needle-point, *B*, was fixed by a paraffin stopper in the bottom of the can, and was connected with the insulated terminal of a Voss electric machine, *C*, and of a vertical electrostatic voltmeter, *D*. A pipe passed from aperture No. 5 in the top of the can to a standard electric filter, *E*, insulated on two paraffin blocks. The filter was of block-tin tube, 5 centims. long and 0·6 centim. bore, and was filled with fine brass filings. It was connected by a short

platinum wire to the insulated terminal of a quadrant electrometer, *F*, and beyond the filter were tubes passing to a mercury gauge, *G*, and air-pump, *H*. The can was connected to the uninsulated terminal of the electric machine and to the sheaths of the voltmeter and electrometer.

74. The experiment was conducted as follows:—Apertures Nos. 1, 2, 3, 4, and 7 in the can were closed, and the electric machine started, the air in the can being charged by a brush discharge from the needle point. The electrometer terminal joined to the filter was insulated, and the pump worked for some time, fresh air filtered through cotton-wool entering the can by a pipe attached to aperture No. 6. The tests were made at

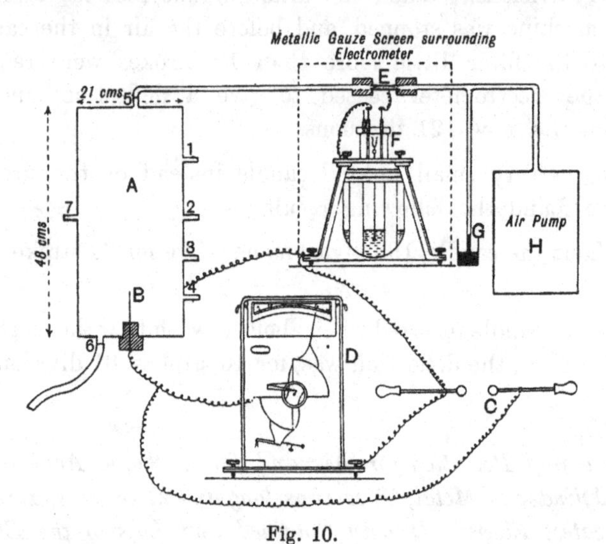

Fig. 10.

potentials ranging from 2500 up to 12,000 volts, and it was found that 3200 volts negative and 4000 volts positive gave about the best results. The speed of the pump was kept constant, and the cubic contents of the cylinders of the pump and the electric capacity of the insulated filter and quadrants of the electrometer being known, the quantity of electricity in absolute measure taken from each cub. centim. of air by the filter could be determined. In experimenting with carbonic acid gas, the procedure adopted was almost exactly the same as that for air, the only difference being that instead of admitting air by the pipe attached to aperture No. 6 the same pipe was attached to the nozzle of an upright pressure cylinder containing carbonic acid gas. The gas was admitted to the can under very slight pressure. For carbonic acid gas, the charging potentials which gave the best results were found to be about 4000 volts negative and 5000 volts positive.

In order to find out the electric density of the electrified air or carbonic acid gas when left in the can for some time after charging had ceased, the electrification was stopped after the machine and pump had been worked for several minutes. The charging wire was removed from the needle and the apertures in the can blocked. The enclosed electrified air or carbonic acid gas was left to itself for different times in different experiments, generally just 1¼ hours. The gas in the can was then drawn from No. 5 aperture through the insulated filter to the pump (aperture No. 6 being opened to admit fresh air), and the pump was stopped when all signs of electrification ceased.

The following results were obtained:—

ELECTRIC Densities in C.G.S. Electrostatic Units.

Gas	Greatest electric density, while working electric machine and pump	Percentage loss in stated time
Air	$0{\cdot}877 \times 10^{-4}$ negative	73 per cent. in 90 minutes
"	$0{\cdot}370 \times 10^{-4}$ positive	92·7 " 120 "
Carbonic acid	$1{\cdot}17 \times 10^{-4}$ "	93·1 " 1¼ hours
"	$0{\cdot}833 \times 10^{-4}$ "	96·1 " 1¼ "
"	$0{\cdot}63 \times 10^{-4}$ negative	98·8 " 1¼ "

Diffusion of Electricity from Carbonic Acid Gas into Air.

·75. We next tried a series of experiments to find if an electric charge given to carbonic acid gas diffused from the gas into air. The method of experimenting finally adopted was as follows:—

Carbonic acid gas was slowly passed from an upright pressure cylinder into the metallic can, *A*, by aperture No. 6; atmospheric air was freely admitted through aperture No. 7; while 12,000 cub. centims. of mixed carbonic acid and air were drawn out per minute from aperture No. 2, and 6100 cub. centims. of air from the top aperture (No. 5); the other openings in the can being kept closed.

In these conditions, it was found that air entered abundantly through No. 7, showing that the volume drawn off from (2) and (5) was much greater than that of the carbonic acid gas entering by (6). Hence there must have been nearly pure carbonic acid gas below the level of aperture No. 7, separated by a very thin transitional stratum from nearly pure air, above the level of No. 7. Nos. 7 and 2 were on the same level. The air drawn off from No. 5 passed through the insulated standard filter, *E* (already described), which was connected to the insulated terminal of the quadrant electrometer, *F*. The Voss electric machine, *C*, was worked to give a brush discharge from the needle point, *B*, the charging potential being indicated by the vertical electrostatic voltmeter, *D*. Within 15 seconds after starting the machine, a decided electrical effect was observed, the reading of the quadrant electrometer almost immediately rising to a maximum rate of deflection of 55 divisions per minute when the needle was charged positively, and 50 divisions per minute when it was charged negatively. The electrification observed was not sensibly affected by stopping the supply of carbonic acid gas. But when the working of the electric machine was stopped and the charging wire to the needle removed, and whether the supply of carbonic acid gas was continued or stopped, the electrical effects noticed on the electrometer rapidly fell, and 3 minutes after the electric machine was stopped, no further electrification was detected. The sensibility of the electrometer was 117 divisions per volt.

In order to test if the carbonic acid gas in the lower half of the can still retained any electrical charge, the connection to the filter and pump was removed from aperture No. 5, at the top of the can, to No. 6 at the bottom, and the gas drawn through the filter, but no electrification could be detected. We were surprised with the results, and we do not see how to explain it: we expected that the stagnant carbonic acid gas in the bottom of the vessel would have retained electricity as in experiments of § 52 and § 74.

76. Further experiments on diffusion of electricity were tried with a porous ball (fig. 11). The mouth of the ball was tightly closed, and through the cork passed two glass tubes, one (*B*)

projecting nearly to the bottom, the other being just through the cork. The ball was suspended in the metallic can, which was filled with carbonic acid gas or air. The gas in the can was electrified from the insulated needle point at the bottom. Meantime, a strong blast of non-electrified air from a large bellows passed into the ball by the tube, *B*, and out again by the other tube to the insulated standard filter of block-tin tubing, 5 centims. long and 0·66 centim. bore, filled with fine brass filings, the filter being connected to the insulated terminal of the quadrant electrometer.

There was thus an air pressure from the inside of the ball towards its outside surface, and under these conditions there was

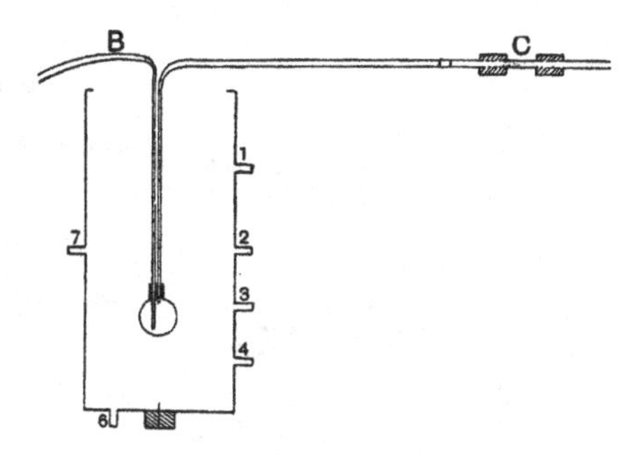

Fig. 11.

no evidence on the electrometer that any part of the electric charge in the carbonic acid gas or air surrounding the ball had made its way against the outward pressure of air, from outside the ball into the interior, and thence to the filter. This experiment was varied by removing the bellows, blocking the tube, *B*, partially or completely, and attaching an air-pump to the insulated filter at *C*. On working the air-pump, some of the electrified carbonic acid gas or air surrounding the ball must have been drawn inside, and thence to the insulated standard filter connected with the electrometer; but, again, no evidence of electrification on the filter could be detected on the electrometer. It thus appears as if the porous ball itself had withdrawn the electric charge from the gas which passed through the ball.

Communication of Electricity from Electrified Steam to Air (§§ 77—81).

77. Steam was generated in a kettle, *A*, and electrified by brush discharge from a needle point, *B*, attached to the lower end of a long copper rod, *CB*. The rod was kept central and insulated in the brass tube, *D*, by two rubber corks. The upper end of the rod was connected to the insulated terminal of a voltmeter, *E*, and of a Voss electric machine, *F* (fig. 12).

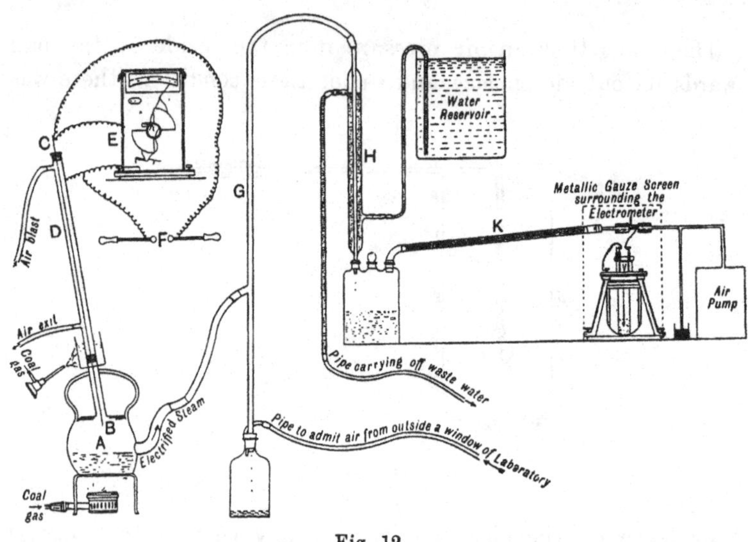

Fig. 12.

78. To preserve the insulating properties of the corks during the experiments it was found necessary to keep a current of air passing in the tube between the corks, and to surround the lower part of the tube with a jacket of oil kept at a temperature of 235° F.

79. The electrified steam from the kettle passed up into a vertical tube, *G*, where it mixed with air drawn, by an air-pump shown on the right-hand side of the drawing, from a bottle into which the lower end of the tube was fitted. Air to take the place of that drawn from the bottle entered by a long pipe from outside a window of the laboratory. The mixed steam and air passed from *G* into a Liebig's condenser, *H*, where the steam was condensed. The water thus formed dropped into a Woulff's bottle,

and the air was drawn from another neck of the bottle through a drying tube, *K*, containing sulphuric pumice. From this it passed direct through an electric filter, *L*, insulated by two paraffin tunnels, and thence to the air-pump. The filter was connected to the insulated terminal of a quadrant electrometer, whose constant was 117 divisions per volt.

80. When the air-pump and the electric machine were worked, with the kettle cold, the electrometer showed no electrification. It also showed no electrification when the kettle was boiling and the air-pump worked, but the electric machine stopped.

81. When the kettle was kept boiling, and the electric machine and air-pump both worked, strong electrification, positive or negative, according as the machine was positive or negative, was observed, 52 divisions per minute being our largest value. This, with the known capacity of the electrometer, corresponds to 0·11 C.G.S. electrostatic unit taken per minute from the air by the filter.

Electrification of Air at Different Air-Pressures and at Different Electric Potentials. Measurement of Current (§§ 82—88).

82. In February, March, and April, 1896, we made experiments on the electrification of air at different air-pressures, using for this purpose the apparatus represented in fig. 13, and electrifying by one needle point joined to the insulated terminal of the Voss electric machine. The air was contained in a glass bell-jar, *A*, which was coated inside with strips of tin-foil kept in metallic connection with one terminal, *G*, of a high-resistance mirror galvanometer, the other terminal of the galvanometer being joined to the sheath of the voltmeter, *V*, and to the disinsulated terminal, *M'*, of the electric machine. The stand of the bell-jar rested on a piece of paraffin. The pressure of the air in the bell-jar was measured by noting the height to which mercury rose in the tube, *B*. The vessel containing the mercury was insulated by a paraffin block. The pressure of the air after it had passed the electric filter (block-tin pipe with fine brass filings) was also measured by means of the rise of mercury in tube, *B'*. A barometer tube not shown in the diagram gave us the atmospheric pressure. The differences of the heights of the mercury in tubes

B, *B′*, and the height of the mercury in the barometer, gave the pressures of the air in the bell-jar and on the exit side of the filter respectively.

All the tubing through which the air passed was block tin.

Throughout each experiment the pressure of the air in the bell-jar was kept constant by regulating the stop-cock, *C*, so that the abstraction of air by the pump was exactly compensated by the gain through *C*.

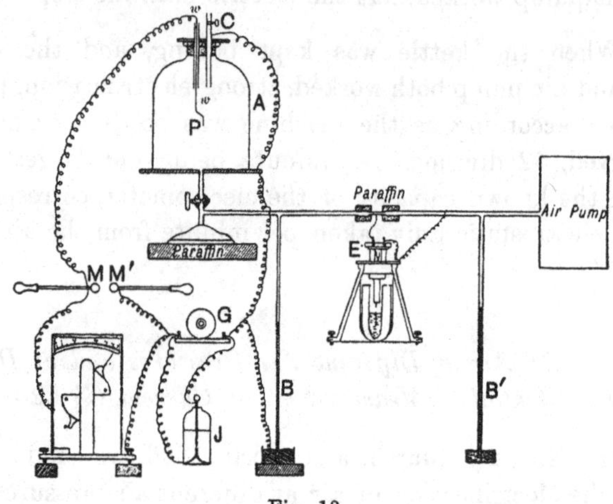

Fig. 13.

The galvanometer measures only that part of the electricity entering the bell-jar by the wire, *ww*, which leaves it by its metal base. This part is *very nearly* the whole. The observed results show that it is enormously great in comparison with the very small part which is carried away in the current of air to the electric filter.

The galvanometer was shunted by a battery of Leyden jars, *J*, to give steady deflections. Its sensitiveness was 1/22·1 of a mikro-ampere per scale division.

83. First, we kept the potential of the needle constant throughout a set of experiments made at different air-pressures, and in this way we found that the current through the air to the metal of the jar became greater as the pressure of the air in the bell-jar became less, down to the lowest pressure to which we

went, which was 40 millims. Curve 10 shows the relation between the current and the air-pressure at a potential of 5000 volts. Similar curves were got for other electric potentials.

84. We found also that as the air became rarer it was not so much electrified. This was shown by the electric filter and electrometer. Thus the electrometer deflection for a pressure of 360 millims. was only about one-sixth of that for 760 millims. with the same number of strokes of the pump, and the same potential of the electric machine.

Curve 10.

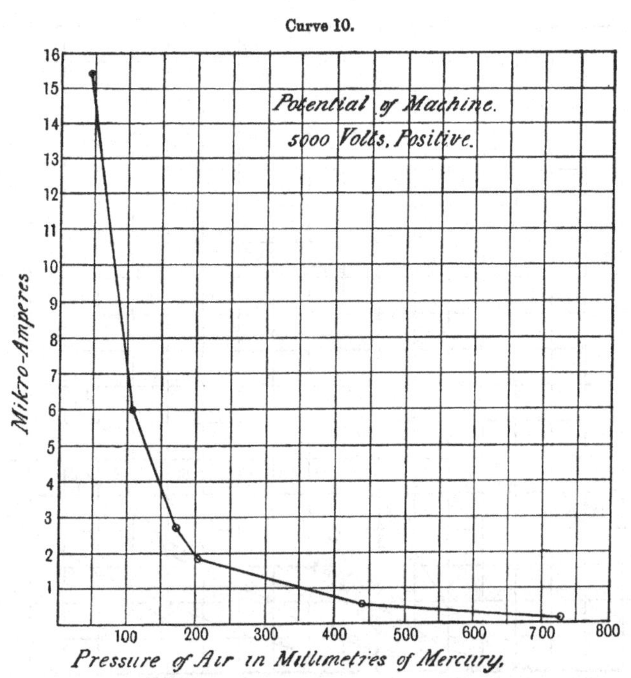

85. We next kept the pressure of the air in the bell-jar constant and varied the electric potential of the electrifying needle. In this manner we found how much the current through the air in the jar was increased by increasing the potential. The curves 11 to 16 represent the relation between the potential of the needle and the current from the needle-point through the air to the metal of the bell-jar, for certain definite air-pressures, and for positive and negative electrification. It will be noticed that for the same air-pressure the current is greater for negative than for positive electrification.

86. At each air-pressure the electrifying needle must be raised to a certain potential before the galvanometer shows any current. Thus for a pressure of 342 millims. no current was shown by the galvanometer at a potential of 2000 volts negative; and for a pressure of 235 millims. no current was shown at a potential of 1000 volts negative or 2000 volts positive.

87. We found also that air at a given pressure was electrified to a greater extent by a certain potential of the needle than by

Curves 11 to 16.

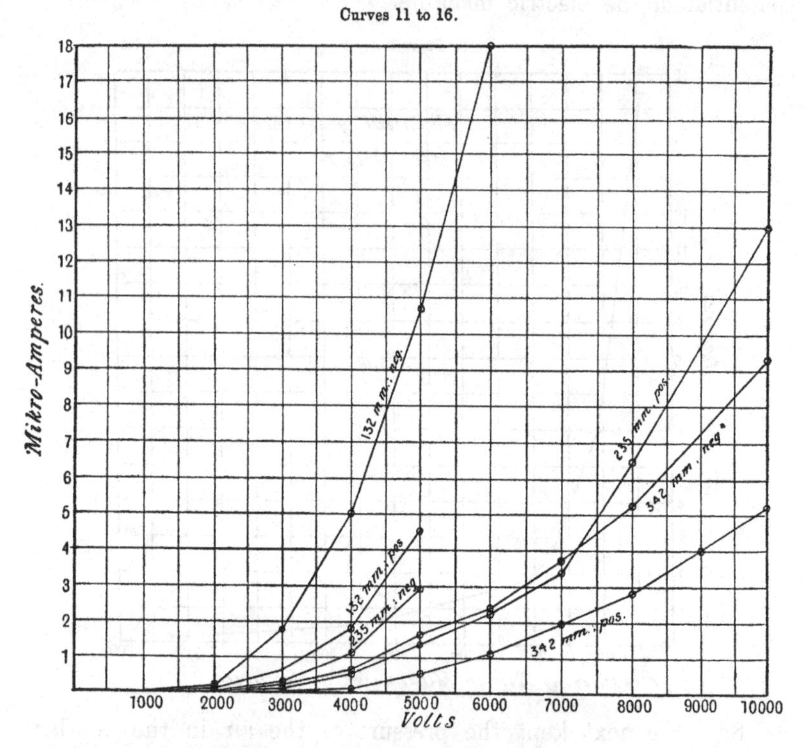

other potentials higher or lower. Thus air at a pressure of 470 millims. was electrified by the needle at 10,000 volts 1·6 times more than at any other potential we tried; while air at a pressure of 340 millims. received maximum electrification, whether positive or negative, when the potential of the needle was 3000 volts.

88. We are indebted for valuable help and co-operation in the carrying out of these experiments to Walter Stewart, M.A., B.Sc., to Vincent J. Blyth, to John F. Henderson, to Wm Craig Henderson, M.A., B.Sc., and to W. S. Templeton, M.A., B.Sc.

253. Leakage from Electrified Metal Plates and Points placed above and below Uninsulated Flames. By Lord Kelvin and Magnus Maclean.

[From *Edinb. Roy. Soc. Proc.* Vol. xxii. [read July 5, 1897], pp. 38—46; *Nature*, Vol. lvi. July 8, 1897, pp. 233—235.]

1. In § 10 of our paper "On Electrical Properties of Fumes proceeding from Flames and Burning Charcoal," communicated to this Society on 5th April, results of observations on the leakage between two parallel metal plates with an initial difference of electric potential of 6·2 volts between them, when the fumes from flames and burnings were allowed to pass between them and round them, were given. The first part (§§ 1—4) of the present short paper gives results of observations on the leakage between two copper plates 1 centimetre apart, when one of them is kept at a constant high positive or negative potential; and the other, after being metallically connected with the electrometer-sheath, is disconnected, and left to receive electricity through fumes between the two.

The method of observation (see fig. 1) was as follows:—Two copper plates were fixed in a block of paraffin at the top of a round tinned iron funnel 96 centimetres long and 15·6 centimetres internal diameter. A spirit-lamp or a Bunsen burner, the only two flames used in these experiments, was placed at the bottom of the funnel, 86 centimetres below the two copper plates. One terminal of a voltaic battery was connected to one plate, *B*, and the other terminal was connected to the sheath of a Kelvin quadrant electrometer. The other copper plate was connected to one of the pair of quadrants of the electrometer in such a way that by pulling a silk cord with a hinged platinum wire at its end, this copper plate and this pair of quadrants could be insulated from the sheath of the electrometer and the rest of the apparatus. On doing so with no flame at the bottom of the funnel, no deflection from metallic zero was observed, even when the other plate was kept at the potential of 94 volts by the voltaic battery; this being the highest we have as yet tried. When the plate was kept at potentials of 2, 4 ... 10 volts, the deflection from metallic zero in three minutes was observed; but for higher potentials, merely the times of attaining to 300 scale divisions from metallic zero were observed.

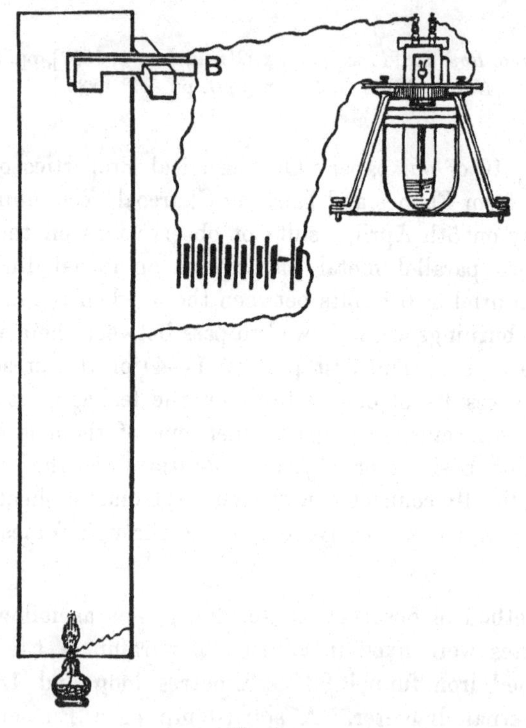

Fig. 1.

2. The results obtained are summarised in the following table. In every case for potentials below 90 volts there was greater leakage when the uninsulated plate was connected to the negative terminal of the battery. This difference depended, partially at all events, on the character of the inner surface of the funnel, which was old tarnished tin-plating.

Spirit Flame.

Sensitiveness of electrometer = 60·7 scale divisions per volt Hence 300 scale divisions corresponds approximately to 5 volts.

Difference of Potentials		+ to plate, B, − to sheath		− to plate, B, + to sheath	
Volts		Deflection	Time	Deflection	Time
		Divisions	Min. Sec.	Divisions	Min. Sec.
2		+ 35	3 0	− 80	3 0
4		+ 92	3 0	−133	3 0
8		+205	3 0	−265	3 0
10		+240	3 0	−311	1 15
Initial	Mean				
12	9·5	+300	0 53	−300	0 38
18	15·5	+300	0 25	−300	0 16
44·5	42·0	+300	0 4·5	−300	0 4
89	86·5	+300	0 2·5	−300	0 2·5

Bunsen Flame.

Sensitiveness of electrometer = 60·7 scale divisions per volt.

Difference of Potentials		+ to plate, B, − to sheath		− to plate, B, + to sheath	
Volts		Deflection	Time	Deflection	Time
		Divisions	Min. Sec.	Divisions	Min. Sec.
2		+ 10	3 0	− 99	3 0
4		+ 73	3 0	−159	3 0
8		+200	3 0	−300	2 20
Initial	Mean				
12	9·5	+300	1 48	−300	0 48
16	13·5	+300	1 12	−300	0 30
19	16·5	+300	0 46	−300	0 18
31	28·5	+300	0 15	−300	0 13
47	44·5	+300	0 11	−300	0 8
75	72·5	+300	0 6·5	−300	0 5
94	91·5	+300	0 5	−300	0 4

3. If the leakage in these experiments were proportional to the difference of potential, then the product of mean difference of potential into time should be constant for the same deflection

from metallic zero. Taking the numbers obtained for the 300 scale divisions of deflection in virtue of the Bunsen flame, we have:—

Positive Charge		Negative Charge	
v. × s.		v. × s.	
9·5 × 108	= 1026	9·5 × 48	= 456
13·5 × 72	= 972	13·5 × 30	= 405
16·5 × 46	= 759	16·5 × 18	= 297
28·5 × 15	= 427	28·5 × 13	= 370
44·5 × 11	= 489	44·5 × 8	= 356
72·5 × 6·5	= 471	72·5 × 5	= 362
91·5 × 5	= 457	91·5 × 4	= 366

Thus it is proved that the leakage between two plates, each 10 square centimetres in area, 1 centimetre apart when the fumes from a Bunsen burner pass between them and round them, is approximately proportional to the difference of potential between them, when that difference is above 20 volts and up to 94 volts, the highest we have tried; but that, below 20, it diminishes with diminishing voltages more than according to simple proportion.

4. To determine the currents which we had in our arrangement, we took a movable plate of a small air condenser charged to a known potential, and applied it to the insulated terminal of the quadrant electrometer. In this way we found that a quantity equal to 0·15 electrostatic unit, gave a deflection of 300 scale divisions. Hence in the experiments with the Bunsen flame and with a potential of -94 volts kept on the uninsulated copper plate, the current to the insulated copper plate opposite to it, when 300 scale divisions was reached in five seconds, was—

$$\frac{0{\cdot}15}{3 \times 10^9} \times \tfrac{1}{5} = 10^{-11} \text{ ampere} = \frac{1}{100000} \text{ mikro-ampere.}$$

5. One of us about the year 1865, when occupied in experimenting with the latest form of portable electrometer, found that if it was held with the top of its insulated wire (which was about 33 centimetres long) a few inches below a gas-burner, a charge of electricity, whether positive or negative, given to this wire was very rapidly lost. The disinsulating power of flames and of hot fumes from flames was well known at that time, but it was surprising to find that cold air flowing up towards the flame did somehow acquire the property of carrying away electricity from a piece of electrified metal immersed in the cold

air*. Circumstances prevented further observations on this very interesting result at that time, but the experiment was repeated with a portable electrometer in December of 1896, and we were made quite sure of the result by searching tests. During April and May of the present year observations were again made by means of (1) a multicellular electrometer reading up to 240 volts, and (2) a vertical electrostatic voltmeter (fig. 3, below) reading up to 12,000 volts. A pointed steel wire 43 centimetres long

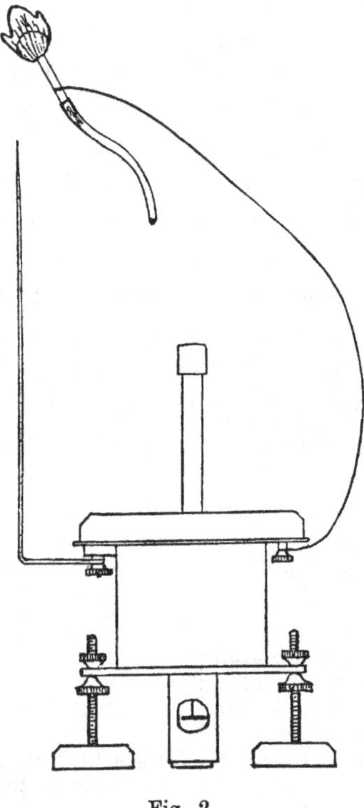

Fig. 2.

* We have recently (June 1897) found the following statement, in Worthington's communication to the British Association (1889, Report, pp. 225, 227) "on the Discharge of Electrification by Flames":—"The observation seems to have been made by Priestley, that the discharge takes place *with apparently equal rapidity*, if the rod be held at the side of, or even below, the flame at the distance of, say, five centimetres." The four words which we have italicised are not verified with the forms and arrangements which we have used, as we find *enormously greater leakage* five centimetres above a flame than five centimetres below it; but it is very interesting to learn that Priestley had found any leakage at all through air five centimetres below a flame.

was fixed to the insulated terminal of the multicellular electrometer, with its point vertically below an ordinary gas-burner, as shown in fig. 2.

6. By means of a small carrier metal plate (a Coulomb's proof plane) a positive or negative charge was given to this wire and the quadrants of the multicellular till the reading on the scale was 240 volts. The leakage was then observed (*a*) with gas not lit, (*b*) with gas lit at different vertical distances above the point of the wire. We found that there was rapid leakage when the flame was one centimetre above the wire; and the times of leakage from 240 volts to about 100 volts increased as the flame was raised to greater distances above the point; or, otherwise, the rate of fall of potential in one minute from 240 volts diminished as the distance of the flame above the point was increased. When the vertical distance of the flame above the point was 15 centimetres, or more, the time of leakage from 240 volts was practically the same as if the flame was not lit at all. A plate of metal, glass, paraffin, or mica, put between the point and the flame, diminished the rate of leakage. The leakage from 200 volts during the first minute is given in the following table, for different distances of the flame, with no intervening plate.

Distance of flame above point	Leakage during one minute	Remarks
Centimetre	Volts	
1·0	200 to 60 = 140	
1·5	200 to 92 = 108	
3·0	200 to 179 = 21	
6·0	200 to 196 = 4	
	200 to 197 = 3	No gas lit, but wire on the electrometer as in the other tests*

7. Similar experiments were made with higher voltages measured by the vertical electrostatic voltmeter, and we found that when the flame was three or four centimetres above the point, there was very rapid discharge; but when the flame was 60 centimetres or more above the point, the leakage from 3500 volts was practically the same as if the flame was not lit.

* We sometimes found the multicellular electrometer to insulate so well that in five minutes there was no readable leakage from 240 volts.

In place of the metal point, a round disc of zinc, 8 centimetres in diameter, was fixed, as shown in fig. 3, to the end of another steel wire of the same length; and leakage from it to the flame above it, observed. For the same distance between the flame and either the point or the metal disc, the rate of leakage through the same difference of potential, was *less for the point than for the disc.* Thus with the flame 25 centimetres above the point the time of drop from 3000 volts to 2000 volts was 1 min. 53 secs., and with the flame the same distance above the disc the time of

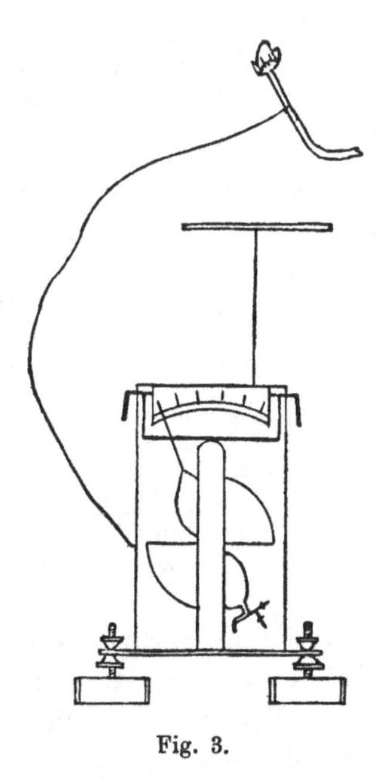

Fig. 3.

drop from 3000 volts to 2000 volts was 1 min. 14 secs. *This is a very important result.*

8. Experiments were next made to find if; and if so, how much; the leakage is diminished by putting non-conducting plates of glass, paraffin, mica, between the point or disc and the flame. At a corner of each plate was pasted a little square of tinfoil, so as to prevent any electrification of the non-conducting substance by handling. These pieces of tinfoil were always kept metallically connected with the sheath of the electrometer. Each plate was fixed with its under surface 1 cm. above the steel point. In preliminary experiments (of which a continuation is deferred until

the insulation of the electrometer is made practically perfect by coating its vulcanite insulators with paraffin) the following numbers were obtained:—

Distance of flame above point	Time of fall from 3000 to 2000 volts		Remarks
I. *Glass Plate* 18 *cms. by* 19 *cms. by* 0·3 *cm.*			
Cms.	Mins.	Secs.	
—	5	30	Insulation test, with no flame
12	2	5	Flame lit: no intervening plate
„	4	7	„ „ glass plate between
II. *Mica Sheet* 18 *cms. by* 9 *cms. by* 0·1 *cm.*			
—	6	46	Insulation test, with no flame
12	1	56	Flame lit: no intervening plate
„	3	50	„ „ mica sheet between
III. *Paraffin Plate* 11 *cms. by* 11 *cms. and* 0·75 *cm. thick*			
—	6	40	No flame. Insulation test
12	1	53	Flame lit: no intervening plate
„	2	20	„ „ paraffin plate between

We hope to return to the investigation with the insulation of the electrometer perfected; and to determine by special experiment, how much of the fall of potential in the electrometer in each case is due to the electricity of opposite kind induced on the uppermost surface of the non-conducting plate, and how much, if any, is due to leakage through the air to the metal disc or point below.

9. To test the quality of the electrification of both sides of the non-conducting plates of glass and paraffin, a thin copper sheet, C, was fixed to one of the terminals of a quadrant electrometer, as represented in fig. 4, where A is the plan of the plate C, and B is the plate of paraffin or glass under test.

In the primary experiment (fig. 3) the non-conducting plate was fixed in a horizontal position one centimetre above the electrified metal (point or disc), and eleven centimetres below the flame. A charge was given to the metal, to raise its potential to about 3500 volts. After some minutes, generally till the potential of the metal fell to 2000 volts, the non-conducting plate was removed and placed, as shown in fig. 4, above the metal plate *C* attached to the quadrant electrometer, and the deflection was observed. For a thin piece of glass (0·3 cm. thick) the whole effect of the two sides was negative when the electrified metal point or disc had been charged positively and *vice versâ*. But on putting two plates of glass above the electrified metal, we found the top plate to be oppositely charged; and the under plate to be charged similarly to the point or disc, but not so highly. We

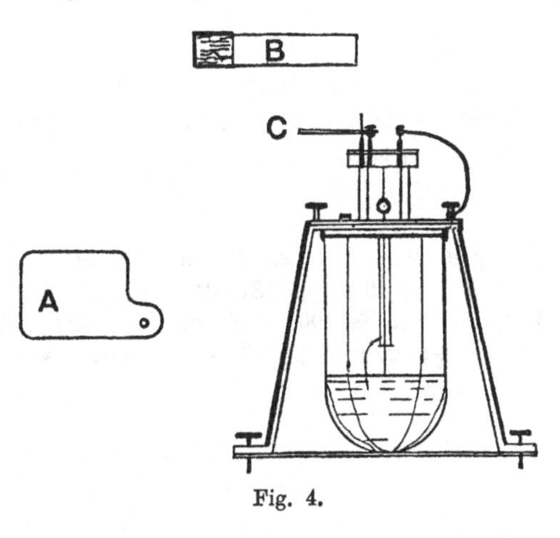

Fig. 4.

found corresponding results with a plate of paraffin 0·75 cm. thick, and with two plates of paraffin, 0·5 cm. and 0·75 cm. thick. When a plate of paraffin 3·25 cms. thick was used, we always found the top face charged oppositely to the charge of the metal, whether disc or needle-point, and the under face charged similarly to the metal below. Thus the apparent total charge of the two faces of a thin non-conducting plate is due to the fact that the face of the plate away from the electrified metal is more highly charged oppositely than the face next the metal is charged similarly.

254. NINETEENTH CENTURY CLOUDS OVER THE DYNAMICAL THEORY OF HEAT AND LIGHT.

[From *Roy. Institution Proc.* Vol. XVI. April 27, 1900, pp. 363—397; *Phil. Mag.* Vol. II. July 1901, pp. 1—40. Reprinted in *Baltimore Lectures*, Appendix B, pp. 486—527. Title also *supra*, Vol. IV. p. 531.]

255. ON THE MOTION PRODUCED IN AN INFINITE ELASTIC SOLID BY THE MOTION THROUGH THE SPACE OCCUPIED BY IT OF A BODY ACTING ON IT ONLY BY ATTRACTION OR REPULSION.

[From *Edin. Roy. Soc. Proc.* Vol. XXIII. July 16, 1900, pp. 218—235; *Phil. Mag.* Vol. L. Aug. 1900, pp. 181—198; *Congrès Internationale de Physique à l'Exposition de* 1900, Vol. II. pp. 1—22. Reprinted in *Baltimore Lectures*, Appendix A, pp. 468—485. Title also *supra*, Vol. IV. p. 552.]

256. ON THE DUTIES OF ETHER FOR ELECTRICITY AND MAGNETISM.

[From *Phil. Mag.* Vol. L. Sept. 1900, pp. 305—307. Reprinted *supra*, as No. 82, Vol. IV. pp. 553—555.]

257. ÆPINUS ATOMIZED.

[From the Jubilee Volume presented to Prof. Bosscha in Nov. 1901. *Phil. Mag.* Vol. III. March 1902, pp. 257—283. Reprinted in *Baltimore Lectures*, Appendix E, 1904, pp. 541—568.]

258. Becquerel Rays and Radio-activity.

[Letter correcting the Report of the Physical Soc. Meeting on Oct. 31. *Nature*, Vol. LXVII. Dec. 4, 1902, p. 103.]

In your report of the meeting of the Physical Society of October 31, I find the following sentence given as having been said by me in the course of some remarks on Mr Ridout's paper on the size of atoms, with the four words which I underline accidentally omitted.

"If the electrions, or atoms of electricity, succeeded in getting out of the atoms of matter, they proceeded with *velocities which might exceed* the velocity of light, and the body was radio-active."

The omission of those four words made it appear that I had considered the velocity of the escaping electrions to be essentially the velocity of light. In reality, the electrions may escape with velocities possibly less or possibly more than the velocity of light, but certainly not all with one definite velocity.

It is probable that the electrification of air produced by the breaking up of liquids into drops*, by a jet of water falling through air†, by water-falls‡, by the bubbling of air through water and other liquids, and by the shaking up of liquids and gases in a bottle§, are all to be explained by the splashing out of electrions in consequence of violent vibrations of molecules of the liquid at surfaces of separation between liquid and gas in rapid relative motion, and at places of disruption between two portions of liquid.

* Holmgren, *Swedish Academy of Sciences*, 1873.

† Maclean and Goto, *Phil. Mag.* August 1890.

‡ Lenard, *Ann. der Phys. und Chem.* 1892.

§ Kelvin, Maclean and Galt, *R. S. Proc.* and *Trans.* 1895.

259. Contribution to Discussion on the Nature of the Emanations from Radium.

[From *Brit. Assoc. Report*, 1903, pp. 535—537; *Nature*, Vol. LXVIII. Oct. 22, 1903, p. 609; *Phil. Mag.* Vol. VII. Feb. 1904, pp. 220—222.]

Let us first consider the mere fact, now known as a result of observation and experiment, that radium has been found to emit three types of rays:—

α. Positively electrified, and largely stopped by solid, liquid, or gaseous screens.

β. More penetrative than *α*, and negatively electrified.

γ. Electrically neutral, and much more penetrative than either *α* or *β*; passing with but little loss through a lead screen 1 centimetre thick, which is an almost perfect screen against *α* and *β* rays.

A simple *prima facie* view is to regard the '*γ* rays' as merely vapour of radium [if they are in truth an emission of matter: but it now appears more probably Röntgen rays, *i.e.* waves of ether. (Signed) K. June 23, 1904]. The '*β* rays' seem certainly to be atoms of resinous electricity—electrions, as I have called them (to specialise Johnstone Stoney's 'electron,' which might be either a vitreous or resinous atom of electricity, or an atom of matter deprived of its natural quantum of electricity). The '*α* rays,' according to my proposed atomic resuscitation of Aepinus's doctrine, are atoms or molecules of matter, probably atoms of radium, or perhaps molecules of bromide of radium; either deprived of electrions, or having less than their neutralising quantum.

The electro-etherial hypothesis, referred to in my communication of last Thursday to Section A*, affords a ready explanation

* *Phil. Mag.* Oct. 1903.

of the relative penetrativities of the three radiations, and of the fact that each one of them makes its existence known to us by conferring electric conductivity on air or any ordinary gas in which it is present.

Taking the γ rays first, we have to explain the free penetration of unelectrified radium molecules through dense liquid or solid matter. An easy assumption suffices: let the Boscovichian mutual forces (that is, the chemical affinities and the repulsions) between an atom of radium and the atoms of lead and other permeable substances be small enough to allow the known permeation.

Taking, next, the α radiation. The apparent great absorption of the vitreous electric emanation from radium is only apparent; it means that an atom shot from radium with less than its neutralising quantum of electrions cannot go far through a solid or liquid without acquiring the neutralising quantum.

The β rays are merely electrions; and their absorption may be regarded as real. Atoms of resinous electricity shot from radium cannot be expected to enter a screen of metal or glass or wood or liquid, and leave at the other side irrespectively of the insulation of the screen and of the radium. The full consideration and experimental investigation of the emission of atoms of resinous electricity from radium hermetically sealed in a glass bulb or tube is forced upon us. It has, I believe, led to surprising and interesting results. As to the γ rays, there is no difficulty in supposing that non-electrified vapour of radium passes very freely through glass or metals without any electric disturbance. It has been published, on authority so far as I know unquestioned, that loss of weight in the course of a few months has been proved. Full information on all that is known on this subject will no doubt be brought forward in the course of the discussion to be opened by Professor Rutherford. I regret much that I am not able to be present, and I shall look forward with eagerness to the earliest published reports of the discussion.

Returning to Becquerel's original discovery in respect to uranium and salts of uranium, the electric conductivity induced in air and other gases by a radio-active substance, we have a ready explanation in my atomic resuscitation of the old doctrine of Aepinus. The ordinary thermal motions within any solid, or liquid, or gas, must cause occasional shootings out of the electrions from the substance; and the motions of these electrions under the influence of electrostatic force must contribute to the electric conductivity of the gas; must, in fact, constitute all of it which is not due to transport of atoms of the gas carrying less than the neutralising quantum of electrions. Thus every substance, solid, liquid, or gas, must possess radio-activity. It is exceedingly interesting to find in Strutt's short paper "On Radio-activity of Ordinary Materials*," that the electric conductivity of dry air contained in a cylinder of solid material differs largely for different materials (1·3 for glass coated with phosphoric acid, 1·4 aluminium, 2 to 3·3 various ordinary metals, 3·9 platinum) It is also exceedingly interesting to be told that radium is 300,000,000 times more active than the most active common material with which he experimented. How are we to explain this enormous radio-activity of radium? I venture to suggest that it may be because it is exceedingly poly-electrionic; that the saturating quantum of electrions in an atom of radium may be hundreds, or thousands, or millions of times as many as those of atoms of 'ordinary material.'

But this leaves THE mystery of radium untouched: Curie's discovery that it (perpetually?) emits heat at a rate of about 90 Centigrade calories per gramme per hour. If emission of heat at this rate goes on for little more than a year, or, say, 10,000 hours ($13\frac{1}{2}$ months), we get as much heat as would raise the temperature of 900,000 grammes of water by 1° C. It seems to me utterly impossible that this can come from a store of energy lost out of the gramme of radium in the 10,000 hours. It seems to me, therefore, absolutely certain, that if emission of heat at the rate of 90 calories per gramme per hour found by Curie at ordinary temperatures, or even at the lower rate of 38 found by Dewar and Curie from a specimen of radium at the temperature of liquid oxygen, can go on month after month, energy must somehow be supplied from without to give the energy of the heat which gets into the material of the calorimetric apparatus.

I venture to suggest that somehow etherial waves may supply energy to the radium while it is giving out heat to the ponderable matter around it. Think of a piece of black cloth hermetically

* *Phil. Mag.* June 1903.

sealed in a glass case, and sunk in a glass vessel of water exposed to the sun; and think of another equal and similar glass case containing white cloth, submerged in an equal and similar glass vessel of water, similarly exposed to the sun. The water in the former glass vessel will be kept very sensibly warmer than the water in the latter. This is analogous to Curie's first experiment, in which he found the temperature of a thermometer, with a little tube containing radium kept beside its bulb, in a little bag of soft material, to be permanently about 2° C. higher than that of another equal and similar thermometer, similarly packed with a little glass tube, not containing radium, beside its bulb.

By changing the water in our two glass vessels, a calorimetric investigation might be made, showing how much heat is given out per hour by the black cloth to the surrounding glass and water. Here we have thermal energy communicated to the black cloth by waves of sunlight, and given out as thermometric heat to the glass and water around it. Thus, through the water, we actually have energy travelling inwards in virtue of waves of light, and outwards through the same space in virtue of thermal conduction.

My suggestion respecting radium may be regarded as utterly unacceptable; but at all events it will be conceded that experiments should be made comparing the thermal emission from radium wholly surrounded with thick lead with that found with the surroundings hitherto used.

260. On the Destruction of Cambric by Radium Emanations. *Editorial Note.*

[From *Phil. Mag.* Vol. VII. Feb. 1904, p. 233.]

LARGS,
Jan. 26, 1904.

I HAVE received from Lord Blythswood a letter of date Jan. 23, with a specimen of cambric rendered thoroughly brittle or rotten by exposure for about three days to radium bromide. He had put a little circle of cambric in place of the circular sheet of mica which is commonly used to cover the cavity containing radium bromide in the little receptacle in which it is usually sold. The cambric is quite broken away, leaving an irregularly shaped hole of about 3 mm. greatest diameter in the place which was directly exposed to the radium. This is certainly a very interesting and, I believe, important discovery. Lord Blythswood found the same result in several other trials with exposures of two or three days.

261. Electrical Insulation in "Vacuum*"

[From *Brit. Assoc. Report*, 1904, p. 472 (title only); *Phil. Mag.* Vol. viii. Oct. 1904, pp. 534—538.]

1. It has long been well known that difference of electric potential between conductors in a high vacuum is maintained without appreciable current, even when the distance between them is a small fraction of a millimetre. Fifty or sixty years ago, when we had no experimental knowledge of what is now called a high vacuum, it was a vexed question whether vacuum is an insulator or a conductor. In a Royal Institution Friday evening lecture of May 18th, 1860†, I find that I made the following statement:—"It has been supposed, indeed, that outside the earth's recognised atmosphere there exists something or nothing in space which constitutes a perfect insulator; but this supposition seems to have no other foundation than a strange idea that electric conductivity is a strength or a power of matter, rather than a mere non-resistance."

2. The labours of many experimenters during the last fifty years, and the comparatively modern atomic theory of electricity, have thoroughly confirmed the view that the space of our best modern vacuum, and interstellar and interplanetary space, and generally, space occupied only by the all-pervading luminiferous ether, is a *very perfect non-resister* of electricity passing through it.

3. Hence we see that the insulation of electricity in "vacuum" is to be explained, not by any resistance of vacant space or of ether, but by a resistance of glass or metal or other solid or liquid against the extraction of electrons from it, or against the tearing away of electrified fragments of its own substance. The kathode torrent of resinously electrified particles, discovered in

* By "vacuum" I mean space occupied only by the luminiferous ether.

† Sir William Thomson's *Electrostatics and Magnetism*, § 281.

1871 by Varley, rediscovered eight years later by Crookes, and generally accepted as a truth some eight years later still, has in many discussions and speculations been attributed to the tearing off of portions of the solid metallic kathode. But I believe the most modern and best experiments* tend rather to show that it consists solely or chiefly of atoms of resinous electricity (electrions as I call them).

4. However this may be, it is quite certain (if we accept the atomic theory of electricity as true) that the extraction of an electrion from the atom is opposed by a definite permanent force which must be overcome before the electrions can be drawn out. But it may be true, and probably is true in many cases of the loss of resinous electricity from a solid, that the forces called into play may be great enough to tear away the atom, with or without its electrion or electrions, out of its place in the solid. This, however, would not contribute to the transference of electricity from the solid: in other words, Varley's torrent may contain non-electrified particles, or vitreously electrified particles, along with his negatively electrified particles which we now believe to be atoms of electricity.

5. It is conceivable also that an atom may, by electrostatic force, be extracted from a solid metallic anode: and its electrion or electrions left behind in the anode. In this case the electric current would consist partially, if not wholly, of vitreously electrified particles: but I believe there is no experimental evidence in support of this supposition. Perhaps there is decisive experimental evidence against it. In the case, however, of a liquid anode, or kathode, whether of a non-conductive substance such as oil, or conductive such as liquid mercury or other melted metal, there is a lifting of spray or spindrift from the liquid surface, if the electrostatic force is strong enough: and this gives something of an electric current of vitreously electrified particles from the anode.

6. To form some idea of the force required to pluck an electrion out of an atom of the metal of the kathode in a very high vacuum: suppose the vacuum so high that no current, nor

* J. J. Thomson, *Conduction of Electricity through Gases*, §§ 50, 279; and *Electricity and Matter*, pp. 86, 87.

torrent of sparks, passes between two blunt-ended electrodes of thick straight wire, $\frac{1}{48}$ mm. asunder, when the difference of potential between them is raised to 200,000 volts. The electrostatic force between the middles of their ends will be approximately uniform through the intervening space: and will amount to 96,000,000 volts per cm.; or 320,000 C.G.S. electrostatic. Taking with this J. J. Thomson's most recent estimate* $e = 3{\cdot}4 \,.\, 10^{-10}$ C.G.S. electrostatic, for the quantity of resinous electricity in an electrion, we find $109 \,.\, 10^{-6}$ dyne as the force which a single electrion would experience in the electrostatic field between the electrodes in these circumstances.

7. Consider now a single mono-electrionic atom having a single electrion within it, in equilibrium in the centre of the field. Let r be the radius of the atom, and x the distance from its centre, at which the electrion rests. The electrostatic force at distance x from the centre is $x/r \,.\, e/r^2$, and therefore if the force of the external field is just sufficient to make $x/r = \frac{1}{9}$, we have $(3{\cdot}4 \,.\, 10^{-10})/9r^2 = 320{,}000$. This gives $r = 1{\cdot}1 \,.\, 10^{-8}$†.

8. Consider next an equal and similar atom in the extreme front of the kathode. Its electrion will certainly be drawn to a considerably greater distance from its centre than $\frac{1}{9}r$; because it is backed by atoms behind it with their electrions pulled forward: it is probable, however, it could not be quite extracted from the atom without a greater electrostatic force than that considered in § 6. But it seems to me certain, from some imperfect mathematical reckonings which I have made, that from two to four or five times that force would suffice to do so. We shall guess it as 1,280,000, being four times that force: though the actual amount required is calculable and would certainly be different for different possible crystalline configurations of the molecules in the kathode. Thus, merely as an illustration of the orders of the magnitudes concerned, we shall assume that, with $2{\cdot}2 \,.\, 10^{-8}$ for the diameter of the atom and $3{\cdot}4 \,.\, 10^{-10}$ C.G.S. for the quantity of vitreous electricity in the atom and of resinous in the electrion, an electrostatic force of 1,280,000 C.G.S. in the ether in front of the kathode would break down the insulation by drawing off electrions from the outlying atoms of the kathode.

* *Electricity and Matter*, p. 78 (1904).

† See *Baltimore Lectures*, Lect. XVII. § 80.

9. Leaving atomic considerations for a moment, remark that, per unit area, the outward attraction experienced by a metallic surface under the influence of electrostatic force R in the air, or the ether, outside is $R^2/8\pi$. This with $R=1{,}280{,}000$ gives 6519×10^7 dynes or approximately 66·4 tons weight per square centimetre. The breaking weight of the strongest steel wire scarcely amounts to 20 tons per square centimetre. Hence the thick straight wire of § 6 would be broken or would have its electrified end shattered and pulled away in fragments by the electrostatic force suggested at the end of § 8. It would, however, bear without breaking, and possibly without any disintegration of its electrified end, the 320,000 C.G.S. of § 6, which would only strain the wire with a force of 16·6 tons weight per square centimetre.

10. Moderate permissible changes in our guess-work assumptions regarding sizes and electric quality of atoms (mono-electrionic or poly-electrionic) might no doubt readily be devised to make the discharge of electrions take place with increasing electrostatic force, before disintegration of either kathode or anode is produced. We have as yet no sure experimental evidence as to what would take place in the perfect vacuum (only ether, no ponderable atoms, in the space between the electrodes) which is our present subject. What has been observed in respect to the highest of modern vacuums (from one one-millionth to one two-hundred-millionth of an atmosphere by the Macleod gauge), shows that a much greater difference of potential than 100,000 volts (which is so far as I know the highest hitherto measured electrostatically) may be maintained between two metallic electrodes without producing a manifest discharge through the "vacuum," even when the electrodes are brought within less than 1 mm. of contact. And when a discharge does take place it is I believe not generally direct between the nearest points of the ends of the electrodes, but in wildly erratic lateral courses, attributable to residual gaseous molecules, according to J. J. Thomson's experimental and theoretical investigations on the passage of electricity through gases.

11. In the experiments by which Varley discovered the kathode torrent of resinously electrified particles, the differences of potential used were those of a Daniell's battery of from 307

to 380 Daniell's elements. The fact that such small electrostatic forces produced luminous discharge, proves that his vacuum was very far from being what is now called a high vacuum: and proves that the molecules of the residual air were largely concerned in all his results. It is exceedingly interesting to learn from J. J. Thomson's experiments of 1897, described in § 50 of his *Conduction of Electricity through Gases*, that, great though the influence of the residual gas (air or hydrogen or carbonic acid gas) was in respect to the results, the virtual mass of the resinously electrified particle in the kathode torrent is the same for the different gases; and is about 1/770 of that of the hydrogen atom; and is so small that he was led to believe it to be an atom of resinous electricity unloaded with ponderable matter. It is also very interesting to know from J. J. Thomson's experiments and from a continuation of them by H. A. Wilson*, that the virtual mass of the particles of the torrent from the kathode was the same whether the metal of the kathode was aluminium, platinum, copper, iron, lead, silver, tin, or zinc. This strongly corroborates Thomson's original conclusion that Varley's† "*attenuated particles of matter projected from the negative pole* by electricity in all directions," are atoms of resinous electricity.

12. It is very much to be desired that careful experiments with the very highest obtainable vacuum should be made to ascertain the greatest steady, measured, difference of potentials, that can be maintained with or without any measurable electric current between two metals separated by a very short length of vacuous space.

* H. A. Wilson, *Proc. Camb. Phil. Soc.* Vol. XI. p. 179 (1901).

† "Some Experiments on the Discharge of Electricity through Rarefied Media and the Atmosphere," by Cromwell Fleetwood Varley, *Proc. R. S.* Oct. 5, 1870.

262. PLAN OF A COMBINATION OF ATOMS HAVING THE PROPERTIES OF POLONIUM OR RADIUM.

[From *Brit. Assoc. Report*, 1904, p. 472 (title only); *Phil. Mag.* Vol. VIII. Oct. 1904, pp. 528—534.]

1. THE properties to be explained are:

(1) To store a large finite amount of energy in a combination having very narrow stability.

(2) To expend this energy in shooting off with very great velocity, vitreously and resinously electrified particles.

2. In the title of the present communication, Polonium means a substance which shoots off vitreously electrified particles abundantly and with very great velocities; but few or no resinously electrified particles. Radium means a substance that shoots off in extraordinary abundance both vitreously and resinously electrified particles. From the kinetic theory of gases, it seems certain that every kind of matter has some radioactivity: that is to say, shoots off both vitreously and resinously electrified particles. Hence it is only in their extraordinarily great abundance and great velocities of shooting, that Polonium and Radium differ from ordinary matter.

3. In the present communication I use the word electrion to signify an atom of resinous electricity, according to a suggestion given in a communication to *Nature*, May 27, 1897: and I use the suggestions regarding atoms of ponderable matter and electrions, which I first proposed in an article under the title "Aepinus Atomized*" in the jubilee volume, presented to Professor Bosscha in November 1901.

* Reproduced *Phil. Mag.* for March 1902, and Appendix E of my recently published volume of *Baltimore Lectures*. This article will be referred to in the text as "Aep." for brevity.

4. A plan for molecular structure of Polonium is represented in fig. 1, and may be shortly described as two void atoms held together against their mutual repulsion by a bond consisting of one electrion. A plan of molecular structure for emission of the β rays of Radium is represented in fig. 2, and may be shortly described as two electrions held together against their mutual repulsion by a bond consisting of one void atom.

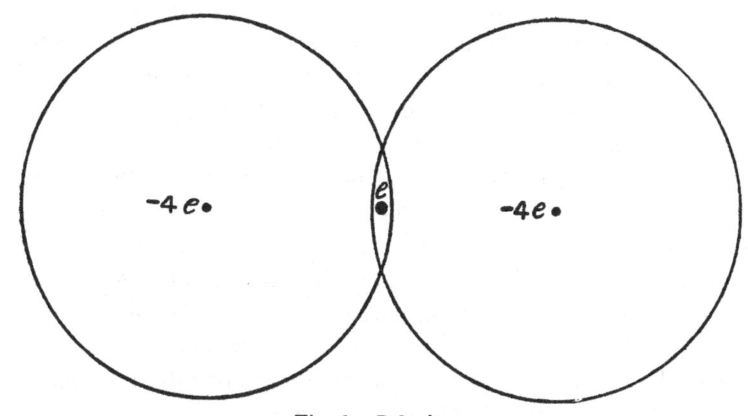

Fig. 1. Polonium.

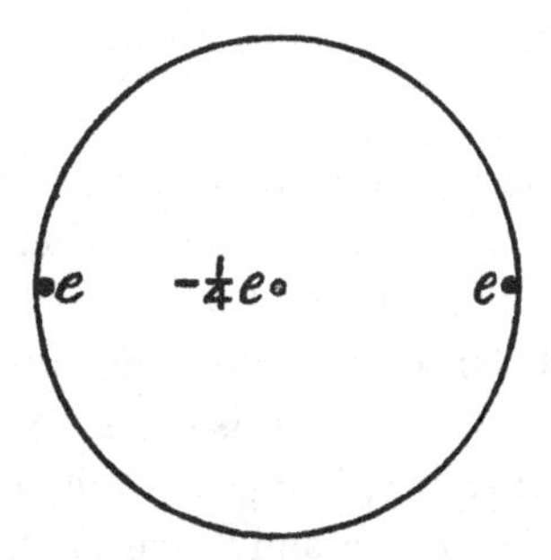

Fig. 2. For β rays of Radium.

5. In fig. 1 the quantity of vitreous electricity belonging to each of the void atoms is four times the quantity of resinous electricity, *e*, belonging to an electrion. The quantity of vitreous electricity belonging to the single void atom of fig. 2 is $\frac{1}{4}e$.

6. Alter fig. 1 slightly to make the two circles touch one another with the centre of the electrion, *e*, at the point of contact. The electrion still experiences equal attractions leftwards and rightwards and is therefore in equilibrium. Each atom experiences

repulsion from its neighbour, which, if r denotes the radius of the atom, is equal to $(4e.4e)/4r^2 = 4e^2/r^2$; and attraction by the electrion equal to $4e.e/r^2 = 4e^2/r^2$. These forces being equal, each atom is in equilibrium. But the equilibrium is unstable: to prove this; separate one atom by a slight distance from the other, leaving the electrion in the other, free to move while the two atoms are held fixed. The electrion will be left vibrating, through a small range, wholly within the last-mentioned atom; and, by sending out waves through ether, will come to rest at a small distance within this atom. The void atom will now experience diminished repulsion from the other atom, and an attraction towards the electrion diminished by a greater difference. Hence repulsion will predominate, and if the system is left free, the two atoms will separate to an infinite distance, the electrion remaining always within one of the two. The whole work done by the excess of repulsion above attraction will be spent in the generation of etherial waves, and uniform motion through ether, of the void atom and of the other atom with the electrion settled at its centre. For brevity, and to keep as nearly as possible in harmony with the language of J. J. Thomson, Rutherford, and other writers on the dynamics of radioactivity, I shall call this action, by which two atoms are sent flying asunder with very great velocity, an explosion.

7. To find the work done in this particular kind of explosion: first separate the two atoms, leaving the electrion in the middle between them. The attraction of the electrion on each atom ($4e^2/r^2$), will exactly balance the repulsion ($16e^2/4r^2$) on it by the other atom; and therefore no work is done. When the two atoms are at a very great distance, bring the electrion slightly nearer to one atom than to the other and leave all free. The electrion will be drawn towards the nearer atom and will ultimately settle at its centre. The work done in this action (Aep., Table* of § 20) will be

$$\tfrac{3}{2} . 4e^2/r = 6e^2/r.$$

Hence this is equal to the work done in the explosion of § 6, because the initial and final configurations of atoms and electrion

* **In this table a denotes the radius of the atom instead of r as at present: and e is the quantity of vitreous electricity belonging to the atom instead of $4e$ as at present. Thus instead of e^2 we have $4e^2$.**

are the same in the two cases. We may make r as small as we please, and so make the energy of the explosion as great as we please.

8. Similar considerations show that if e and e are placed on the circumference of the circle in fig. 2 instead of slightly within it, the configuration is unstable and is liable to an explosion in which one of the electrions e is shot off to an infinite distance, while the other settles at the centre of the atom. And just as in § 7, we find that the work done in this explosion is equal to the work required to extract the electrion from the centre of the atom and carry it off to an infinite distance: which is

$$\tfrac{3}{2} \cdot (\tfrac{1}{4}e \cdot e)/r = \tfrac{3}{8} \cdot e^2/r$$

(Aep., Table* of § 20).

9. In fig. 1 the total quantity of the two electricities is $8e$ of vitreous and e of resinous. Hence to make a neutral or unelectrified combination of atoms and electrions we must add a combination electrically equivalent to 7 electrions. If we simply placed seven electrions in the neighbourhood of the combination shown in fig. 1, they would instantly explode into the atoms: and the thus augmented combination might ultimately settle in two tetraelectrionic atoms moving from one another with some finite velocity, and each having its quartet in one of the stable configurations of equilibrium of four electrions within it (Aep. § 17). Or it might settle into any of a great number of possible configurations of two overlapping tetraelectrionic atoms with 8 electrions in some configuration of stable equilibrium within them. In any of these results the explosive energy for which we are planning is lost. We must therefore find another plan for supplying the $7e$ of resinous electricity. Any such plan involves essentially the addition of 8 or more electrions. We might try one atom containing vitreous electricity equal in amount to one electrion, and try to charge it with 8 electrions: which we should almost certainly find impossible. The simplest plan really is to take fourteen atoms each possessing vitreous electricity equal to $\tfrac{1}{2}e$, and place within it one electrion. This would add to our vitreously electrified explosive combination represented in fig. 1, $14e$ of resinous electricity and $7e$ of vitreous;

* In this table the quantity of vitreous electricity belonging to the atom is e instead of $\tfrac{1}{2}e$ as at present. Thus instead of e^2 we have $\tfrac{1}{2}e^2$.

and would so add an electrical equivalent of the required $7e$ of resinous electricity to make up a non-electrified explosive combination.

10. These fourteen atoms may be first put together in two groups of seven as shown in fig. 3 and then applied symmetrically on the right-hand and the left-hand sides of fig. 1, in planes perpendicular to the axis. By making these atoms very large in comparison with the two atoms of fig. 1, we avoid any great interference with the forces described in §§ 6, 7, 8;

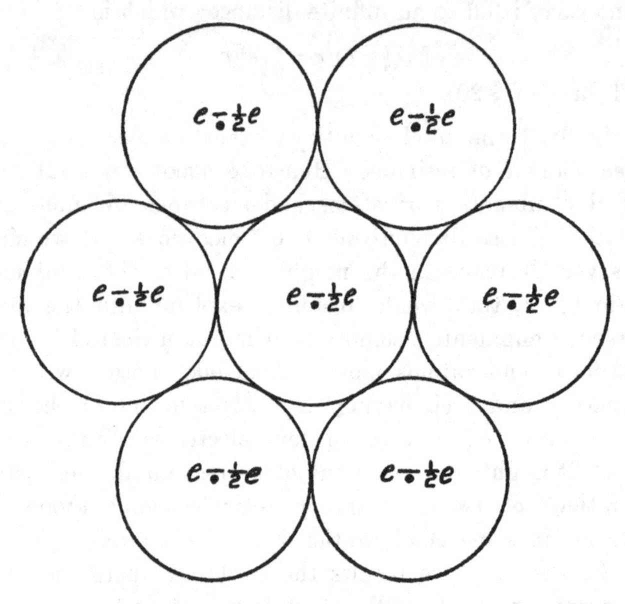

Fig. 3. Neutralizing preservative guard for Polonium molecules.

and by placing them so as to overlap one another slightly and to make the central atoms of the two groups of seven overlap slightly the atoms of fig. 1, we can, according to the last four lines of Aep. § 4, give any mutual forces we please between the atoms in order to secure stability of the group of sixteen. The fourteen electrions will find places of stable equilibrium within them, not disturbing at all the electrion shown in fig. 1: because they repel it equally in opposite directions, and with such small forces that they do not render its equilibrium unstable. Thus we have a beautifully symmetrical explosive group of fourteen large atoms and two small atoms containing in all fifteen electrions in positions of stable equilibrium within them.

11. The limits of stability of the equilibrium of the central electrion and the two small overlapping atoms which contain it as shown in fig. 1, are so narrow that a shock of a very slight but suitable kind, will produce an explosion shooting out these two atoms in opposite directions with prodigious velocities, one of them carrying the central electrion with it. Each of them will probably shoot through the neighbouring guard atom, without carrying its electrion away. Thus both of the atoms shot away will be found vitreously electrified: one with a quantity $4e$ of vitreous electricity, the other only $3e$, because it carries an electrion (e of resinous electricity) with it. This agrees perfectly with the behaviour which experiment has proved for Polonium.

12. Going back to fig. 1, alter to make the two circles touch one another in e; and for simplicity suppose the two atoms held fixed in this position; e is unstable in the central position, but if disturbed ever so little to either side it will vibrate between the two; and by giving out etherial waves will come to rest in either of the atoms at the point in the line joining the two centres at which the attractions on the electrion are equal and opposite. The distance x of this point from either centre is given by the equation

$$x/r \,.\, m/r^2 = m/(2r - x)^2,$$

where m denotes the quantity of vitreous electricity belonging to each atom. This is a cubic equation of which one root is $x = r$: its greatest root does not belong to this problem: and its least root is $x = r(3 - \sqrt{5})/2 = \cdot 382r$, which is the required distance of the position of stable equilibrium from the centre of atom in which the electrion rests.

13. Considering now the β rays of Radium, look to fig. 2. We have $2e$ of resinous electricity and $\frac{1}{4}e$ of vitreous: requiring $\frac{7}{4}e$ of vitreous for neutralization. The simplest way of applying this, which has also the advantage of converting into stability the instability of the electrions of fig. 2, is to take an atom of very large radius endowed with vitreous electricity to an amount $\frac{7}{4}e$, and place it concentric with the smaller atom shown in fig. 2. This addition gives us a non-electrified combination of two atoms

and two electrions all in stable positions with the electrions slightly inside the boundary of the smaller atom as shown in fig. 2. For brevity denote by A the smaller atom and by B the larger: and by R the radius of the larger.

14. The larger is R/r, the smaller are the distances of the two electrions from the boundary of A inwards; and the smaller and the narrower is the range of their stability; and the more nearly correct is the estimate $3e^2/8r$, in § 8 for the energy of the explosion, when an explosion takes place. By making r small enough we may make the explosive energy as great as observation shows it to be.

15. Thus in §§ 10, 13 we have two un-electrified molecules, which, if put together in any substance, would give it the α, β properties of Radium. There are many other plans, some no doubt very much simpler than the combination of these two now suggested, for a combination of atoms to give the properties of Radium. It is indeed easy enough to design a single atom possessing vitreous electricity in a fixed distribution of equal density at equal distances from the centre, into which a neutralising quantum of electrions could be shot and come to rest in such a configuration that, in the presence of other atoms or disturbing electrions, it would act as Radium does. This is in fact done for the β rays of Radium in § 13 above: because the two atoms A, B there put together concentrically may be supposed fixed relatively to one another and called one atom.

[From *Nature*, Vol. LXX. Sept. 22, 1904, p. 516.]

LORD KELVIN described his models of radium atoms to give out α and β rays respectively. The former consisted of an "electrion" e placed at the point of contact of two spheres, through the volumes of which charges $-4e$ are uniformly distributed. When equilibrium is destroyed and the spheres move apart the electrion accompanies one sphere and we have the α particle. In the same way if two electrions e are in equilibrium at opposite extremities of a diameter of a sphere through the volume of which a charge $-\frac{1}{4}e$ is uniformly distributed, and equilibrium is destroyed, one of the electrions moves away from the sphere and gives the β ray.

263. On the Statistical Kinetic Equilibrium of Ether in Ponderable Matter at any Temperature.

[From *Brit. Assoc. Report*, 1905, pp. 346, 347; *Phil. Mag.* Vol. x. Sept. 1905, pp. 285—290.]

1. Consider first the simplest possible case, a piece of solid matter of a few millimetres or a few centimetres greatest diameter, placed in space at the earth's distance from the sun, say 150 million kilometres; for particular example, let us suppose two globes of metal, or rock, or glass, or the bulbs of two thermometers, of a centimetre diameter, one of them coated with black cloth and the other with white cloth, side by side, at a distance of a few centimetres or metres asunder. For the most extreme simplification, suppose no other matter in the universe than ether, the sun, and our test globes. From our knowledge of the properties of matter, it is obvious that each of our test globes will, in a few minutes of time, come to a steady temperature. In these circumstances, each globe sends out by radiation as much energy of waves travelling out through ether, as it takes in from the sun; after it has been long enough exposed to come to a steady temperature.

2. The internal mechanism in each globe consists of atoms of ponderable matter, with ether permeating through the whole volume of the globe, and locally condensed and rarified in the space around the centre of each atom; as I have assumed, with explanations, in §§ 162, 163, 164 of pp. 412, 413, and in § 3 of pp. 487, 488 of my volume of *Baltimore Lectures.*

3. The action of this mechanism in our case under consideration, involves the communication of energy from incident waves of sunlight to the atoms of the solid, in the surface of the hemisphere illuminated by the sun; and the communication of energy from the atoms to ether outside the globe, in the form of waves travelling out *in all* directions from the surface of the

globe. The travelling of this energy through the volume of the globe is carried on according to the laws of the conduction of heat through solids; modified, but scarcely perceptibly modified, by convection currents in the case in which the globe is the bulb of a mercury thermometer.

4. Our present knowledge of the radiational properties of matter does not quite suffice to let us pronounce for certain, which of the two globes will have the higher steady temperature: as this depends, not only on the well-known higher receptivity of the black surface than of the white for sun-heat; but also, on the difference of radiational emissivities of all parts of the two surfaces to surrounding space. It seems most probable that the black globe will be steadily warmer than the white; but we cannot say with certainty that this is true. Suppose for a moment that the steady temperatures of the two are the same; and now whiten the hemisphere of the black globe remote from the sun. This will cause the globe which is now black and white, to be warmer than it was, because it will radiate less into void ether than it did when it was all black.

5. Now blacken the hemisphere facing from the sun, of the globe originally all white. Its temperature obviously will be lowered. Thus we have, side by side, two globes each with a white hemisphere and a black hemisphere; facing respectively towards and from the sun. The globe of which the black hemisphere is towards the sun, will certainly be warmer than the other, when a few minutes of time has been given for the temperature of each to become steady.

6. It is not possible for a human experimenter to attain to the extreme simplicity ideally prescribed in §§ 1—5 above. But it has occurred to me (and probably to many others) that instructive experiments might be made by observing the temperatures of two equal and similar thermometers, placed beside one another on a wooden table (or on two similar tables of the same materials) or on a cushion or layer of very fine cotton wool: each thermometer between the folds of a doubled sheet; one of white cloth and the other of black; both exposed in the open air under sunlight, or under the light of a more or less cloudy sky, or under moonlight or starlight, or in the darkest attainable cellar.

7. Not being at present able to undertake experiments of the kind, I asked Dr Glazebrook a few weeks ago if he could conveniently allow some such experiments to be made under the auspices of the National Physical Laboratory. He kindly consented, and asked Dr Chree to commence an investigation of the kind. I have to-day (28th July, 1905) received the annexed description of his work, and statement of results [omitted].

8. It is very interesting to see in Dr Chree's results how large are the differences in the temperatures of the thermometers under black and under white cloth, ranging from ·5° to ·6° Cent., even at times when the sky is covered with dark clouds; and how comparatively moderate are the differences ranging from 1°·1 to 3°·6 Cent., at times of exposure to direct sunshine.

9. Returning to § 4 with one of the globes black over its whole surface and the other white: suppose the two to be taken to 1000 times the earth's distance from the sun; and suppose, all at about the same distance (for simplicity of calculation), 999 stars, each equal to our own sun, to be scattered through space, round the place of our ideal experiment. The total of radiational energy coming from all these suns to the place of observation per unit area, will be one one-thousandth of the amount coming from our own sun in the case of §§ 1, 2, 3; and the difference of steady temperature between the white globe and the black globe may be about one one-thousandth of that which it would be in §§ 1, 2, 3. This last arrangement would be somewhat similar to an exposure to starlight on a cloudless night, at the top of a high mountain of our earth, with two or three polished silver screens between the tested globes and the mountain top. It does not, however, seem probable that any differences of temperature will be perceptible on the two thermometers exposed only to stellar radiation from the sky. Even less of difference may be expected when the two thermometers are placed in the darkest attainable cellar. The bolometric method would of course be much more sensitive than the comparison of two ordinary thermometers: even of the most extreme sensibility: and it will, I think, be worth while to try it in cases in which the thermometric method fails, or almost fails, to show any difference between the temperatures in the two cases.

AIX-LES-BAINS, SAVOIE,
Aug. 16, 1905.

P.S.—I have made some rough experiments in this place, about 250 metres above the sea-level; with two small dismantled bath-thermometers hung side by side from a horizontal bar in an open window about 12 metres above the ground. The thermometers were double coated; one with black silk, the other with white cotton, round the bulb and up to about 17° Cent. of the scale. The black was always warmer during daylight. The greatest difference which I have hitherto observed in the course of eight days was this morning, 37°·6 – 30° = 7°·6 in bright sunshine. This was with air freely circulating round the two bulbs. In a special experiment with the two thermometers laid side by side on a slab of red blotting-paper, in bright sunshine, the black-coated one ran rapidly up to above 40° (the end of its scale), and had to be removed to escape breakage, as it had no safety space above the top of its tube. The white-coated thermometer did not rise as high as 40°.

264. Plan of an Atom to be capable of Storing an Electrion with Enormous Energy for Radio-activity.

[From *Phil. Mag.* Vol. x. Dec. 1905, pp. 695—698.]

1. In a communication to the *Philosophical Magazine* of October 1904, I described combinations of atoms and electrions having certain definite qualities of radio-activity; holding myself, for the time, bound to the precise description of the electric property of an atom of ponderable matter, which I had suggested in § 4 of "Aepinus Atomised" (*Baltimore Lectures*, Appendix E). In that description each atom of ponderable matter is supposed to have ideal electric matter of the vitreous kind distributed uniformly through it. No longer binding myself by this limitation to uniformity of vitreous electric-density, I now propose to consider an atom of ponderable matter intrinsically charged with concentric strata of electricity, vitreous and resinous, of equal electric density at equal distances from the centre; and with an excess of the total quantity of the vitreous above the total quantity of the resinous. I still suppose (with, I believe, all at present concerned with radio-activity) that free resinous electricity consists of equal atoms. I assume, and I believe there is general agreement in this assumption, that each of these atoms of resinous electricity, which I am calling electrions, has, besides its ordinarily defined property of electric attraction or repulsion, a property of somehow acting upon ether, and in virtue of this action having quasi-inertia. The nature of this action I believe to be attraction and consequent enormous condensation of ether around the centre of the electrion. (*Baltimore Lectures*, Lecture xx. §§ 238, 239.)

2. My present assumption is Boscovichianism pure and simple. It merely declares that there is, between a single electrion and a single atom of ponderable matter void of electrions, a definite force in the line of their centres varying according to the distance;

which for distances greater than the radius of the atom is attraction according to the inverse square of distance between the electrion and the centre of the atom. It leaves us absolutely free to assume any law of force whatever that suits our purpose, when the electrion is within the atom. To give capacity to the atom for storing enormous electric energy by placing a single electrion at its centre, or at a very small distance from its centre, I assume, as indicated in the accompanying diagram, fig. 1, that the force on the electrion at distances less than a certain distance CM from the centre is towards the centre; and that at all distances between CM, and CN (the latter being slightly less than the radius of the atom), there is repulsion from the centre, rising to a maximum of

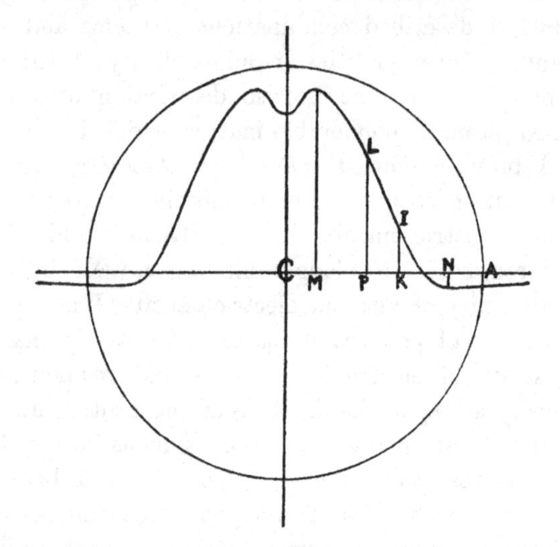

Fig. 1. Work-curve.

On the right side of the diagram, slope up to right implies attraction; slope down implies repulsion.

enormously great amount at some distance CK between CM and CN, and coming to zero at the distance CN. Between CN, and CA the radius of the atom the force becomes again attractive, and continues so, varying inversely as the square of the distance for all the distances of the electrion from the centre of the atom, greater than CA. The curve shown in the diagram may for brevity be called the *work-curve*. It shows by the ordinate PL the work, positive or negative, required to move an electrion from an infinite distance to any point P within or without the atom; which I denote by w. Thus if we denote CP by r, we have

$$F = dw/dr \quad \text{..........................(1)},$$

where F denotes the force (positive when attractive, negative when repulsive) of the atom on an electrion in the position P. Hence this force is indicated in the diagram by the tangent of the inclination of the curve at any point L, to the line of centres, CA.

3. For all points outside the atom we have

$$F = \alpha e/r^2 \quad \text{..........................(2)},$$

where α denotes the excess of the vitreous over the resinous electricity permanently belonging to the atom; and e denotes, according to the general usage of scientific writers on radio-activity, the quantity of resinous electricity in any one electrion. Thus for the equation of the curve outside the atom, we have

$$w = -\alpha e/r \quad \text{..........................(3)}.$$

By this we see that our curve outside the atom consists of portions of two rectangular hyperbolas.

The maximum ordinate through M, and the minimum ordinates through C and N, show that the point M is a position of unstable equilibrium, and that the points C and N are positions of stable equilibrium; for a single electrion placed within the atom. The point I on the curve, being a point of inflexion in the branch sloping downwards to the right, indicates that K is a position in which the atom experiences a maximum of repulsive force. Considering a spherical surface through any point P within the atom, we see that if Q denote the excess of vitreous over resinous electricity of the portion of the atom within this sphere, we have

$$F = Qe/r^2 \quad \text{..........................(4)},$$

because the resultant force of all the electricity of the atom in the shell of outer and inner radii CA and CP is zero for every point inside its hollow.

4. We may vary the work-curve within the atom as we please, with a view to trying to explain the different radio-activities of different atoms, and the different modes of radio-activity which seem to be presented by one and the same atom at different times. Thus for example we may draw the curve

with four or six or eight or any even number of minimums, instead of the two minimums at C and N. There will of course be in every case an odd number of maximums, being less by one than the number of minimums. Thus we may arrange for any even number of stable positions of equilibrium within the atom. The work-minimum for the stable position nearest to the boundary of the atom is essentially negative, and somewhat less (somewhat more negative) than the negative work required to carry an electrion to the surface of the atom from an infinite distance outside. All the other minimums may be as large positive quantities as we please. The magnitude of any one of them is the explosive energy which will be spent in shooting the electrion outwards or inwards by any shock or any kind of influence, if it is displaced away from its position of equilibrium far enough to reach an unstable position on either side of it. Look for example to the diagram. An electrion placed at C has stability, but only through a narrow range. If it is shaken away farther from the centre than M, the electric force of the atom upon it will shoot it out of the atom, with prodigious velocity, which will be but slightly diminished by the attraction of the whole atom when it gets outside. If it gets quite out of the atom, it will be shot through the ether outside with a velocity whose kinetic energy is something greater than the value of w at the unstable position from which it was shot, provided of course we can neglect its loss of energy by motions which, while it is in the atom, it gives to the ether in the atom and outside it.

265. An Attempt to Explain the Radioactivity of Radium.

[From *Phil. Mag.* Vol. XIII. March 1907, pp. 313—316.]

1. One chief action concerned in radioactivity is the shooting out of electrions from a non-electrified solid or liquid body. In the equilibrium of kinetic averages in any solid or liquid, every individual electrion must occasionally have so high a velocity that it is shot out of the body. Hence every solid or liquid body has something of radioactivity.

2. The Radium atom must, so far as we can at present judge, be assumed to have a special property of being adapted to store enormously more energy, by an electrion within it, than the atom of any other substance hitherto known to us. In a short article, published in the *Phil. Mag.* of Dec 1905, I explained the plan of an atom by the purely Boscovichian assumption of mutual force in the line between the centre of the atom and an electrion anywhere within it, according to which there is for the electrion one position of stable equilibrium, near the boundary of the atom, with very small potential energy; and another position of stable equilibrium at the centre of the atom, with very great potential energy. For brevity I shall call the atom "loaded," when there is an electrion at its centre, or anywhere within the range of the stability of the central position: and I shall call the atom "unloaded," when there is no electrion within the central range of stability.

3. In a solid crystal of Bromide or Chloride of Radium we may suppose the Bromine or Chlorine, and the Helium which Ramsay and Soddy produced from it, to be not directly concerned in the marvellous radioactivity, which the crystal presents. For brevity at present I shall assume that the radioactivity depends primarily on the Radium atoms in the compound. Suppose now

the crystal to be given with every Radium atom in it unloaded. In a very short time of progress towards equilibrium of kinetic averages, perhaps the millionth of a second, or the millionth of the millionth of a second, some of the atoms will become loaded. As time advances a greater and greater proportion of the Radium atoms will become loaded until, perhaps in the course of a few months, a permanent average of loadings and unloadings will be reached. The energy of the work done in loading the Radium atoms is taken from the energy of thermometric heat in the crystal. A cooling is thus experienced by the crystal, until heat conducted *and radiated* in from the surrounding matter compensates the cooling effect of the loadings, and a permanent equilibrium of temperature is reached.

4. A certain definite proportion of all the loaded atoms, probably a very small proportion, will, according to the equilibrium of kinetic averages, become unloaded every second of time. Electrions will be projected out of these atoms with enormous velocities; sufficient no doubt to shoot them through the substance of the crystal into space outside. It will be convenient for us to call this action a discharge, or explosion; just as the immediate result of igniting the cordite in a loaded great gun is called a discharge, or explosion. The exciting cause of our supposed atomic explosion is a shaking of the electrion from its stable equilibrium in the centre, far enough out to get beyond the range of the stability, and to be expelled away by repulsion. This repulsion increases to a very high maximum, and thence diminishes to zero, and changes continuously to the relatively small amount of the attraction between the resinous electrion and the vitreous atom, experienced when the electrion passes out of the atom.

5. Those of the discharges of loaded atoms which send the electrion inwards, relatively to the crystal, would, by the recoil, force the unloaded atom outwards: and, if it is near enough to the surface of the crystal, would send it out into the surrounding space, with comparatively small velocity and energy. The unloaded atoms, being vitreously electric, constitute the "α radiations," when they are sent outside the crystal by the force of the explosion. They would generate comparatively little heat in being brought to rest by the resistance of the matter outside the crystal.

The generation of heat by the electrions which are shot out, would be much greater because of their much greater kinetic energy.

6. It seems to me that it must be chiefly the electrions (the "β particles") which do the work of producing heat at the rate of one hundred gråm water Centigrade calories, per gram of Radium, per hour: but the "α particles" must also contribute to it.

7. This process can go on for ever, without violating the law of conservation of energy, and without any *monstrous*, or *infinite*, store of potential energy in the loaded Radium atom. The shooting out of electrions with prodigious velocities generates heat

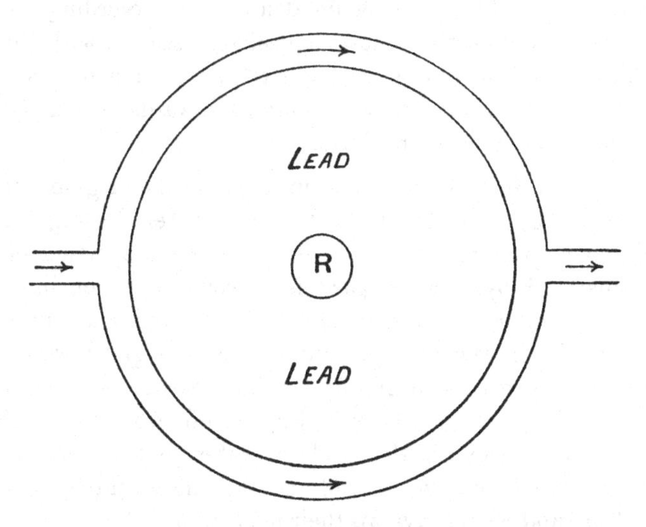

locally in the material around the Radium crystal; while heat is taken into the crystal by conduction, *and radiation*, to supply the energy for the continued loadings of unloaded Radium atoms.

8. Go back to the "equilibrium of temperature" reached at the end of § 3. Suppose the matter around the Radium to be a globe of lead of 50 cms. radius, surrounded by a spherical sheath so arranged as to let a stream of water flow steadily along meridional semicircles of the globe. Every "α particle," and "β particle," shot from the Radium into the lead, will generate the heat-equivalent of its kinetic energy in the lead. A certain proportion of the heat will pass inwards into the Radium to *cooperate with radiation in* supplying energy for the continued

loadings. The remainder will be conducted outwards and will be carried away by the water. The arrangement suggested will form a calorimeter to measure the steady, permanent, thermal, effect of Radium.

9. If the outer surface of the lead were coated with an ideal impermeable varnish, the whole heat generated in the lead by the electrions and atoms shot out from the crystal, would return into the crystal by conduction and would supply the energy for the loadings.

10. It has occurred to many radioactive minds, that by making the lead globe large enough the heating effect might possibly be all lost. This is no doubt true, according to the explanation of radioactivity now suggested: because, with thicker and thicker lead, a greater and greater proportion of the heat generated in the lead will be conducted inwards to supply the energy for the loadings in the crystal.

11. If the Radium is sealed in a glass tube or globe, and if its heating effect is tested by a stream of water, flowing around it through a glass tube of small bore, a large proportion of the heat generated by the "β particles" would not get into the calorimetric water, but would be lost in heating the outer air around it. If, in the arrangement of § 8, the globe were only half a centimetre instead of fifty centimetres radius, a considerable proportion of the "β particles" would get away without leaving their energy in the lead or water, but practically all the "α particles" that get out of the Radium would be stopped by the lead, and would give up their energy to it.

12. The dimensions stated in § 8 are chosen merely for the sake of illustration, and are not suggested as suitable for any practical experiment.

13. The words "*radiated*" and "*radiation*," printed in italics in §§ 3, 7, and 8, above, indicate what seems to me the only possible way of escaping the conclusion that Radium contains a quasi-infinite supply of energy, which can be drawn upon for hundreds of years, without any compensating extraneous source. It seems to me not absolutely impossible that Radium may be, as it were, an exceedingly black body, relatively to waves of ether so short that lead and other solid and liquid substances are transparent for them.

266. On the Motions of Ether produced by Collisions of Atoms or Molecules, containing or not containing Electrions.

[From *Brit. Assoc. Report*, 1907, p. 439 (title only); *Electrician*, Vol. LIX. Aug. 16, 1907, pp. 714—716; *Phil. Mag.* Vol. XIV. Sept. 1907, pp. 317—324; *Nature*, Vol. LXXVI. Aug. 29, 1907, p. 457; *Écl. Élec.* Vol. LII. Sept. 21, 1907, pp. 415—417; Vol. LIII. Oct. 5, 1907, pp. 14—16.]

1. By atom is meant an indivisible element of ponderable (gravitational) matter, or of electricity; by molecule, an assemblage of two or more ponderable atoms, held together by mutual attractions balanced by mutual repulsions.

2. In the atomic theory of electricity, electrion means an atom of resinous electricity, commonly hitherto called negative electricity. It is at present commonly assumed, and I believe in all probability rightly assumed, that all electrions are equal and similar.

3. An ancient hypothesis, which has had large consideration among philosophers in all times assumes that there is only one kind of atom, and that groups of equal and similar atoms constitute the chemical elements, with all their marvellous variety of quality. But, though no doubt some important and interesting differences of quality, such as the difference between ordinary, and red, phosphorus, are due to differences of grouping in assemblages of one kind of atom, it seems extremely improbable that differences of grouping of atoms all equal and similar suffice to explain all the different chemical and other properties of the great number of substances now commonly called chemical elements. It seems indeed almost absolutely certain that there are many different kinds of atom, each eternally invariable in its own specific quality; and that different substances, such as gold, silver, lead, iron, copper, oxygen, nitrogen, hydrogen, consist each of them of atoms

of one invariable quality; and that every one of them is incapable of being transmuted into any other.

4. The sole properties of an atom are:—(1) its mass (being the measure of the inertia of its translatory motion), (2) its law of mutual force between itself and every other gravitational or electrical atom in the universe, varying according to the distance between them. As to the mutual force between ponderable atoms, we have strong reason to believe that this law is practically the Newtonian Law of universal gravitation, for all distances exceeding the millionth of a centimetre. For distances considerably less than the millionth of a centimetre, the Newtonian Law of attraction according to the inverse square of distance merges into repulsions resulting in mutual pressure of two bodies resisting joint occupation of space. For smaller distances, we have attraction again, in the inevitable theory of Boscovich, constituting cohesions and chemical affinities.

5. The assumption, that the mutual force between two atoms depends merely on the distance between their centres, implies that each atom is utterly isotropic. An æolotropic atom, that is to say, an atom having different attractive and repulsive forces in different directions, is conceivable; and may possibly come in future to have a place in atomic theory. Hitherto it has been universally assumed that every atom, whether gravitational or electrical, is thoroughly isotropic, and I do not propose at present to enter upon any theoretical consideration of æolotropic atoms.

6. I do not propose to enter on any atomic theory of ether. It seems to me indeed most probable that in reality ether is structureless; which means that every portion of ether however small has the same elastic properties as any portion however great. There is no difficulty in this conception of an utterly homogeneous elastic solid, occupying the whole of space from infinity to infinity in every direction. We sometimes hear the "luminiferous ether" spoken of as a fluid. More than thirty years ago I abandoned, for reasons which seem to me thoroughly cogent, the idea that ether is a fluid presenting appearances of elasticity due to motion, as in collisions between Helmholtz vortex rings. Abandoning this idea, we are driven to the conclusion that ether is an elastic solid, capable of equi-voluminal waves in which the motive force is elastic resistance against change of shape.

7. We now meet the question:—Is ether incompressible? We should be compelled to answer—Yes, it is incompressible, if it is subject to the law of universal gravitation. But, presently, when we try to account for motion produced in ether, by ponderable or electrical atoms moving through it, we shall feel ourselves persuaded that ether is compressible*. Believing this, we are forced to believe that it is non-gravitational. Thus we find ourselves settled in the conviction that ether is compressible and that ether experiences no gravitational forces between its parts.

8. Suppose now that an atom, whether ponderable or electrical, disturbs ether solely by attracting it or repelling it with a force varying according to distance; and that, with no other mutual influence than this, the atom and the ether jointly occupy the same space. If ether were incompressible, this attraction or repulsion would be utterly ineffective. The atom would move through the space occupied by the ether, without giving any motion to the ether, and without itself experiencing any influence of force due to the ether. Hence, in order that atoms may take energy from motions of ether, and that ether may take energy from motions of matter, we must suppose the ether to be compressible and dilatable; and to be compressed, or to be dilated, or compressed at some distances, and dilated at other distances, in virtue of the force exerted on it by the atom.

9. While assuming ether to be compressible, we suppose its resistance to compression (positive or negative) to be so very great that the velocity of condensational-rarefactional waves in pure ether is practically infinite, and that the energy of whatever of such waves may be produced by collisions of atoms or electrions is practically *nil* in comparison with the energy of the equivoluminal waves, constituting radiant heat and light, which are actually produced by these collisions. It is only under the *enormous* forces of attraction or repulsion exerted by atoms on ether that augmentation or diminution of its density is practically influential.

10. By purely dynamical reasoning, it may be proved to follow from the hypotheses of §§ 4, 6, 8, and 9, that an atom, (supposed for a moment to be infinitely small,) kept moving

* *Baltimore Lectures*, Appendix A; Appendix B, § 3.

through ether at any velocity, q, greater than v, the velocity of light, produces no disturbance in the ether in front of a cone having its vertex at the atom and semi-vertical angle equal to $\sin^{-1} v/q$*; but that the moving atom produces, in its rear, wholly within the cone, an ever growing disturbance of ether, and therefore requires the application of a continual pull forward to keep it moving uniformly at any constant velocity exceeding the velocity of light. In 1888, Oliver Heaviside† arrived at a corresponding conclusion by purely mathematical work, from Maxwell's electromagnetic formulas, without any dynamical foundation: and in 1897‡, still without assuming any dynamical or chemical properties of ether and atoms, he corrected an erroneous hypothesis, that no force however great could give an atom a velocity equal to the velocity of light, which has been somewhat extensively adopted within the last ten years in speculations and reckonings regarding radioactivity.

11. Purely dynamical reasoning§ on our physical assumptions of §§ 4, 6, 8, and 9, teaches us further that:—

(*a*) No force is required to keep an atom moving uniformly through ether, at any velocity less than the velocity of light.

(*b*) To start an atom suddenly into motion from rest, causes a spherical pulse to travel outwards with the velocity of light, from the place in which the atom was when it was receiving the supposed velocity.

(*c*) The magnitude of this spherical pulse is a maximum in the plane through the centre perpendicular to the line of motion, and is zero at the two points in which the spherical surface is cut by that line||.

(*d*) This spherical pulse carries outwards through infinite space a finite quantity of energy, l, due to a part of the work, w, done by the force which was applied to the atom to start it in motion. The sharper the suddenness of the stopping, the greater is l.

* *Baltimore Lectures*, Appendix B, §§ 6, 7.

† Heaviside's *Electrical Papers*, Vol. II. pp. 494, 516.

‡ Heaviside's *Electromagnetic Theory*, Vol. II Appendix G.

§ *Baltimore Lectures*, Appendix B, §§ 4—7.

|| *Baltimore Lectures*, Lect. VIII. p. 88; Lect. XIV. p. 197.

(*e*) If at any time a resisting force suddenly stops the atom, work is done on the ether, in virtue of which another pulse carries away an amount of energy, l'; and work is done on the stopping agent amounting to $w-l-l'$.

(*f*) If the suddenness of the stopping is equal and similar to the suddenness of the starting, the second pulse is equal and similar to the first, and l' is equal to l.

12. To understand clearly the meaning of (*e*), take an example. Let three equal and similar ideal non-electric atoms, *A*, *B*, *C*, be given in a straight line; *B* at rest, *A* moving with velocity q towards *B*, and *C* moving towards *B* in the contrary direction with a velocity just great enough that *B* is left at rest after its collision with *C*. The initial distances must be such that the collision between *A* and *B* precedes the collision between *B* and *C*. Amounts of energy equal to l and l' are carried away into infinite space in the pulses produced by the two collisions. In the arrangement now described, the suddenness of the starting and stopping of *B* are not precisely equal and similar; and, because of their difference, l' might generally be somewhat less than l; but the law of force between the atoms might be such as to render l' equal to, or greater than, l, for certain ranges of values of q.

Take an analogous case of collisions between three ideal billiard balls, each perfectly elastic. The clicks of *A* on *B*, and of *B* on *C*, cause losses of energy, l and l', to be carried off through air by sound-waves.

13. Consider now the collisions in a non-electrified monatomic gas, that is to say, an assemblage of single atoms, each having within it its neutralizing quantum of electrions; except a small proportion, from or to which electrions may have been temporarily taken or given. For simplicity we shall first take the case in which a single electrion is the electric neutralizing quantum for each ponderable atom. The collisions will keep the electrions continually in a state of vibration within the atoms; except, in the comparatively rare case of an electrion being knocked out of an atom, or in the infinitely rare case of the relative motion of an atom and electrion being reduced exactly to zero, by a collision.

14. The law of force between the electrion and atom may be such that the centre of the atom is the only position of stable equilibrium for an electrion within it.

15. Or the law of force may be such that there are any number, i, of concentric spherical surfaces within the atom, on each of which an electrion may rest in equilibrium radially stable; and others, on each of which an electrion would be in equilibrium radially unstable*. In the statistical average of collisions, the electrion may, immediately after a particular collision, be ranging, in non-sinusoidal vibration, across several spherical surfaces of stable and unstable equilibrium, and losing energy by sending out irregularly reciprocating waves through ether. Before the next collision, the electrion may probably have settled down into very approximately sinusoidal vibrations in and out across any one of the surfaces of radial stability.

16. This last condition we may suppose to be generally prevalent during the greater part of the free path between successive collisions. We may indeed suppose it to be more frequently the immediate result of a collision than the wilder vibration described in § 15, which, however, must undoubtedly be an occasional, though probably a rare, condition immediately after a collision.

17. We are not bound to assume that a single electrion is the saturating quantum of any particular ponderable atom: nor are we bound to suppose that it is electrically neutralized by any integral number of electrions†. The most general supposition we can make is that, with j electrions to each atom, the atom and electrions act externally as a vitreously electrified body, and, with $j+1$ electrions, the atom and electrions act as a resinously electrified body.

18. It seems to me indeed exceedingly probable that the persistence of the two-atom molecule in the common diatomic gases, O_2, N_2, H_2, Cl_2, is due to the impossibility of electrically neutralizing the ponderable atom by any integral number of electrions. Suppose for example that one electrion suffices to electrically neutralize two atoms of Nitrogen. A monatomic

* "Plan of an Atom to be capable of storing an Electrion with Enormous Energy for Radioactivity," by Lord Kelvin, *Phil. Mag.* Dec. 1905.

† "Aepinus Atomized," §§ 5, 6; *Baltimore Lectures*, Appendix E.

Nitrogen gas (N), if non-electrified as a whole, would have half of its atoms without electrions, and therefore vitreously electric, with electric quantity equal and opposite to half that of a single electrion. Each of the other half of its whole number of atoms would have one electrion within it, and therefore its external action would be resinous, with half the potency of a single electrion. Thus there would be a strong electric attraction between the atoms destitute of electrions and the atoms each containing one electrion, within it. This attraction would tend to bring the atoms together in pairs, N_2, each pair containing one electrion, of which one position of equilibrium would be at the middle of the line joining the centres of the two ponderable atoms. It seems quite probable that this is the real condition of ponderable atoms and electrions, in the ordinary diatomic gases.

19. The dissociation of a considerable number of such pairs of atoms would be exactly the "ionization" by which, following Schuster's and J. J. Thomson's theory of the conduction of electricity through gases, the latest developed theories of radio-activity explain the specially induced electric conductivity of diatomic gases, such as Lenard found to be produced in air by ultra-violet light traversing it, and Becquerel found in air all round an apparently inert piece of metallic Uranium, or a Uranium salt.

20. But, to give electric conductivity to a monatomic gas, the "ionization" could not be anything else than dissociation of electrions from ponderable atoms. This kind of dissociation might be produced in a very hot gas by mere impacts between the atoms of the gas itself, with the large translational velocities to which high temperatures are due. Or it might be produced by extraneous bodies, such as the "α" or "β" particles shot out with high velocities from radioactive substances. We are now however chiefly concerned with the motions of ether produced by collisions of atoms, in circumstances less abnormal than those in which dissociations and recombinations are largely influential.

21. The pulses described in §§ 11, 12, as due merely to mutual collisions between ponderable atoms (without consideration of electrions whether present or not), constitute a kind of motion in the ether, which, if intense enough to produce visible light,

would, when analysed by the spectroscope, show a continuous spectrum without the bright lines, which, when seen, prove the existence of long-continued trains of sinusoidal vibrations of particles of ether, in the eye perceiving them, and therefore also in the source, and in all the ether between the source and the eye. On the other hand, the vibrations of electrions referred to in § 13 would, if intense enough, produce bright lines in the spectrum.

22. There is another kind of vibration in the source, which might produce, and which probably does produce, bright lines in the spectrum. If there are two or more ponderable atoms in the molecule of a glowing gas, not dissociated by the violence of the collisions, each atom of the molecule must have a vibratory motion, of which an isolated ponderable atom is incapable; and these vibratory motions of the atoms of a group must give rise to bright lines in the spectrum, when the frequency of the vibrations in any one, or in all, of the vibrating modes, is between four hundred million million and eight hundred million million per second, if we take this as the range of frequencies for visible light.

23. The spectroscopic phenomena to be accounted for in a dynamical theory of light include continuous spectrums, with large numbers of bright lines superimposed on the more or less bright background of continuous spectrum. Even when every care has been taken, in artificial sources of light, to eliminate influence of more than one of the substances commonly called chemical elements, the number of bright lines is generally very large: indeed we are not sure that we have been able to count the whole number of those which are presumably due to any single element.

24. In a glowing monatomic gas, with just one electrion to each atom, and only the central position of stable equilibrium for the electrion in the atom, there could be only one bright line in the spectrum. But in reality, every one of the known monatomic gases, Mercury vapour, Argon, Helium, Neon, Krypton, Xenon, gives a highly complicated spectrum with a large number of bright lines. We infer; that, if there is just one electrion to each atom, it has many positions of stable equilibrium; or that there are many electrions, with only the central position

of equilibrium for one of them alone; or that there are several electrions, and several stable positions for one of them alone in the atom.

25. It seems as if only on the third supposition—several electrions and several positions of stable equilibrium—we can imagine the great number of bright lines, and the great complexity of their arrangement in the spectrums of the monatomic gases.

26. But we can feel little satisfaction in this, or any other, attempt to discover details of dynamical theory, unless it gives some reasonably acceptable explanation of the laws of arrangement of trains of bright lines in the spectrums of different chemical elements, which have been experimentally discovered by Runge, Kayser, Rydberg, Schuster, and others.

NAVIGATION AND TIDES.

267. ON THE DETERMINATION OF A SHIP'S PLACE FROM OBSERVATIONS OF ALTITUDE.

[From the *Proceedings of the Royal Society*, No. 125, 1871, pp. 259—266.]

THE ingenious and excellent idea of calculating the longitude from two different assumed latitudes with one altitude, marking off on a chart the points thus found, drawing a line through them, and concluding that the ship was somewhere on that line at the time of the observation, is due to Captain T. H. Sumner*. It is now well known to practical navigators. It is described in good books on navigation, as, for instance, Raper's (§§ 1009—1014). Were it not for the additional trouble of calculating a second triangle, this method ought to be universally used, instead of the ordinary practice of calculating a single position, with the most probable latitude taken as if it were the true latitude. I believe, however, that even when in a channel, or off a coast trending north-east and south-west, or north-west and south-east, where Sumner's method is obviously of great practical value, some navigators do not take advantage of it; although no doubt the most skilful use it habitually in all circumstances in which it is advantageous. I learned it first in 1858, from Captain Moriarty, R.N., on board H.M.S. 'Agamemnon.' He used it regularly in the Atlantic Telegraph expeditions of that year and of 1865 and 1866, not merely at the more critical times, but in connexion

* *A new and accurate method of finding a Ship's Position at Sea*, by Capt. T. H. Sumner. Boston, 1843. "In 1843, Commander Sullivan, R.N., not having heard of this work, found the line of equal altitude on entering the River Plate; and identifying the ship's place on it in 12 fathoms by means of the chart, shaped his course up the river. The idea may thus have suggested itself to others; but the credit of having reduced it to a method and made it public belongs to Capt. Sumner." (Raper's *Navigation*, edition 1857.)

with each day's sights. Instead of solving two triangles, as directed by Captain Sumner, the same result may be obviously obtained by finding a second angle (Z) of the one triangle (PZS) ordinarily solved (P being the earth's pole, Z the ship's zenith, and S the sun or star). The angle ordinarily calculated is P, the hour-angle. By calculating Z, the sun's azimuth also, from the same triangle, the locus on which the ship must be is of course found by drawing on the chart, through the point which would be the ship's place were the assumed latitude exactly correct, a line inclined to the east and west at an angle equal to Z. But, as Captain Moriarty pointed out to me, the calculation of the second angle would involve about as much work as solving for P a second triangle with a slightly different latitude; and Capt. Sumner's own method has practical advantages in affording a check on the accuracy of the calculation by repetition with varied data.

A little experience at sea suggests that it would be very desirable to dispense with the morning and evening spherical triangles altogether, and to abolish calculation as far as possible in the ordinary day's work. When we consider the thousands of triangles daily calculated among all the ships at sea, we might be led for a moment to imagine that every one has been already solved, and that each new calculation is merely a repetition of one already made; but this would be a prodigious error; for nothing short of accuracy to the nearest minute in the use of the data would thoroughly suffice for practical purposes. Now, there are 5400 minutes in 90°, and therefore there are 5400^3 or 157,464,000,000 triangles to be solved each for a single angle. This, at 1000 fresh triangles per day, would occupy above 400,000 years. Even with an artifice, such as that to be described below, for utilizing solutions of triangles with their sides integral numbers of degrees, the number to be solved (being 90^3 or 729,000) would be too great, and the tabulation of the solutions would be too complicated (on account of the trouble of entering for the three sides) to be convenient for practice; and Tables of this kind which have been actually calculated and published (as, for instance, Lynn's Horary Tables*) have not come into general use.

* *Horary Tables for finding the time by inspection, &c.*, by Thomas Lynn, late Commander in the sea-service of the East-India Company. London, 1827, 4to.

It has occurred to me, however, that by dividing the problem into the solution of two right-angled triangles, it may be practically worked out so as to give the ship's place as accurately as it can be deduced from the observations, without any calculation at all, by aid of a table of the solution of the 8100 right-angled spherical triangles of which the legs are integral numbers of degrees.

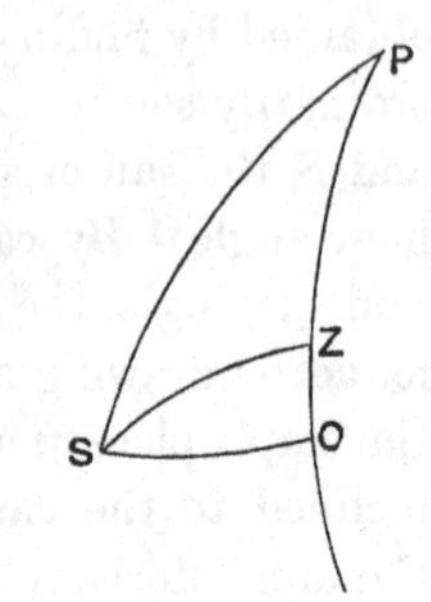

Let O be the point in which the arc of a great circle less than 90° through S, perpendicular to PZ, meets PZ or PZ produced.

If the data were SP, PZ, and the hour-angle P, the solution of the right-angled triangle SPO would give PO and SO. Subtracting PZ from PO, we have ZO; and this, with SO in the triangle SZO, gives the zenith distance, SZ, and the azimuth, SZO, of the body observed.

Suppose, now, that the solution of the right-angled spherical triangle SPO for PO and SO to the nearest integral numbers of degrees could suffice. Further, suppose PZ to be the integral number of degrees closest to the estimated co-latitude, then ZO will be also an integral number of degrees. Thus the two right-angled spherical triangles SPO and SZO have each arcs of integral numbers of degrees for legs. Now I find that the two steps which I have just indicated can be so managed as to give, with all attainable accuracy, the whole information deducible from them regarding the ship's place. Thus the necessity for calculating the solutions of spherical triangles in the ordinary day's work at sea is altogether done away with, provided a convenient Table of the solutions of the 8100 triangles is available. I have accordingly, with the cooperation of Mr E. Roberts, of the *Nautical Almanac* Office, put the calculation in hand; and I hope soon to be able to publish a Table of solutions of right-angled spherical triangles, showing co-hypotenuse* and one angle,

* It is more convenient that the complements of the hypotenuses should be shown than the hypotenuses, as the trouble of taking the complements of the declination and the observed altitude is so saved. [The tables were published in book form with the title, *Tables for facilitating the use of Sumner's Method at Sea*: a specimen page is here omitted from the text.]

to the nearest minute, for every pair of values of the legs from 0° to 90°. The rule to be presently given for using the Tables will be readily understood when it is considered that the data for the two triangles are their co-hypotenuses, the difference between a leg of one and a leg of the other, and the condition that the other leg is common to the two triangles. The Table is arranged with all the 90 values for one leg (b) in a vertical column, at the head of which is written the value of the other leg (a). Although this value is really not wanted for the particular nautical problem in question, there are other applications of the Table for which it may be useful. On the same level with the value of b, in the column corresponding to a, the Table shows the value of the co-hypotenuse and of the angle A opposite to the leg a. I take first the case in which latitude and declination are of the same name, the latitude is greater than the declination, and the azimuth (reckoned from south or north, according as the sun crosses the meridian to the south or north of the zenith of the ship's place) is less than 90°. The hypotenuses, legs, and angles P and Z of the two right-angled triangles of the preceding diagram are each of them positive and less than 90°, and the two co-hypotenuses are the sun's declination and altitude respectively. We have then the following rule:—

(1) Estimate the latitude to the nearest integral number of degrees by dead reckoning.

(2) Look from one vertical column to another, until one is found in which co-hypotenuses approximately agreeing with the declination and altitude are found opposite to values of b which differ by the complement of the assumed latitude.

(3) The exact values of the co-hypotenuse of the angle A corresponding to these values of b are to be taken as approximate declination, hour-angle, altitude, and azimuth.

(4) Either in the same or in a contiguous vertical column find similarly another set of four approximate values, the two sets being such that one of the declinations is a little less and the other a little greater than the true declination.

(5) On the assumed parallel of latitude mark off the points for which the actual hour-angles at the time of observation were exactly equal to the approximate hour-angles thus taken from

the Table. With these points as centres, and with radii equal (miles for minutes) to the differences of the approximate altitude from the observed altitude, describe circles. By aid of a parallel ruler and protractor*, draw tangents to these circles, inclined to the parallel of latitude, at angles equal to the approximate azimuths taken from the Table. These angles, if taken on the side of the parallel away from the sun, must be measured from the easterly direction, or the westerly direction, according as the observation was made before or after noon. The tangent must be taken on the side of the circle towards the sun, or from the sun, according as the observed altitude was greater or less than the approximate altitude taken from the Tables in each case. The two tangents thus drawn will be found very nearly parallel. Draw a line dividing the space between them into parts proportional to the differences of the true declination, from the two approximate values taken from the Tables. *The ship's place at the time of the observation was somewhere on the line thus found.*

To facilitate the execution of clause (2) of the rule, a narrow slip of card should be prepared with numbers 0 to 90 printed or written upon it at equal intervals, in a vertical column, equal to the intervals in the vertical column of the Table, 0 being at the top and 90 at the bottom of the column as in the Table. Place number 90 of the card abreast of a value of co-hypotenuse in the Table approximately equal to the declination, and look for the other co-hypotenuse abreast of the number on the card equal to the assumed latitude. Shift the card from column to column according to this condition until the co-hypotenuse abreast of the number on the card equal to the assumed latitude is found to agree approximately enough with the observed altitude.

When the declination and latitude are of contrary names and the azimuth less than 90°, or when they are of the same names, but the declination greater than the latitude, the sum, instead of the difference, of the legs b of the two triangles will be equal to the complement of the assumed latitude; and clause (2) of the

* A circle divided to degrees, and having its centre at the centre of the chart, ought to be printed on every chart. This, rendering in all cases the use of a separate protractor unnecessary, would be useful for many purposes.

rule must be altered accordingly. The slip of card in this case cannot be used; but the following scarcely less easy process is to be practised. Put one point of a pair of compasses on a position in one of the vertical columns of the hypotenuse abreast of that point of the column of values of b corresponding to half the complement of the assumed latitude; this point will be on a level with one of the numbers, or midway between that of two consecutive numbers, according as the assumed latitude is even or odd: then use the compasses to indicate pairs of co-hypotenuses equidistant in the vertical column from the fixed point of the compasses, and try from one column to another until co-hypotenuses approximately agreeing with the observed altitude and the correct declination are found. It is easy to modify the rule so as to suit cases in which the azimuth is an obtuse angle; but it is not worth while to do so at present, as such cases are rarely used in practice.

The following examples will sufficiently illustrate the method of using the Tables:—

(1) On 1870, May 16, afternoon, at 5 h. 42 m. Greenwich *apparent* time, the Sun's altitude was observed to be 32° 4′: to find the ship's place, the assumed latitude being 54° North.

The *Nautical Almanac* gives at 1870, May 16, 5 h. 42 m. Greenwich *apparent* time, the Sun's apparent declination N. 19° 10′. On looking at the annexed Table (which is a portion of the solutions of the 8100 right-angled spherical triangles) under the heading $a = 56°$, and opposite $b = 54°$, the co-hypotenuse (representing the Sun's declination) is 19° 11′, and opposite $b = 18°$ (differing from 54° by the complement of the assumed latitude), the co-hypotenuse (representing the Sun's altitude) is 32° 8′, which are sufficiently near the actual values; we therefore select our sets of values from these columns as follows:—

		b (°)	Co-hyp. (° ′)	A (° ′)	
$a = 56°$	1.	$b = 54$	19 11	61 23	= Sun's hour-angle
		$b = 18$	32 8	78 14	= Sun's azimuth (S. towards W.)
	2.	$b = 55$	18 42	61 5	= Sun's hour-angle
		$b = 19$	31 55	77 37	= Sun's azimuth (S. towards W.)

from which we have the following:—

Greenwich apparent time (in arc)		85° 30′		85° 30′
Sun's hour-angle	(1)	61 23	(2)	61 5
Diff. = Longitude		24 7 W.		24 25 W

Sun's altitude (observed)		32° 4′		32° 4′
Sun's altitudes (auxiliary)	(1)	32 8	(2)	31 55
Diff. =		− 4		+ 9

Sun's declination from N. A.		19° 10′		19° 10′
Sun's declinations (auxiliary)	(1)	19 11	(2)	18 42
Diff. =		− 1		+ 28

This example is represented graphically in the first diagram annexed [p. 252]. The second set of values could have been selected equally well from the contiguous columns ($a=57°$), which on trial will be found to give an almost identical result.

Again, (2), on 1870, May 16, afternoon, at 5 h. 42 m. Greenwich *apparent* time, the Sun's altitude was observed to be 30° 30′: to find the ship's place, the assumed latitude being 10° North.

The Sun's declination from N. A. is N. 19° 10′, and the half complement of the assumed latitude 40°. By a few successive trials, $a=56°$ will be found to contain values of co-hypotenuses approximately equal to the Sun's declination and altitude at the time, and which are equidistant from 40°; we therefore select the following sets of values from this column as follows:—

			Co-hyp.	A	
$a=56°$	1.	$b=54°$	19° 11′	61° 23′	= Sun's hour-angle
		$b=26$	30 10	73 32	= Sun's azimuth (N. towards W.)
	2.	$b=55$	18 42	61 5	= Sun's hour-angle
		$b=27$	29 53	72 58	= Sun's azimuth (N. towards W.)

from which we have the following:—

Greenwich apparent time (in arc)		85° 30′		85° 30′
Sun's hour-angle	(1)	61 23	(2)	61 5
Diff. = Longitude		24 7 W.		24 25 W.

Sun's altitude (observed)		30° 30′		30° 30′
Sun's altitudes (auxiliary)	(1)	30 10	(2)	29 53
Diff. =		+ 20		+ 37

		° ′		° ′
Sun's declination from N. A.		19 10	....	19 10
Sun's declinations (auxiliary)	(1)	19 11	.........(2)	18 42
	Diff. =	− 1		+ 28

In this case the sun passes the meridian to the north of the ship's zenith, the azimuth, from the Tables being less than 90°, is measured from the north towards the west. In this case also the second set of values might have been taken from $a = 57°$, which will be found on trial to give a position nearly identical with the above.

This example is represented in the second diagram annexed.

Again, (3), on 1870, May 16, afternoon, at 5 h. 42 m. Greenwich *apparent* time, the Sun's altitude was observed to be 18° 35′: to find the ship's place, the assumed latitude being 20° South.

The Sun's declination from N. A. is N. 19° 10′, and the half complement of the assumed latitude is 55°, to be used because the Sun's declination and the assumed latitude are of different names. Proceeding as in the previous example, we find the column $a = 56°$ again to contain values of co-hypotenuses approximately equal to the given values; and therefore have

			Co-hyp.	A	
		°	° ′	° ′	
$a = 56°$	1.	$b = 54$	19 11	61 23	= Sun's hour-angle
		$b = 56$	18 13	60 47	= Sun's azimuth (N. towards W.)
	2.	$b = 55$	18 42	61 5	= Sun's hour-angle
		$b = 57$	17 44	60 30	= Sun's azimuth (N. towards W.)

which give

		° ′		° ′
Greenwich apparent time (in arc)		85 30	....	85 30
Sun's hour-angle	(1)	61 23	.........(2)	61 5
	Diff. = Longitude	24 7 W.		24 25 W.

		° ′		° ′
Sun's altitude (observed)		18 35	....	18 35
Sun's altitudes (auxiliary)	(1)	18 13	.........(2)	17 44
	Diff. =	+ 22		+ 51

		° ′		°
Sun's declination from N. A.		19 10	....	19 10
Sun's declinations (auxiliary)	(1)	19 11	.........(2)	18 42
	Diff. =	− 1		+ 28

This example is represented in the third diagram annexed.

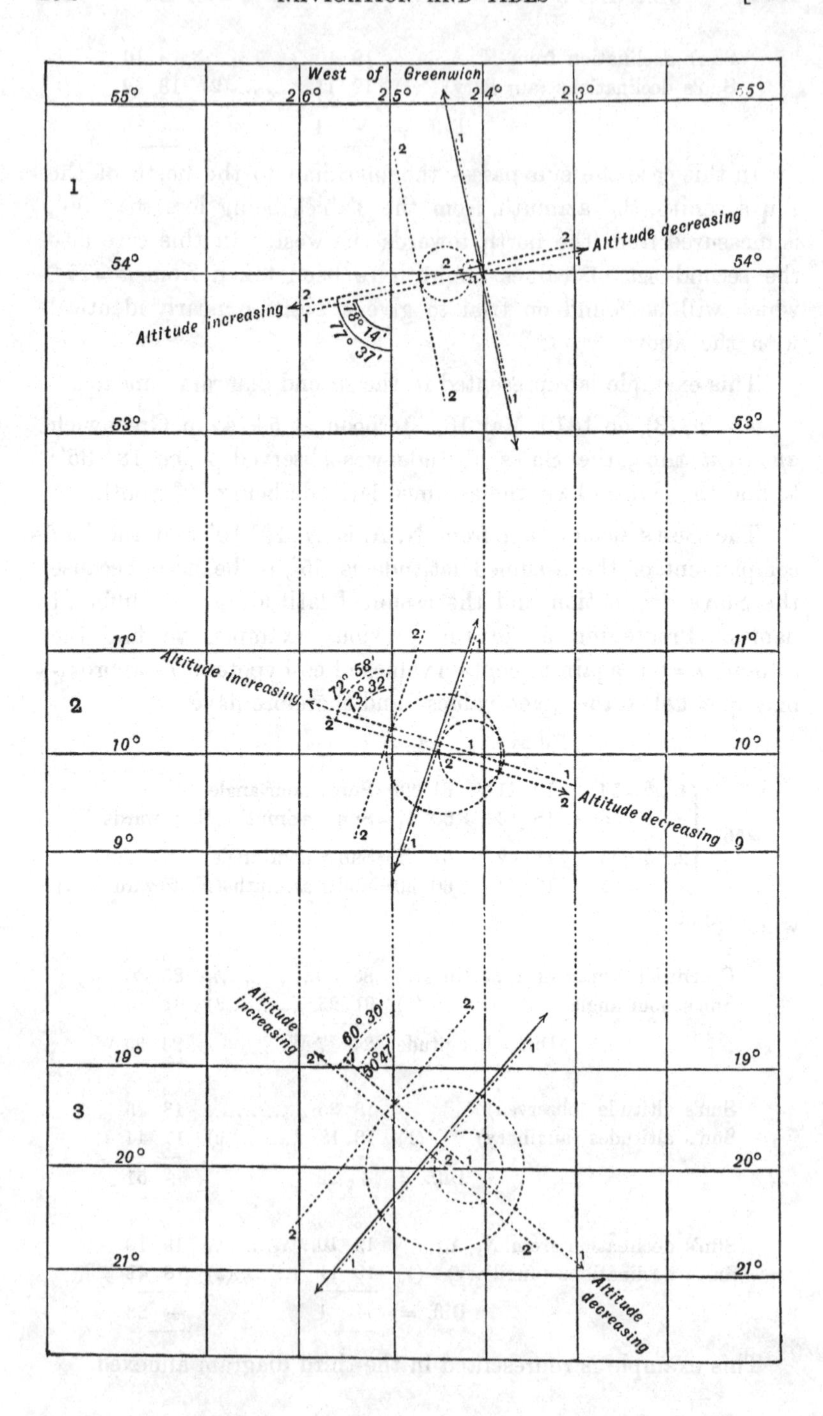
West of Greenwich
55°
26°
25°
24°
23°
55°
1
Altitude decreasing
54°
54°
Altitude increasing
78° 14'
77° 37'
53°
53°
11°
11°
Altitude increasing
72° 58'
73° 32'
2
10°
10°
Altitude decreasing
9°
9
Altitude increasing
60° 30'
60° 47'
19°
19°
3
20°
20°
21°
21°
Altitude decreasing

268. Amended Rule for working out Sumner's Method of finding a Ship's Place.

[From the *Proceedings of the Royal Society*, No. 129, 1871, pp. 524—526.]

In my previous communication on this subject [*anteà*, p. 244] I described a plan according to which, in the first place, two auxiliary lines were to be drawn on the chart, from two sets of numbers taken out of a proposed Table, and then Sumner's line (the line on which the observation shows the ship to be) was to be interpolated, dividing the space between them in the proportion of the differences of the sun's declination from two of the tabular numbers. I find a better plan in practice to be as follows:—

(1) Take two solutions out of the Table as directed in my previous paper.

(2) Taking the two hour-angles and the two altitudes from these two solutions, interpolate to the nearest minute the hour-angle and the altitude corresponding to the correct declination, according to the simple proportion of its differences from the declinations of the two solutions; and estimate, by inspection, the proper azimuth to the nearest half degree, from the azimuths shown in the two solutions.

(3) Using the interpolated hour-angle, azimuth, and altitude found by clause (2), find on the chart, in the assumed parallel of latitude, the point whose longitude is the difference between the interpolated hour-angle and the Greenwich hour-angle at the time of the observation; through this point draw, by aid of a protractor, a line inclined to the north and south at an angle equal to the azimuth, and on the proper side according to whether

the observation was made before or after noon; on this azimuthal line* measure off towards the sun a length (miles for minutes) equal to the correct altitude of observation above the interpolated altitude of clause (2); and through the point thus reached draw a perpendicular to the azimuthal line. This perpendicular is Sumner's line.

The Table (of which a specimen page was shown in my former communication) has now been completed by Mr Roberts, and has been in my hands long enough to allow me to test its use in actual practice. I find the assistance of compasses for measuring off the assumed colatitude preferable to the slip of card with numbers which I first suggested; and I find the process to be altogether very easy and *unfatiguing* (in respect to fatigue a great contrast to the ordinary method). I find that all the cases (as azimuth and hour-angle both acute, azimuth acute and hour-angle obtuse, or azimuth and hour-angle both obtuse, or, again, declination greater than latitude, but of same name, and declination of opposite name to latitude) work out without ambiguity or perplexity. Still the mere fact of there being different cases may possibly deter practical navigators from leaving the ordinary method, which, though considerably longer and *much more laborious*, has the excellent quality of presenting no variety of cases. I intend, however, to push forward the preparation of a short paper of practical directions, illustrated by examples of all ordinary and critical cases, and to publish it with the Table; so that practical men may have an opportunity of judging from actual experience whether the plan of working Sumner's method which I have proposed will be useful to them or not.

I thought it unnecessary in my former communication to remark that every determination of longitude at sea (except from soundings or sights of land interpreted in connexion with observations for latitude) involves the unknown error of the chronometer, and makes the ship 1′ West or East of the true place for every four seconds of time that the chronometer's

* It is unnecessary to mark this azimuthal on the chart. By holding one side of a "set square" (or other proper drawing instrument for making right angles) along the azimuthal line, the Sumner line perpendicular to it is readily drawn, and this "Sumner line," or line of equal altitude, is the only mark which need be actually made on the paper.

indication is in advance of or behind correct Greenwich time. Although I believe that every man who uses a chronometer at sea knows this perfectly well, I shall not omit to state it in the practical directions which I propose to publish, as the Astronomer Royal, Professor Stokes (*Proceedings*, April 27, 1871), and Mr Gordon (writing in the *Mercantile and Shipping Gazette*) are of opinion that an explicit warning of the kind might be desirable in connexion with any publication tending to bring Sumner's method into more general use than it has been hitherto.

269. On a Septum permeable to Water and impermeable to Air, with practical applications to a Navigational Depth-gauge.

[From the *Report of the British Association*, 1880, pp. 488, 489.]

A small quantity of water in a capillary tube, with both ends in air, acts as a perfectly air-tight plug against difference of pressure of air at its two ends, equal to the hydrostatic pressure corresponding to the height at which water stands in the same capillary tube when it is held upright, with one end under water and the other in air. And if the same capillary tube be held completely under water, it is perfectly permeable to the water, opposing no resistance except that due to viscidity, and permitting a current of water to flow through it with any difference of pressure at its two ends, however small. In passing it may be remarked that the same capillary tube is, when not plugged by liquid, perfectly permeable to air.

A plate of glass, or other solid, capable of being perfectly wet by water, with a hole bored through it, acts similarly in letting air pass freely through it when there is no water in the hole; and letting water pass freely through it when it is held under water; and resisting a difference of air-pressures at the two sides of it when the hole is plugged by water. The difference of air-pressures on the two sides which it resists is equal to the hydrostatic pressure corresponding to the rise of water in a capillary tube of the same diameter as the narrowest part of the hole. Thus a metal plate with a great many fine perforations, like a very fine rose for a watering-can for flowers, fulfils the conditions stated in the title to this communication. So does very fine wire cloth. The finer the holes, the greater is the difference of air-pressures balanced, when they are plugged with water. The shorter the length of each hole the less it resists the passage of water when completely submerged; and the greater the number of holes, the less is the whole resistance to the permeation of water through the membrane.

Hence, clearly, the object indicated in the title is more perfectly attained the thinner the plate and the smaller and more numerous the holes. Very fine wire cloth would answer the purpose better than any metal plate with holes drilled through it; and very fine closely-woven cotton cloth, or cambric, answers better than the finest wire cloth. The impenetrability of wet cloth to air is well known to laundresses, and to every naturalist who has ever chanced to watch their operations. The quality of dry cloth to let air through with considerable freedom, and wet cloth to resist it, is well known to sailors, wet sails being sensibly more effective than dry sails (and particularly so in the case of old sails, and of sails of thin and light material).

An illustration was shown to the meeting by taking an Argand lamp-funnel, with a piece of very fine closely-woven cotton cloth tied over one end of it. When the cloth was dry, and the other end dipped under water, the water rose with perfect freedom inside, showing exceedingly little resistance to the passage of air through the dry cloth. When it was inverted, and the end guarded by the cloth was held under water, the water rose with very great freedom, showing exceedingly little resistance to the permeation of water through the cloth. The cloth being now wet, and the glass once more held with its other end under water, the cloth now seemed perfectly air-tight, even when pressed with air-pressure corresponding to nine inches of water, by forcing down the funnel, which was about nine inches long, till the upper end was nearly submerged. When it was wholly submerged, so that there was air on one side and water on the other the resistance to permeation of air was as decided as it was when the cloth, very perfectly wet, had air on each side of it.

Once more, putting the cloth end under water; holding the tube nearly horizontal, and blowing by the mouth applied to the other end:—the water which had risen into the funnel before the mouth was applied, was expelled. After that no air escaped until the air-pressure within exceeded the water pressure on the outside of the cloth by the equivalent of a little more than nine inches of water; and when blown with a pressure just a very little more than that which sufficed to produce a bubble from any part of the cloth, bubbles escaped in a copious torrent from the whole area of the cloth.

The accompanying sketch represents the application to the Navigational Depth Gauge. The wider of the two communicating tubes, shown uppermost in the sketch, has its open mouth guarded by very fine cotton cloth tied across it. The tube shown lower in the diagram is closed for the time of use by a stopper

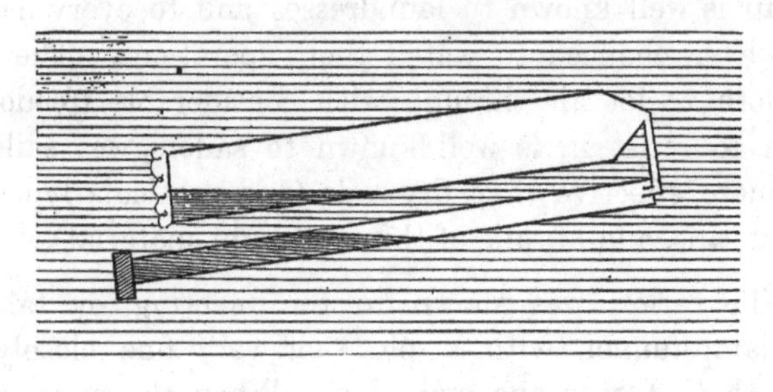

Water indicated by horizontal shading; air by white paper.

at its lower end. A certain quantity of water (which had been forced into it during the descent of the gauge to the bottom of the sea) is retained in it while the gauge is being towed up to the surface in some such oblique position as that shown in the sketch. While this is being done the water in the wide tube is expelled by the expanding air. The object of the cloth guard is to secure that this water is expelled to the last drop before any air escapes; and that afterwards, while the gauge is being towed wildly along the surface from wave to wave by a steamer running at fourteen or sixteen knots, not a drop of water shall re-enter the instrument.

270. ON THE NEW NAVIGATIONAL SOUNDING MACHINE AND DEPTH-GAUGE*.

[From the *Journal of the United Service Institution*, Vol. XXV. pp. 374—381; read March 4, 1881.]

THE objects sought to be attained by this new instrument are twofold; first, to protect the wire from rust more completely and more conveniently than was done in the old machine; and secondly, to supersede the necessity for using the chemically-prepared tubes, one of which had to be used every time a cast was taken.

A depth-gauge, which is in some respects satisfactory, already exists in the form of the Massey or Walker log modified to measure distance travelled vertically downwards by the sinker going down to the bottom, and has given good results in Sir Cooper Key's method of flying soundings, which is not so generally known, and not so fully appreciated, as it ought to be. This method, which has been used with perfect success on board a ship running at 12 knots in 40 fathoms of water, as I have myself witnessed on board Her Majesty's ship "Northampton," consists in letting the cord run direct *and unresisted* from a reel near the stern; instead of from coils held by hands stationed along the whole length of the ship's side, as usual in soundings by the old method from a ship under way.

The great labour required to haul in the hemp line and the heavy lead sinker used in this method is an objection seriously telling against its *very* frequent use, and a great advantage will be gained if we can see our way to obtain the same result with less labour. This we do by my method of using fine steel wire. Then, again, the greatest depth at which it is possible to obtain a

* [The previous papers on this subject are reprinted in *Popular Lectures and Addresses*, Vol. III. "Navigation," 1891, pp. 337—388.]

sounding from a ship running at anything from 8 to 16 knots is probably not equal to that at which we can successfully use the wire. But the Massey or Walker depth-gauge does not give correct indications when used for flying soundings with wire, because the *pull* which must be kept on the wire to prevent it from slackening and kinking prevents the sinker from going vertically down to the bottom. Looking upon the wire as a labour-saving appliance, with the further advantage of being available at speeds and in depths for which it would be impossible to use the lead and hemp line, even by Sir Cooper Key's method, we may consider what may be done to supply the want of a depth-gauge suitable for use with it.

A depth-gauge was patented by Ericsson in the year 1837, in which the required indication was derived from the compression of air in a hollow vessel having two chambers, an outer one and an inner one, connected by a serpentine passage, through which water was allowed to flow into the top of the inner chamber, when the sinker goes down to a greater depth than that which compresses into the volume of the inner chamber, the air that occupied the whole volume of the two chambers and serpentine passage, when the apparatus is first plunged into the water. Many air-compression depth-gauges have been subsequently constructed, but all on the same general principle as Ericsson's. For an up-and-down cast the action of Ericsson's gauge is perfectly satisfactory. If it is let straight down and brought straight up, it will bring up, without spilling any of it, the water which falls into the inner chamber during the descent; but my problem is to secure that no water shall be spilt out of the inner chamber when the gauge is hauled up from the bottom to the surface, and then dashed about from wave to wave in the wake of a ship running at 16 knots. That problem I despair of solving if the inner chamber, which is as it were an uncorked bottle, is more than half-full. If it is less than half-full, I attain the object by guarding the neck of the bottle by a piece of fine tube, projecting downwards like the guard tube of a non-emptying ink-bottle. You may then shake it about as much as you like without spilling or shaking out any of the water, provided it is not more than half-full. If it is anything more than half-full, it is liable to let water be expelled from it by the expanding air within it when, in the course of ascent, the depth-gauge is dragged into a horizontal, or nearly horizontal, position, as it is continually when the ship is going at any speed exceeding 5 or 6 knots. I have found, by actual experience, at a speed of about 12 knots, that when the depth has been great enough to cause the inner chamber to be more than half-filled with water during the descent, water is, during the ascent, irregularly expelled from it in such quantities as to thoroughly vitiate the desired indication of the depth.

In my apparatus there is an outer chamber of brass, and the inner chamber, as in Ericsson's, is a glass measuring tube, and the proportion of the volume of this chamber to the volume of the glass is so arranged that the overflow shall take place when a certain depth has been reached. But I arrange it so that I never use a gauge when the reservoir is more than half-full. When the bottle is *not* more than half-full of water, I am quite satisfied, by much and varied experience, that none is ever thrown out in the practical use of the instrument, whether at high or low speed. I have taken several hundred casts, and never had evidence of any of the depth-measuring water being rejected through knocking about as the gauge is brought up to the surface. Fortified by this experience, I made the *Triple Depth Gauge*, which I will now explain to you. It is composed of three independent gauges, of which the first measures depths of from 11 fathoms to $27\frac{1}{2}$; the second, from $27\frac{1}{2}$ fathoms to $60\frac{1}{2}$; and the third from $60\frac{1}{2}$ to $126\frac{1}{2}$. The outer chamber in each is a brass tube, the inner chamber, or "receiver," as I will call it, is a glass tube. In each the communication between the outer chamber and the receiver is through a short and very fine aperture bored in the brass mounting by which the tops of the six tubes (three of brass and three of glass) are held together. In gauge No. I the outer chamber is of double the capacity of the receiver; in No. II it is five times; and in No. III it is eleven times. Thus, before the overflow takes place in the No. I gauge, the air must be condensed to one-third of its original bulk, in No. II to one-ninth, and in No. III to one-twelfth. Now the law of the compression of air is, that the density is in simple proportion to the pressure. The pressure of the air in which we live is, when the barometer stands at 29·9, equal to the pressure of $5\frac{1}{2}$ fathoms of sea-water. The weight of the whole column of air standing on a horizontal square inch is 14·7 lbs., which is equal to the weight of, say, a column of water of 1 square inch base and 33 feet (or 5½ fathoms) length. Hence at the depth of 5½ fathoms below the surface of the water the pressure is doubled, an amount equal to the atmospheric pressure being added by the water to the pressure of the atmosphere. From this and from Boyle's law of the compression of air, which I have already quoted, it follows that in a diving-bell, or in the hollow of a diver's helmet, with the interior water-level 5½ fathoms below the surface of the sea, the air is condensed to half its natural bulk. At the depth of twice 5½ fathoms, air is compressed to one-third; at five times 5½ (or 27½) fathoms it is compressed to one-sixth; at eleven times 5½ (or 60½) fathoms air is compressed to one-twelfth; and at twenty-three times 5½ (or 126½) fathoms air is compressed to one twenty-fourth of its natural bulk (or bulk at ordinary atmospheric pressure).

Hence, in No. I gauge, the whole air originally distributed through the outer chamber and receiver becomes compressed into the bulk of the receiver, and at greater depths than 11 fathoms the water overflows into the receiver.

At a depth of 27½ fathoms the air is condensed to one-sixth of its natural bulk, and then No. I reservoir is half-full.

For greater depths, therefore, I distrust No. I gauge, and pass on to No. II. In No. II the capacity of the brass outer chamber is five times the capacity of the glass receiver: thus, No. II begins to overflow at the depth of 27½ fathoms, when No. I is half-full.

At a depth of 60½ fathoms the air is condensed to one-twelfth of its original volume. No. II reservoir is then half-full, and I distrust it for the greater depths, and use instead No. III, which begins to overflow when No. II is at the end of its limit of safe use. No. III has its outer chamber of eleven times the capacity of its glass receiver, and therefore its receiver is half-filled, when the air is condensed to one twenty-fourth part of its bulk; that is, at the depth of 126½ fathoms.

I must now explain the cloth guards which you see across the open mouths of the outer chambers. When I first commenced trials of the triple gauge at sea last June, I sometimes got very puzzlingly vicious indications, which, after much anxious consideration, I traced to the continual wash of water backwards and forwards in the outer chambers, by which drop after drop got shot into the reservoirs viciously, while the gauge was being brought up to the surface and dragged from wave to wave in the wake of the ship. The remedy I found for this was simply to tie a piece of very fine strong cotton cloth across the open mouths of the brass tubes. The principle is:—very fine cloth becomes air-tight when the interstices between the threads and the fibres of each thread are closed by water. Water blocks the passage through such fine holes effectively against air striving to burst through, unless the pressure exceeds that due to a head of 5 or 6 inches of water. Thus wet cloth is absolutely impermeable to air if not forced too much. The pressure required to force air through wet cloth depends on the size of the holes, and the smaller they are, the greater is the amount of pressure it will bear. If you blow through a piece of dry cloth you find that the air goes through with perfect freedom. This is shown by the lamp funnel with dry cotton cloth tied across its mouth, which I hold in my hand. You see I can blow through it freely by the very slightest effort. But now, after I have dipped the cloth in water, I cannot blow air through it at all without very considerable effort; and when I plunge it with the open end downwards in this deep glass basin of water, you see that absolutely no air passes through the wet cloth till the level of the water inside the tube is 5 inches below the level in the basin outside the tube. But when I place the funnel with the cloth-guarded end downwards in the basin of water you see that water passes through the wet cloth with perfect ease. The precise object of the cloth guards in the depth-gauge is, to allow the pressure of the expanding air to expel the last drop of water from the outer chamber before any air escapes at all. After the outer chamber is thus perfectly drained of water, a quantity of the air equal to that displaced by the water which has fallen into the inner chamber is expelled during the last part of the ascent of the gauge to the surface. Thus the gauge comes up to the surface without any water in it at all except that which remains in the receiver, and which marks the depth. I have frequently tested the efficiency of the cloth guard by letting the gauge and sinker slowly down to the surface of the sea by proper manipulation of the sounding machine, and towing them along the surface from wave to wave for ten minutes in rough water behind a vessel running at 8 or 9 knots, with various lengths of wire, from 10 to 30 fathoms; and I have never found any water to be lodged in the outer chamber, or any change in the quantities of water in the inner chambers (or receivers) to be produced, by the violent shocks which the instrument experienced in the circumstances. This severe test proves the action of the cloth guard to be practically perfect.

The three glass tubes (receivers) are all closed by one india-rubber pad, a quarter of an inch thick, screwed to a flat brass disc. The more continually india-rubber, in such an application, is kept in use, the less liability there is of its going wrong. If it does go wrong in the course of years, it can easily be repaired.

The new sounding machine is kept, in intervals between casts, in a box filled with lime water, under the surface of which the wheel and wire are immersed. This is a great improvement on the old machine, described in this Institution two years ago, in which the wire could only be immersed in protective liquid, by taking the wheel off its bearings and carrying to a separate tank of lime water, or of oil, or of solution of caustic soda. Thus, in the intervals between casts, in making the English Channel for example by a homeward bound ship, or in going down St George's Channel by an outward bound ship from the Clyde, the wire of the old sounding machine was often left for twenty-four or forty-eight hours unprotected on the wheel placed over the taffrail, and so became weakened by rust.

In the new machine, the wheel and wire are dipped in lime water instantly after every cast, even when casts are being taken every five minutes with not a minute's interval. From six months' experience of the working of the new machine, and of the condition of the wire which I have observed from day to day and from month to month in the course of practical work with it, I am able to say with much confidence regarding the wire itself, that if it is *always* kept in lime water at all times between the acts of taking casts, it will be as good after twenty years' service as it is when first supplied for use, and will remain bright and absolutely free from rust. The machine and the depth gauge, and the whole apparatus in position for use, with a fair-lead pulley fixed to the taffrail, are represented in the following drawings (Figs. 1, 2, 3, 4).

Description of Diagrams.

Fig. 1 shows the machine with a small weight *A* (3 lbs. of lead) resting on the long weight *W* (56 lbs. of iron), and the brake cord *B* slack as it is when the wire is being wound in.

To put on the brake:—Lift the long weight *W* by the hand-rope *E*, and place the small weight *A* in the recess in the large weight, and then slack the hand-rope again. While a sounding is being taken, the long weight *W* is held up by the rope *E*, so as to allow the small weight *A* to hang freely. As soon as the sinker reaches the bottom the brake is put on by easing the rope *E*, and allowing the weight *W* to be supported, by means of its jaws, on the small weight *A*. The whole weight of *W* should not be allowed to come suddenly on *A*, but it should be eased down gradually. If when the whole weight of *W* is resting on *A*, the wire still continues to run out, the brakesman should press his hand down on the top of *W*, until he stops the wheel.

Figs. 2 and 3 are drawings of the depth-gauge. Fig. 2 shows the complete gauge, and Fig. 3 the inside tubes.

a, Fig. 2, is a screw for screwing up the valve *d* against the ends of the three glass tubes, g^1, g^2, g^3. Before the gauge is allowed to go down it should be examined to see that there is no water in any of the tubes, and then the valve *d* should be closed by means of the screw *a*.

The inside tubes can be taken out of the case by unscrewing the top of the case *c*. At *f* a small piece of the case is cut away to facilitate the getting of the tubes out.

At the bottom of the brass tubes b^1, b^2, Fig. 3, there is a piece of very fine strong cotton or linen cloth. If it becomes necessary to renew this, the brass tube must be heated gently to soften the wax on the screw of the tube, and the cap then screwed off. A fresh piece of cloth can then be tied on, and the cap screwed back to its original position, and fixed with a little beeswax or lard. If there is a difficulty in starting the cap, it may be done by binding a piece of twine twice around it, and tightening it by pulling on the ends and pressing the brass cap with a finger or thumb applied to it where the twine is round it.

Fig. 4 shows the sounding machine fixed on board a ship with the stern pulley on the taffrail. Between the sinker and the depth-gauge there is a two-fathom length of *plaited* rope, and

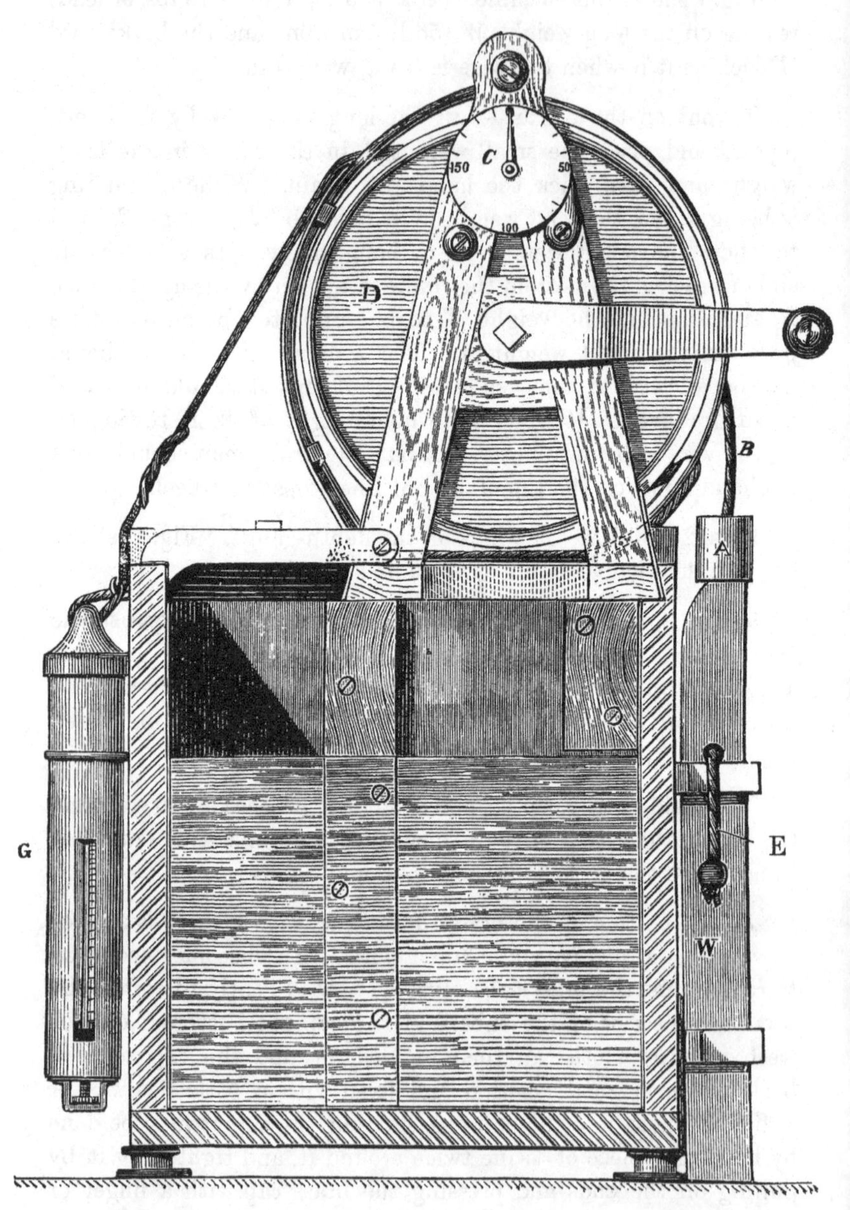

Fig. 1.

between the depth-gauge and the link at the end of the wire there are other two fathoms of *plaited* rope. It is very important that this rope should be plaited, *not twisted.* If twisted rope is used, the turns from the rope go into the wire, and it becomes very liable to kink.

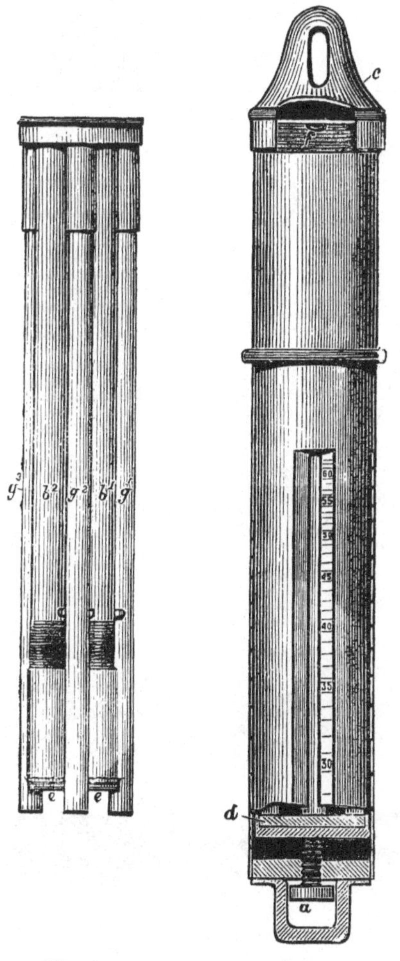

Fig. 3. Fig. 2.

In reply to questions Sir William Thomson said: The depth-gauge is graduated to 126 fathoms; but from 90 to 100 fathoms is about the greatest depth that can be reached with my present arrangements when you are going at 16 knots. At the speed of

11 or 12 knots, a cast can be taken at 110 or 120 fathoms with ease.

Sir George Nares has asked if some other method may be found than the use of lime water for preserving the wire. I most

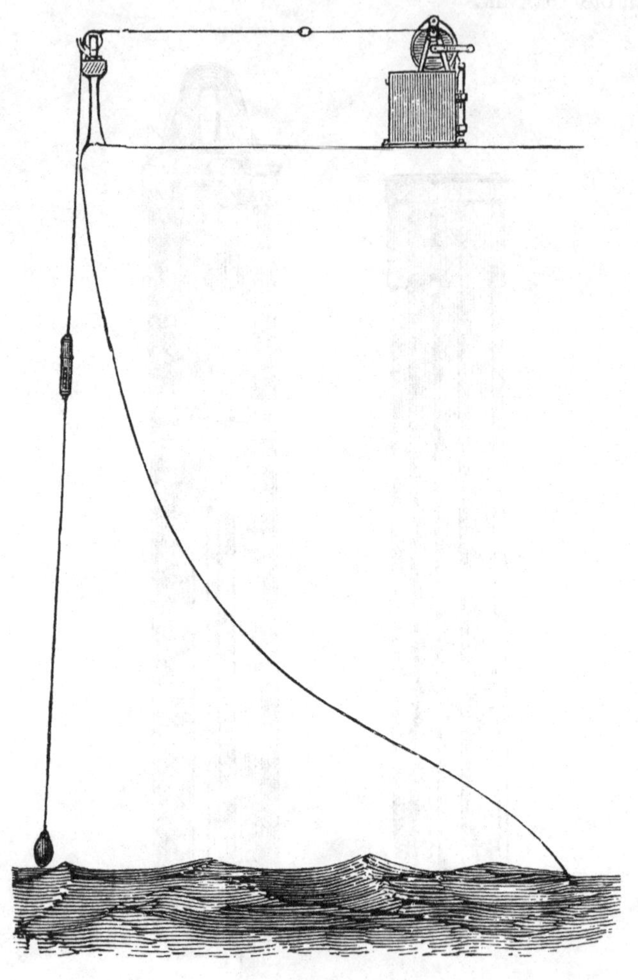

Fig. 4.

cordially agree with him in this desire previously to the introduction of the new machine, but I now hope that Sir George Nares will agree with me that lime water is satisfactory. In the old machine the trouble of taking the wheel off its bearings, putting it into the lime water, and bringing it out again for a fresh cast, was very great. The separation of the lime water tank and the machine was an annoyance all round. That, however, is now obviated; the machine is conveniently kept in a box in any case, and that is all the apparatus now needed for the protection against rust. Look at it on the ship's deck; you need not know that the lime water is there at all, and you simply see a box convenient for keeping the machine in when out of use. Besides keeping the wire bright and free from rust, the arrangement for placing the machine in the box of lime water in the intervals between casts has the great advantage of allowing the iron to be kept at a convenient temperature in very cold weather, by occasionally pouring a kettleful of hot water into the box, and by that means the whole of the machine will be kept at such a temperature as to be capable of being used. That will be found a very great advantage in winter time in the North Atlantic, the German Ocean, and anywhere else where there is severe frost, reducing the iron to such a temperature that it could not, unless artificially warmed, be touched or handled without blistering the hands. Sir George Nares also suggested the use of thicker wire, and that it should be galvanized. No doubt galvanizing would work very well, and for some surveying work a stronger wire galvanized would be more convenient. The greatest strength of galvanized wire, however, is only 52 tons to the square inch, whereas the strength of the special steel wire which I use is 130 tons; so that the steel wire for my sounding machine is more than twice as strong as the galvanized steel wire hitherto made. For wire-rope dredging the galvanized wire may no doubt be found more convenient than ungalvanized, all things considered in practice: but the high quality of the ungalvanized steel pianoforte-wire is of vital importance in deep-sea survey soundings in depths of from 1,000 to 5,000 fathoms, and in flying soundings at all depths; so that I think the lime water is to be regarded not as a disadvantage but as an advantage in this machine. As to the very important notes of warning Sir George Nares has given, I feel them most deeply. I may say, however, that it is not the machine, but the want of the machine, or the non-use of the machine, or the omission to take soundings by the older methods, in ninety-nine out of a hundred cases of vessels running ashore in moderate weather, whether foggy or thick or clear, which is to blame. I knew a case where with a sounding machine on board the ship was as much out of her course as 2½ points of the compass. They heard a sound which really was the fog-siren of Sanda, but they assumed it was the fog-horn of a steamer. The Captain thought he had one of the great Transatlantic steamers on his port bow, and that he was therefore safe in mid-channel, and though he had my machine on board he did not take any sounding, and went on shore on Cantyre. I really think this machine can never be a source of danger, but will, on the other hand, when used with proper diligence, be a help to navigation, by telling the ship's place long before she could be in danger. With regard to my old chemically prepared glass gauges, the trouble and expense of preparation prevented as many soundings from being taken as ought to be taken, because the question of expense comes in. In many cases soundings ought to be taken every ten minutes for five or six hours, and then the cost of the tubes would be too much; thus the necessity for the chemical preparation of the tube on my old system diminished the number of soundings that could (practically speaking) be taken: but with my new machine there is not the slightest difficulty in taking a sounding every five or six minutes. As to the time occupied, you must remember that there is, in almost all localities, a good hour, and generally much more, between the hundred fathoms' sounding and danger. Off Ushant, for example, for a ship approaching the British Channel there are 35 miles from the hundred fathoms' line to the nearest danger ("The Saints," or "Chaussée de Sein"); and before the danger-line of 70 fathoms is reached half a dozen good casts can be taken. The right principle is not to wait until you are in danger and then to take the cast, but to take a cast before you run into danger. I know there used to be a feeling against taking soundings when the ship had to be stopped or rounded-to every time, especially on passenger ships, because the Captains knew very well that the passengers would say, "Oh, he is taking soundings, he does not know where he is," and would become alarmed. What ought to be done, is to do as the blind man does, who feels his way with a stick: that is to say, when you do not *see* where you are, use the lead incessantly to *feel* your way in any water of less than 120 fathoms. This can only be done practically when you can take soundings without reducing speed. Captain Curtis remarked that the stern-pulley could be put on a swivel. I have a special pulley such as he has suggested, which I use in cases in which it is suitable. His suggestion as to using wet cloth for floating seems to me to be an extremely valuable one. No doubt women are often saved from sinking by their dress and men by their shirt-sleeves. It must be remembered, however, that wet cloth does not provide a very *hardy* floater. If the difference of depth between highest and lowest point of a completely submerged bag, of such cloth as you have seen in my depth-gauge and in the demonstrative experiments which I showed, be greater than six inches, air will escape through the cloth in its uppermost parts.

271. THE TIDE GAUGE, TIDAL HARMONIC ANALYSER, AND TIDE PREDICTER*.

[From the *Minutes of the Proceedings of the Institution of Civil Engineers*, March 1, 1882.]

I. *The Tide Gauge.*

THE self-registering tide gauge is a well-known instrument for automatically recording, by a curve traced on paper, the height of the sea level at every instant above or below some assumed datum line. The first essential of the instrument is a floater, which rises and falls with the water. The practical annulment of wave disturbance, so that the floater at each instant may be nearly enough in the position corresponding to the mean of the water level for several minutes, is an important detail. The next thing is mechanism to cause a marking pencil to move in a straight line in simple, but much reduced, proportion to the motion of the floater. The instrument is completed by clockwork, carrying paper with a uniform motion perpendicularly across the line of motion of the pencil, with proper arrangements to cause the pencil to press with sufficient force on the paper to make its mark. An ink marker, as in the Tide Predicter, to be explained later, has been tried for tide gauges both by the Author and by others, but has hitherto been found unsuccessful, on account of the slowness of the motion, and the long time

* [See also *Popular Lectures and Addresses*, Vol. III. "Navigation," 1891, pp. 139—227, especially pp. 171—190. The papers on the dynamical theory of the tides are reprinted in *Math. and Phys. Papers*, Vol. IV. pp. 231—269, and later additions are to be found in Part II. of the Second Edition of Thomson and Tait's *Natural Philosophy*. For the long series of reports to the British Association on the harmonic analysis of the tides at various localities on the Earth, which were inspired mainly by Lord Kelvin, references are given in *Math. and Phys. Papers*, Vol. IV. pp. 235, 269, to which should be added Sir G. Darwin's *Collected Papers*, and his book on the *Tides*.]

through which the action has to be continued; and as there is ample driving power in the tide gauge, there is not the strong reason that there is in the tide predicter for preferring the ink-marker to the pencil; so, for the present at all events, a pencil is by general consent the marker of an automatic tide gauge.

The Tide Gauge now to be described differs from other tide gauges only in certain dynamical and geometrical details, designed for giving, in more convenient form, results of greater, or of better assured, accuracy; with a smaller floater and finer and smaller, but not less hardy, mechanism. The leading idea for the design of every machine ought to be the work to be done by it. With this idea properly kept in view, the force in each part of the mechanism, essentially involved in the work which the machine has to do, is the force to be designed for. The strength in all the parts of the machine ought to be designed to suit the force thus calculated, and nothing more except what may be entailed by the massiveness required for the hardiness of the machine, according to the circumstances in which it is to be used. In the tide gauge, the work done is the moving of the pencil across the paper, subject to the pressure required to produce the mark. For this pressure 50 grammes weight is sufficient, and about 10 grammes may therefore be the force required to move the pencil. The motion of the pencil is made to be from one-tenth to one-hundredth of the motion of the floater, according to the place where the tide gauge is to be used. For the Mediterranean, one-tenth is a convenient scale; for Bristol, St Malo, or the Bay of Fundy, one-fiftieth to one-hundredth would be suitable. A convenient general scale for English tide gauges is $\frac{4}{10}$ inch to the foot, or a ratio of $\frac{1}{30}$. With even the lowest ratio, say $\frac{1}{10}$, the force at the floater corresponding to 10 grammes at the pencil is only 1 gramme, and therefore, so far as merely moving the pencil is concerned, the area of the floater's water-line need not be more than 10 square centimetres to avoid any error on this account of as much as 1 millimetre above or below its correct position. But in ordinary tide gauges the frictional character of the slide is such as to entail the need for a force to move the pencil-carriage scores of times the force required to do the essential work of moving the pencil across the paper. This entails a much larger floater than need be, and heavier and more frictional mechanism all through the instrument. The Author's first object, therefore, was to minimise the friction of the pencil-carriage. Considerable progress had been made towards attaining this object in the ink-marker of the South Kensington tide predicter: and in the Author's first tide gauges, exactly the same geometrical slide was used as in this instrument. In it the motion of the marker is vertical, the axis of the paper cylinder being vertical, instead of both being horizontal, as had been the case in nearly all previous tide gauges. The marker is weighted to the amount of about 600 grammes; so as, in the Clyde or other tide gauges with ratios of motions of 1 to 30, to produce (effect of frictions of the mechanism in either direction largely allowed for) an upward pull on the floater of from 16 to 24 grammes. This pull is transmitted through a fine platinum wire, which it keeps tight enough, and which is strong enough to bear a pull of 110 grammes. The platinum wire is wound round a wheel of 6 inches diameter. The shaft of this wheel carries a pinion, working into a wheel on a second shaft. This second shaft has a fine flat grooved drum, of 3 inches diameter. On this drum is wound a wire cord, strong enough to bear 1,200 grammes, on which is hung the marker, of 600 grammes weight. Fig. 1 represents the design for the second of three tide gauges erected, or to be erected, on the Clyde by the Trustees of the Clyde Navigation.

A large drawing, which is exhibited, shows the first of the tide gauges now at work, having been set in action immediately after the cessation of the frost in January. In it the marker, as in the Author's two tide predicters, and all his tide gauges hitherto made, travels vertically up and down, and is hung by a vertical bearing thread attached to it at a point between the vertical through its centre of gravity and the paper. The couple constituted by gravity downwards through its centre of gravity and the upward pull of the bearing wire, is balanced by the reaction of the paper on the marker, in a horizontal line, near the bottom of the mass, and an equilibrating horizontal force constituting the reaction of the plane front of a fixed guiding flat bar against a round-ended pin fixed to the back of the marker near the top, as shown in section in Fig. 2. Thus the equilibrium of the marker is a very simple and direct application of Poinsot's theory of the equilibrium of couples. Fig. 2 also shows an upper front pin projecting towards, but not touching, the paper. The fixed guiding frame is completed by four upright rectangular bars, one on each side of the pencil tube and upper front pin, and one on each side of the upper back pin; these two last being joined by the flat bar already mentioned, which constitutes a back-plate for the whole frame. The five sliding points of a geometrical slide* are thus provided for the marker as follows:

1. The end of the pencil tube slipping on the paper.

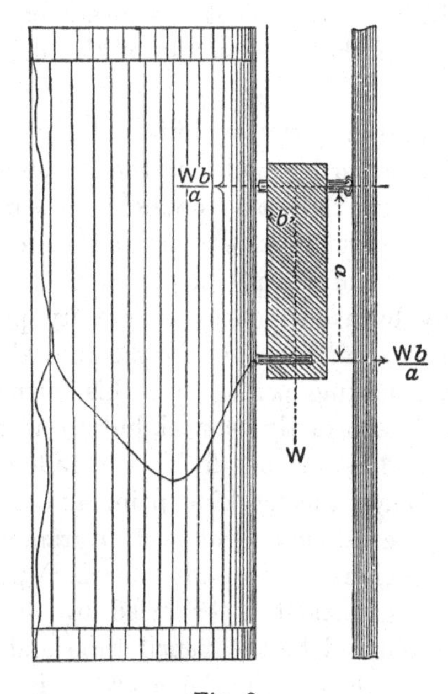

Fig. 2.

2. The point of contact of the round end or knob of the upper back pin on the plane surface of the back-plate.

3. The contact of this knob on one or other of the two side guide-rods.

4. The contact of the upper front pin on one or other of the two front guide-rods.

* *Vide* Thomson and Tait's *Natural Philosophy*, sec. 198; or *Elements of Natural Philosophy*, sec. 168.

5. The contact of the pencil tube, on that one of the front guide-rods on which it is pressed by the frictional force of the paper.

The distance between the two front guide-rods is so small that even if, through any disturbance, the pencil tube is brought into contact with the other guide-rod, the corresponding error in the apparent time of the mark is insensible. This plan has worked well in the several tide gauges to which it has been hitherto applied, and in the British Association and India Office tide predicters. But the Author has now returned to a much better plan, which he tried, not resolutely enough, to realise during the course of the construction of the first tide predicter eight years ago.

It is shown in the model exhibited, and it is to be used in the second tide gauge to be made for the Trustees of the Clyde Navigation, and in the third tide predicter now nearly finished*. In it the back-plate and its side guards are done away with, and the requisite pressure of the pencil or ink tube on the paper is produced directly by a component of gravity perpendicular to the cylindric surface of the paper, in the line along which it is traversed by the marking point. For this purpose the axis of the cylinder is not set exactly vertical, but to an incline of 1 in 5 (angle of about $11\frac{1}{2}°$, or one-fifth of a radian) between the tangent plane through the marking point and the true vertical. This inclination gives a component of 120 grammes (being one-fifth of the whole weight of the loaded marker) perpendicular to the paper. This component is balanced by the paper at the place where it is touched by the pencil tube and the contained lead, or by the tube alone when the marking is by ink. Fig. 3 [omitted] shows the equilibrium of four forces in this plane, consisting of two components of gravity ($W \sin i$ and $W \cos i$), the normal component of the pressure of the paper on the pencil, and the tension of the bearing cord. It will be seen from the figure that the centre of gravity is intended to be in the very point of the pen or pencil. This condition is easily attained, with all needful accuracy, as follows:—Place the marker with its point resting on one side of a horizontal bar of wood, and adjust till it balances on the point; then, holding it by its bearing

* [This has been most successful. It constitutes a real and great simplification. MS. note signed W. T.]

thread, adjust so that when the bar is inclined at any angle to the horizontal, the marker still balancing on its point, keeps the bearing thread as nearly as may be parallel to the bar, at a distance of not more than 1 or 2 millimetres from it. The centre of gravity of the loaded marker being thus adjusted to be nearly enough in the marking point of the pen or pencil, it is clear that when the marker is hung in its place in the instrument, the pen or pencil will press on the paper with a force equal to the whole of the component ($W \cos i$) of the weight of the marker (W) perpendicular to the tangent plane of the cylinder at the point touched. Hence with the weight and inclination stated above, the pressure on the paper is 120 grammes, which is more than twice the amount required for marking by blacklead. The part of it used for pressing the lead on the paper is applied by means of a spring on the end of the little blacklead bar remote from the paper, and the remainder of the 120 grammes is spent in pressing the end of the marking tube on the paper.

To complete the equilibrium, the component of the frictional force of the paper, pulling the pen or pencil point in the direction of the paper's motion, must be balanced. This is done by a directly opposing force—the pressure of a guide-rod—pressing against the wing of the marker, at a point in the tangent plane through the marking point, 6 centimetres from it in the horizontal line of the paper's motion, on the side towards which the paper moves. A second guide-rod is fixed symmetrically on the other side, and the two wings of the marker are quite symmetrical. Thus, if it is desired to turn the paper cylinder, sometimes in one direction and sometimes in the other, the marker works undisturbedly, experiencing just an infinitesimal motion to annul the pressure on one wing, and give the requisite pressure on the other wing when the motion is reversed. The fulfilment of this condition is valuable in the tide predicter, though not so in the tide gauge; but independently of it, the symmetrical form is simpler and more easily made than any other. The two guide-rods serve to keep the marker in a convenient position—very near its true working position—when at any time the paper cylinder is removed for fresh paper. The model constructed for No. 2 Clyde gauge (Fig. 1) shows these details already realised.

This plan is a great improvement on the one described as hitherto used in the tide gauges, and in the two tide predicters already completed, which essentially involves a sum of normal pressures equal to double the pressure on the marking point, and is besides much less simple and less easily constructed.

The sum of normal pressures of the marker on pen and guide-rod in the newly-finished plan, about 135 grammes in all, may produce at most about 30 grammes of resisting frictional force against the upward and downward motion of the marker. This, with the ratio one-thirtieth adopted in the Clyde tide gauges, would give alternate augmentations and diminutions of the pull of the platinum wire on the floater to the extent of a gramme, which is but a small contribution to the ample allowance of ± 4 grammes made above for variation of amount of pull on the floater by friction in all parts of the mechanism. A specimen of the floater used in the first Clyde gauges, and for other tide gauges made by Mr White, which have already been sent to Australia, Italy, and Madeira, is exhibited. The area of the water-line in the floater, which is a circle of 6 centimetres diameter, is about 28 square centimetres, and therefore the supposed extreme vertical disturbing force of ± 4 grammes would disturb its position by no more than $\frac{1}{7}$ centimetre, which is insensible. A much smaller floater might have been taken; the actual size was chosen for the sake of hardiness, so that no slight obstruction should check its motion up and down with the water-level in the tube, and yet its weight be small enough for safety in case of accidental stoppage. Its weight, being only 35 grammes, is less than one-fifth of the breaking weight of the platinum wire attached to it. Thus if the motion of the wire wheel of the tide gauge is stopped while the water-level is still sinking, or if at any time it be turned so as to lift the floater, the platinum wire will not be broken. This floater is of thin sheet copper. Those to be made for Nos. 2 and 3 Clyde gauges are to be of greenheart*, and of the shape and size of the specimen exhibited, which weighs 56 grammes. This wood sinks in water, and an upward pull is thus required to keep it afloat. In the actual specimen a pull of 4 grammes suffices to just prevent it from sinking, and it floats well with the standard upward pull of 20

* [The greenheart has been given up and a copper floater (as shown in Fig. 1, Pl. 1) used instead. W. T.]

grammes. The difference between the smallest pull of 4 grammes and the greatest, 56, which it can give to the wire, constitutes an ample margin for working the wheelwork and marker of the gauge. Its property of sinking in water, unless held up, gives an important security in rendering it impossible for the wire to slack and kink when the water rises, if by any accident the motion of the wheelwork has been arrested, whether by the downward motion of the marker being stopped, or from its suspending wire being broken, or otherwise. Besides remarking that the breaking weight of the wire is 110 grammes, and that forces, whether by jerks or otherwise, which could put more of a pull on it than about 70 grammes are to be avoided, no other caution for safety is required than:—"Beware of kinks."

The Author's tide gauges have hitherto been provided with the means of using continuously a long enough band of paper to allow the curve to be traced on it simply, and without break, through the whole length required for a year. Thus, if it be not desired to take away any tracings of the machine for examination or reduction through the course of a year, no other work is necessary than to wind the clock once a week or once a fortnight, and to put in a fresh lead for the pencil as often as one is needed. The two cylinders for the paper supply and for the haul-off may, however, be omitted or left unused in any case in which it is desired to mark a week's or a fortnight's curves on a single twenty four hours' length of paper, as has been hitherto usually done in other tide gauges. This involves the trouble of taking off a paper of curves, and putting on a fresh piece of blank paper, once a week or once a fortnight; and has the considerable disadvantage of giving a more confused appearance, and involving some practical difficulty and inconvenience in picking out the right curve for a particular day, as will be seen by looking at some of the specimens of tide-gauge work exhibited. On the other hand, the old plan has the great advantage of using less paper, and giving the results in a more compact form, and also some advantage in respect to facility of application to the Tidal Harmonic Analyser. Specimens of both the old and the new plans are exhibited (Figs. 4 and 5); and engineers must choose according to the circumstances which plan is to be preferred in any particular case. Whatever plan be adopted, to secure accuracy in respect both of time and datum-height, the

Author uses rows of fine pins at the top and bottom of the main cylinder; with grooved rollers above and below, pressed on the paper by springs, so as to cause it to be perforated by them when the long continuous band of paper is used. By these pins the hours and half-hours are indelibly marked on the paper, and the two rows which they form give two absolutely fixed datum lines from either of which the water-level indicated by the curve may be measured. By double and triple pins the noon and midnight and six hours are distinguished from the others. In setting up the instrument it is adjusted once for all, so that the marker shall be at the desired distance from either row of pins when the water is at the desired datum-level, or at a measured distance above or below it. The paper drum is set so that the wire bearing the markers is seen to cover the corresponding time on a scale of hours divided down to five minutes, which is engraved on the upper brass rim of the drum.

II. *The Tidal Harmonic Analyser.*

The object of this machine is to substitute brass for brain in the great mechanical labour of calculating the elementary constituents of the whole tidal rise and fall, according to the harmonic analysis inaugurated for the tides by a Committee of the British Association appointed for this purpose in 1867, and carried on from year to year till 1876, with the aid of grants of money from the British Association and the Royal Society, and recently adopted by the Government of India. The machine consists of an application of Professor James Thomson's Disk-Globe-and-Cylinder Integrator to the evaluation of the integrals required for the harmonic analysis. The principle of the machine and the essential details are fully described and explained in Papers communicated by Professor James Thomson and the Author to the Royal Society, in 1876 and 1878, and published in the Proceedings for those years*, and reprinted, with a Postscript dated April 1879, in Thomson and Tait's *Natural Philosophy*, 2nd Edition, Appendix B. It remains now to describe and explain the actual machine referred to in the last of these communications, which is the only Tidal Harmonic Analyser hitherto made. It may be mentioned, however, in passing, that

* *Vide* Vol. XXIV. p. 262, and Vol. XXVII. p. 371.

the same instrument, with the simpler construction wanted for the simpler harmonic analysis of ordinary meteorological phenomena, has been constructed for the Meteorological Committee, and is now regularly at work at their office, harmonically analysing the results of meteorological observations, under the superintendence of Mr R. H. Scott.

Figs. 6 and 7 represent the Tidal Harmonic Analyser, constructed under the Author's direction, with the assistance of a grant from the Government Grant Fund of the Royal Society. The eleven cranks of this instrument are allotted as follows:

Cranks	Object	Distinguishing Letter	Speed
1 and 2	To find the mean lunar semi-diurnal tide .	M	$2(\gamma-\sigma)$
3 " 4	" mean solar " " .	S	$2(\gamma-\eta)$
5 " 6	" luni-solar declinational diurnal tide	K_1	γ
7 " 8	" slower lunar " " .	O	$(\gamma-2\sigma)$
9 " 10	" slower solar " " .	P	$(\gamma-2\eta)$
11	" mean water-level . . .	A_0	—

The two cranks of each of the five pairs are fixed at right angles to one another on one shaft. The speeds of revolution of the five shafts are in simple proportions to the speeds of the respective tidal constituents (S, M, K_1, O, P) which they serve to extract from the given compound result of observation. To give to each crank shaft its proper speed, the Author applied in this first machine an intermediate or "idle shaft," because he was under the impression that the required accuracy could only so be conveniently obtained in practice. He thought that the numbers of teeth in the wheels to give good enough approximations to the true speeds would be inconveniently great without the mechanical complexity of four idle shafts carrying the eight toothed wheels upon them.

Application was made to Mr Edward Roberts, of the Nautical Almanac Office, who had shown much ability in performing the arithmetical work of the British Association Committee, to find convenient factors of numbers expressing the ratio of the speeds of the four shafts M, K_1, O, P, to that of the mean solar shaft (S). He kindly took the thing in hand and found the following solution, which the Author received in a letter dated October 27th, 1878:

Mean solar 12 hours : mean lunar 12 hours : : $184 \times 256 : 199 \times 245$. M
Mean solar 12 hours : sidereal 24 hours : : $119 \times 317 : 209 \times 360$. K_1
12 solar hours : 24 hours of ideal star O : : $58 \times 92 : 89 \times 129$. O
12 solar hours : 24 hours of ideal star P : : $178 \times 221 : 242 \times 326$. P

The following Table shows how very close an approximation to absolute truth is given by these numbers:

SPEEDS of the SEVERAL TIDAL CONSTITUENTS in DEGREES per HOUR.

—	Numerical Approximation	True Speeds	Differences
M	$30° \times \frac{184 \times 256}{199 \times 245} = 28°·9841042$	28°·9841042	0·0000000
K_1	$30° \times \frac{119 \times 317}{209 \times 360} = 15°·0410686$	15°·0410686	0·0000000
O	$30° \times \frac{58 \times 92}{89 \times 129} = 13°·9430363$	13°·9430356	0·0000007
P	$30° \times \frac{178 \times 221}{242 \times 326} = 14°·9589312$	14°·9589314	0·0000002

Thus it will be seen that Mr Roberts' figures give the speeds for two of the constituents correct to the 7th decimal of a degree; and in the other two the larger difference, that for O is scarcely more than $\frac{1}{1500000}$ of a degree per hour. An error of $\frac{1}{1500000}$ of a degree per hour would only amount to one degree in about one hundred and sixty years, so that the approximations may be taken as practically perfect.

The Author afterwards found, in designing the No. 3 Tide Predicter, that his supposition of need for two pairs of toothed wheels to get the true speed for each shaft nearly enough for practical purposes was a mistake, and that all the needful accuracy is readily obtainable by a single pair of toothed wheels for each speed. The numbers of the teeth in the four pairs of wheels thus required will, for the analyser, be given in the description of the No. 3 Tide Predicter.

No. 2 Tidal Harmonic Analyser, whenever it comes to be made, will not be allowed any idle shafts. It will have four toothed wheels on the main shaft, of period twelve solar hours, and four separate crank shafts, carrying one toothed wheel, driven by one of the toothed wheels of the main shaft. It is probable that in No. 2 machine, with this great simplification, at least three more crank shafts will be allowed, so as to include the lunar six-hourly and three-hourly "overtides" or "shallow-water tides" (M_4, M_6), and the luni-solar shallow-water tide (MS) of approximately six-hour period (the harmonic mean of six lunar hours and six solar hours). The toothed gearing for two of these three proposed additional shafts (M_4, M_6) is, of course, very simple; the ratios to be dealt with being 2 to 1 and 3 to 1 (or 3 to 2), as in the meteorological analyser. All this, however, is prospective. It need only be further remarked now, that the first extension beyond the scope of No. 1 machine, in the way of constituents to be analysed for, ought to include the shallow-water tides, because they are of great practical importance, and they cannot be estimated theoretically in any case, and can only be determined by accurate observation and a rigorous analysis of the results; while the whole series of the astronomical constituents N, L, K (semi-diurnal), λ, ν, μ, S, T, R, can each be estimated with all needful accuracy for practical purposes from the analysed values of S and M, by judgment, enlightened by examination of the corresponding results of the analysis already performed by the British Association and by the Indian Government, and to be found in the British Association Reports*; and in the Tide Tables for the Indian ports, published by authority of the Secretary of State for India in Council.

Returning to No. 1 Tidal Harmonic Analyser (Figs. 6 and 7), the actual machine (which is about 20 feet long) is now in a room in the University of Glasgow, where, unhappily idle, it waits for hands to work it. Signor Capello has kindly communicated a series of tidal curves for Lisbon; and Loanda, on the west coast of Africa, lat. 8° 48′ S., long. 13° 8′ E.: which the Author hopes soon to be able to analyse by passing through the machine.

The general arrangement of the several parts may be seen

* *Vide* Reports, 1868, 1870, 1871, 1872, 1876.

from Figs. 6 and 7. The large circle at the back, near the centre, is merely a counter to count the days, months, and years for four years, being the leap year period. It is driven by a worm carried on an intermediate shaft, with a toothed wheel geared on another on the solar shaft. In front of the centre is the paper drum, which is on the solar shaft, and goes round in the period corresponding to twelve mean solar hours. On the extreme left, the first pair of disks, with globes and cylinders, and crank shafts with cranks at right angles between them, driving their two cross-heads, corresponds to the K_1, or luni-solar diurnal tide. The next pair of disk-globe-and-cylinders corresponds to M, or the mean lunar semi-diurnal tide, the chief of all the tides. The next pair lie on the two sides of the main shaft carrying the paper drum, and correspond to S, the mean solar semi-diurnal tide. The first pair on the right correspond to O, or the lunar diurnal tide. The second pair on the right correspond to P, the solar diurnal tide. The last disk on the extreme right is simply Professor James Thomson's disk-globe-and-cylinder integrator, applied to measure the area of the curve as it passes through the machine.

The idle shafts for the M and the O tides are seen in front respectively on the left and right of the centre. The two other longer idle shafts for the K and the P tides are behind, and therefore not seen. That for the P tide serves also for the simple integrator on the extreme right.

The large hollow square brass bar, stretching from end to end along the top of the instrument, and carrying the eleven forks rigidly attached to it, projecting downwards, is moved to and fro through the requisite range by a rack and pinion, worked by a handle and crank in front above the paper cylinder, a little to the right of its centre. Each of these eleven forks moves one of the eleven globes of the eleven disk-globe-and-cylinder integrators of which the machine is composed. The other handle and crank in front, lower down and a little to the left of the centre, drives by a worm, at a conveniently slow speed, the solar shaft and through it, and the four idle shafts, the four other tidal shafts.

To work the machine the operator turns with his left hand the driving crank, and with his right hand the tracing crank, by which the fork-bar is moved. His left hand he turns always in one direction, and at as nearly constant a speed as is convenient to allow his right hand, alternately in contrary directions, to trace exactly with the steel pointer the tidal curve on the paper, which is carried across the line of to-and-fro motion of the pointer by the revolution of the paper drum, of which the speed is in simple proportion to the speed of the operator's left hand.

The eleven little counters of the cylinders in front of the disks are to be set each at zero at the commencement of an operation, and to be read off from time to time during the operation, so as to give the value of the eleven integrals for as many particular values of the time as it is desired to have them.

A first working model harmonic analyser, which served for model for the meteorological analyser, now at work in the Meteorological Office, is exhibited. It has five disk-globe-and-cylinders, and shafting geared for the ratio 1 : 2. Thus it serves to determine, from the deviation curve, the celebrated "*ABCDE*" of the *Admiralty Compass Manual*, that is to say, the coefficients in the harmonic expression

$$A + B \sin \theta + C \cos \theta + D \sin 2\theta + E \cos 2\theta,$$

for the deviation of the compass in an iron ship.

III. *The Tide Predicter.*

After having worked for six years at the Tidal Harmonic Analysis, the Author designed an instrument for performing the mechanical work of adding together the heights (positive or negative) above the mean level, due to the several simple harmonic constituents, determined by the analysis, from observations or from the curves of a self-recording tide gauge for any particular port, so as to predict for the same port for future years, not merely the times of high and low water, but the position of the water-level at any instant of any day of the year. To produce a single simple harmonic motion is one of the best known elementary problems of mechanism, as indicated in the following passage of Thomson and Tait's *Natural Philosophy* (1st ed. 1867, p. 37, or 2nd ed. 1879, p. 39), sec. 55*, "Those common kinds of mechanism

* Or *Elements of Natural Philosophy*, by the same authors, sec. 72.

for producing rectilineal from circular motion, or *vice versâ*, in which a crank, moving in a circle, works in a straight slot belonging to a body which can only move in a straight line, fulfil strictly the definition of a simple harmonic motion in the part of which the motion is rectilineal, if the motion of the rotating part is uniform."

There are many known ways of combining two or more motions in the same or in parallel lines by levers and otherwise. The number of separate motions to be combined made it, however, not easy to see any very acceptable details for the application of levers to produce the combination which the Author desired for the Tide Predicter. On his way to attend the British Association in 1872, with Mr Tower for a fellow-passenger, the Author was deeply engaged in trying to find a practical solution for the problem. Having shown his plans and attempts to Mr Tower, whose great inventiveness is well known, Mr Tower suggested, "Why not use Wheatstone's plan of the chain passing round a number of pulleys, as in his alphabetic telegraphic instrument?" This proved the very thing wanted. The plan was completed on the spot; with a fine steel hair-spring, or wire, instead of the chain which was obviously too frictional for the tide predicter. Everything but the precise mode of combining the several simple harmonic motions had, in fact, been settled long before. At the Brighton Meeting, in presenting the Report of the Tidal Committee to Section A, the Author described minutely the tide-predicting machine thus completed in idea, and obtained the sanction of the Tidal Committee to spend part of the funds then granted to it on the construction of mechanism to realise the design for tidal investigation by the British Association.

Before the end of the meeting he wrote from Brighton to Mr White at Glasgow, ordering the construction of a model to help in the designing of the finished mechanism for the projected machine (Fig. 8). The instrument, which is exhibited, has eight pulleys on cranks, and a cord, passing over and under them alternately, is fixed at one end, and carries a weight representing the marker at the other. The Author will have to return to it presently to explain one part of the original design, the counter-poising, which he did not succeed in having carried out in the first working Tide Predicter;—this instrument belonging to the South Kensington Museum.

The following statement, taken from the third edition of the "Catalogue of the Special Loan Collection of Scientific Apparatus at the South Kensington Museum, 1876," page 11, describes the general object of the tide predicter, and some of the details of the first instrument, exhibited here this evening by the permission of the authorities of the Science and Art Department, under whose care it has been permanently placed by the British Association.

"The object is to predict the tides for any port for which the tidal constituents have been found by the harmonic analysis from

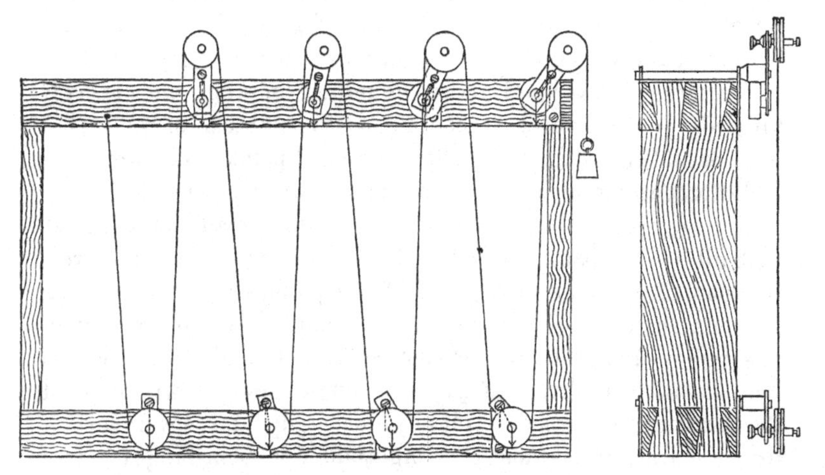

Fig. 8. First Model for Tide Predicter. Scale $\frac{1}{6}$ real size.

tide-gauge observations: not merely to predict the times and heights of high water, but the depth of water at any and every instant, showing it by a continuous curve, for a year, or for any number of years in advance.

"This object requires the summation of the simple harmonic functions representing the several tidal constituents to be taken into account, which is performed by the machine in the following manner:—For each tidal constituent to be taken into account the machine has a shaft, with an overhanging crank, which carries a pulley pivoted on a parallel axis adjustable to a greater or less distance from the shaft's axis, according to the greater or less range of the particular tidal constituent for the different ports for which the machine is to be used. The several shafts, with their axes all parallel, are geared together so that their periods are to a sufficient degree of approximation proportional to the periods of the tidal constituents. The crank on each shaft can be turned round on the shaft and clamped in any position; thus it is set to the proper position for the epoch of the particular tide which it is to produce. The axes of the several shafts are horizontal, and their vertical planes are at successive distances one from another, each equal to the diameter of one of the pulleys (the diameters of these being equal). The shafts are in two rows, an upper and a lower, and the grooves of the pulleys are all in one plane perpendicular to their axes. Suppose, now, the axes of the pulleys to be set each at zero distance from the axis of its shaft, and let a fine wire or chain, with one end hanging down and carrying a weight, pass alternately over and under the pulleys in order, and vertically upwards or downwards (according as the number of pulleys is even or odd) from the last pulley to a fixed point. The weight is to be properly guided for vertical motion by a geometrical slide. Turn the machine now, and the wire will remain undisturbed, with all its free parts vertical, and the hanging weight unmoved. But now set the axis of any one of the pulleys to a distance $\frac{1}{2}T$ from its shaft's axis, and turn the machine. If the distance of this pulley from the two on each side of it in the other row is a considerable multiple of $\frac{1}{2}T$, the hanging weight will now (if the machine is turned uniformly) move up and down with a simple harmonic motion of amplitude (or semi-range) equal to T, in the period of its shaft. If, next, a second pulley is displaced to a distance $\frac{1}{2}T'$, a third to a distance $\frac{1}{2}T''$, and so on, the hanging weight will now perform a complex harmonic motion equal to the sum of the several harmonic motions, each in its proper period which would be produced separately by the displacements $\frac{1}{2}T$, $\frac{1}{2}T'$, $\frac{1}{2}T''$. Thus, if the machine was made on a large scale, with T, T'... equal respectively to the actual semi-ranges of the several constituent tides, and if it is turned round slowly (by clockwork, for example), so that each shaft goes once round in the actual period of the tide which it represents, the hanging weight would rise and fall exactly with the water-level as affected by the whole tidal action. This, of course, could be of no use, and is only suggested by way of illustration. The actual machine is made of such magnitude that it can be set to give a motion to the hanging weight equal to the actual motion of the water-level reduced to any convenient scale; and provided the whole range does not exceed about 30 centimetres, the geometrical error due to the deviation from perfect parallelism in the successive free parts of the wire is not so great as to be practically objectionable....In the actual machine there are ten shafts, which, taken in order from the hanging weight, give respectively the following tidal constituents:—

1. The mean lunar semi-diurnal . . . . M
2. The mean solar semi-diurnal . . . . S
3. The larger elliptic semi-diurnal . . . . . . N
4. The luni-solar diurnal declinational . . . . . K_1
5. The lunar diurnal declinational . . . . . . O
6. The luni-solar semi-diurnal declinational . . . K_2
7. The smaller elliptic semi-diurnal L
8. The solar diurnal declinational . . . . . . P
9. The lunar quarter-diurnal, or first shallow-water overtide of mean lunar semi-diurnal . . . . M_2
10. The luni-solar quarter-diurnal, shallow-water tide . . MS

"The hanging weight consists of an ink-bottle with a glass tubular pen, which marks the tide level in a continuous curve on a long band of paper moved horizontally across the line of motion of the pen, by a vertical cylinder geared to the revolving shafts of the machine. One of the five sliding points of the geometrical slide is the point of the pen sliding on the paper stretched on the cylinder, and the couple formed by the normal pressure on this point, and on another of the five, which is about 4 centimetres above its level and 1½ centimetres from the paper, balances the couple due to gravity of the ink-bottle and the vertical component of the pull of the bearing wire, which is in a line about a millimetre or two farther from the paper than that in which the centre of gravity moves. Thus is ensured, notwithstanding small inequalities of the paper, a pressure of the pen on the paper very approximately constant, and as small as is desired.

"Hour marks are made on the curve by a small horizontal movement of the ink-bottle's lateral guides, made once an hour; a somewhat greater movement, giving a deeper notch, to mark the noon of every day.

"The machine may be turned so rapidly as to run off a year's tides for any port in about four hours.

* * * * * * *

"The general plan of the screw gearing for the motions of the different shafts is due to Mr Légé, the maker of the machine. The construction has been superintended throughout by Mr Roberts, and to him is due the whole arithmetical design of the gearing to give with sufficient approximation the proper periods to the several shafts."

Specimens of the working of this machine, executed by Mr Roberts for the undermentioned places, for which, in his work for the British Association Tidal Committee, and for the Author, he had calculated the harmonic analysis, are exhibited: Ramsgate, Liverpool, West Hartlepool, Portland, San Francisco, Cat Island, San Diego, Fort Clinch, Beechy Island, Malta, Brest, Mauritius, Port Leopold. Most of them were drawn by Mr Roberts for sending with the instrument to the Paris Exhibition of 1878.

It was intended that each crank should carry an adjustable counterpoise, to be adjusted so that when the crank is not vertical the pulls of the approximately vertical portions of wire acting on it through the pulley which it carries shall, as exactly as may be, balance on the axis of the shaft, and that the motion of the shaft shall be resisted by a slight weight hanging on a thread wrapped once round it and attached at its other end to a fixed point. This part of the design, planned to secure against "lost time" or "back-lash" in the gearings of the shafts, and to preserve uniformity of pressures between teeth and teeth, teeth and screws, and ends of axles and "end-plates," was not carried out, but can easily be applied to the machine now exhibited.

The way to realise the counterpoising specified in the preceding statement is shown in the original wooden model (Fig. 8). Each pulley with its central stud is equal to twice the weight of the marker hung on one end of the wire. The slotted crank arm of each shaft of the lower row is permanently balanced by a counterpoise rigidly connected with it in its prolongation on the other end of the shaft (this obvious counterpoise is not executed in the model). Each pulley of the upper row is overcounterpoised by an adjustable counterpoise to such a degree that if the shafts were all loosed from the gearing so as to be each free to turn round its axis, every one of them would rest in any position. Thus any one of them may be turned round its axis in any position and it rests there. This condition is clearly not vitiated by shifting out or in the stud of any of the pulleys of the lower row in its slot; but if any of the pulleys of the upper row be shifted, its counterpoise must also be shifted a corresponding distance out or in on the other side.

It will be seen that the plan of this first tide predicter involves a great simplification in attaching the bearings of each pulley direct to its crank arm, in a proper position adjustable to be either at the centre, in which case its contribution to the resultant motion will be zero, or at any distance from the centre to correspond to the range of the harmonic constituent which it is to represent, instead of having a crank-pin adjustable to different distances from the centre, and causing this crank-pin to produce simple harmonic motion in the manner described in Thomson and Tait's *Natural Philosophy*. Thus the more obvious plan has the advantage of imparting simple harmonic motion to the centre of each pulley. On the other hand the simpler plan gives circular motion to the centre of each pulley, which is equivalent to simple harmonic vertical motion compounded with an equally simple harmonic horizontal motion. The deviation from verticality which the horizontal motion gives to the straight intermediate parts of the thread is a derogation from perfect accuracy in the desired composition of simple harmonic motions, and is a serious drawback to be weighed against the advantage of its great simplicity of mechanism. The model produced, which was made in the winter of 1872–73 for the Tidal Committee of the British Association, by Mr Légé, under the superintendence of Mr Roberts, and which was exhibited at the Bradford Meeting of the Association, shows the composition of two simple harmonic motions on the rigorous plan. The Author did not, however, for the first working instrument to be made for the British Association, venture on the great expenditure which would have been required to carry out the rigorous plan for a sufficient number of tidal constituents to be practically useful, even with all the improvements anticipated in the way of proper geometrical and dynamical designs for the slide, and vertical motion for the marker; and the simple plan of the original wooden model was adhered to, which was accordingly realised in the British Association Tide Predicter.

The second Tide Predicter was made by the Indian Government for the purpose of predicting the tides for the Indian ports, for which, in consequence of the large diurnal tides, the ordinary plan of tide-tables, showing the time and height of high water on the days of full and change, or on every day of the year, does not afford information enough for practical purposes. A design by Mr Roberts and Mr Légé was submitted for the Author's approval, involving the true simple harmonic motions for the centres of the pulleys instead of the circular motions of the first machine. This modification, though making the instrument less simple, was rendered in fact necessary for the large range which it was proposed to give for the resultant curve, and which would have required inconveniently great lengths for the straight parts of wire between the upper and lower rows of pulleys to nearly enough annul the geometrical error of the simpler plan. The Author approved generally of the plan, and recommended to the India Office that it should be carried out by Mr Légé as maker, under the superintendence of Mr Roberts. The details of the slides had not been laid before him in the first instance; and after some progress had been made, Mr Roberts proposed instead of slides to introduce for converting circular into rectilineal motion a system of link-work, of which the suggestion had come from France in some of the numerous pieces of ingenious mechanism which followed the celebrated "Peaucellier's cell" in rapid succession. The Author pointed out that the simplest mechanism sufficed, and that no advantage could be gained by abandoning the elementary slide; and as a suggestion towards details he sent a working drawing of the slide which he had then designed for his harmonic analyser (Figs. 6 and 7). In other respects the India Office machine was a repetition of the British Association tide predicter, with twice as many tidal constituents, greatly improved arithmetical exactness in respect to the periods of the several shafts, and on an enlarged scale. In it, as in the British Association instrument, the number of teeth in the toothed wheels was calculated by Mr Roberts.

For the India Office instrument the Author gave the same principle of counterpoising as that of the original wooden model, but carried out, not by counterpoises fixed oppositely to the crank-arms on the shafts, but by cords passing vertically upwards from the slides over fixed pulleys, and stretched by proper weights hung on their other ends. The condition to be fulfilled is the same, being that, if each shaft was loosed from its gearing and left free to turn without friction, it would remain in whatever position it was placed. The precise way to carry out this condition will be described with No. 3 Tide Predicter, now nearly finished.

The same general plan of gearing as that devised by Mr Légé for the first tide predicter was also adopted in the second instrument. The motions are transmitted from a main driving shaft by intermediate shafts bearing endless screws to toothed wheels on the tidal constituent shafts. In thinking of the mechanism to give the proper speeds to the several shafts, it did not occur to the Author till he began to design practical details for the harmonic analyser, after the tide predicter for the Indian Government was more than half finished, that the simpler plan of merely toothed wheels, gearing into one another, was preferable to any use of intermediate shafts with endless screws. Besides the great unnecessary complication which it gives to the machine, the method by endless screws involves the practical disadvantage that speed is got up very high and run down again between the driving and the driven shafts. Thus, in the India Office machine, when working at such a rate as to trace a year's curves in four hours, some of the wheels and screws turn at from 1,100 to 1,600 revolutions per minute. In the British Association tide predicter the high speed of the intermediate screw-shafts had not been noticed as a fault, because the mode of marking time on the curve in that instrument had limited the speed of working to something less than the greatest speed that, so far as wheel-work was concerned, could have been easily attained. But the jigging motion of the ink-bottle for marking time had been discarded from his tide gauges before the India Office tide predicter was designed, and never entered into the design of this instrument; and he was disappointed to find that its rate of working was limited to one year per four hours through the great speed that this required in the screw-shafts.

To move the ink-bottle marker up and down through the range of the semi-diurnal tide in two seconds is a very moderate speed of working, and this would produce a year's curve in twenty-four minutes. But this is just ten times the speed to which the method of mechanism chosen limits the practical working of the India Office instrument.

The Author has therefore designed and nearly made a third Tide Predicter, in which there is no getting up of speed and running down again, and the proper speeds for the several tidal shafts are obtained by the simple and obvious method of toothed wheels. It is even unnecessary to have the intermediate idle shafts of the harmonic analyser, and thus in the new tide predicter there is only one main shaft carrying eleven toothed wheels; and separate tidal shafts, each carrying one toothed wheel, gearing into one of the wheels on the main shaft except in four instances (M_4, M_6, K_2, MS), in each of which the toothed wheel on the tidal shaft gears into another toothed wheel on another of the tidal shafts. A complete plan of the whole gearing between the main shaft and the tidal shafts is shown in Fig. 9, with the numbers of teeth on each wheel worked, and with pins projecting from the wheel or one of the wheels on each tidal shaft to indicate a crank-pin with its distinguishing letter according to the British Association schedule of tidal constituents.

The following Table shows how close an approximation to

Tidal Constituents	Speeds in Degrees per Mean Solar Hour		Losses of Angle in Machine	
	Accurate	As given by Machine	Per Mean Solar Hour	Per Half Year
M_2	28°·9841042	$15 \times \frac{485}{251} =$ 28°·9840630	+0°·0000412	0°·180
K_1	15°·0410686	$15 \times \frac{366}{365} =$ 15°·0410959	−0°·0000267	0°·117
O	13°·9430356	$15 \times \frac{343}{369} =$ 13°·9430894	−0°·0000538	0°·237
P	14°·9589314	$15 \times \frac{364}{365} =$ 14°·9589040	+0°·0000242	0°·119
N	28°·4397296	$15 \times \frac{802}{423} =$ 28°·4397163	+0°·0000133	0°·059
L	29°·5284788	$15 \times \frac{313}{159} =$ 29°·5283018	+0°·000177	0°·78
ν	28°·5125830	$15 \times \frac{230}{121} =$ 28°·5123966	+0°·0001864	0°·82
MS	58°·9841042	$15 \times \frac{230}{121} \times \frac{271}{131} =$ 58°·9836600	+0°·000444	1°·95
μ	27°·9682084	$15 \times \frac{468}{251} =$ 27°·9681275	+0°·0000809	0°·36
λ	29°·4556254	$15 \times \frac{487}{248} =$ 29°·4556451	−0°·0000197	0°·087
Q	13°·3986609	$15 \times \frac{410}{459} =$ 13°·3986928	−0°·0000318	0°·14

astronomical accuracy is given by the numbers chosen for the teeth of the several wheels. These numbers the Author found by the ordinary arithmetical process of converging fractions.

The main shaft goes once round in the period corresponding to twenty-four solar hours, and a crank-pin, marked S_1, is supplied to allow the meteorological tide of that period to be taken into account in the tide prediction for any port for which it has been found to exist in sufficient amount to be of practical importance. Each of the other fifteen shafts carries a crank-arm and pin, giving simple harmonic motion to the centre of one of the pulleys by means of a crosshead and slide, as shown in Fig. 10, which represents the instrument in the stage up to which it is already completed, and in which the whole working of the composition of fifteen simple harmonic motions is in action, pulling up and letting down a weight, which in the completed instrument will be replaced by an ink-bottle. The cylinder for carrying the paper has not yet been made, and is therefore not shown in the drawing.

In this machine there is no idle shaft; every shaft carries a crank contributing to the general result, and the greatest speed of any one is that of M_6, corresponding to a revolution in four mean lunar hours.

The several slides, pulley-frames, and pulleys in the upper row are not all of the same weight, those for the small tidal constituents being lighter. Those of the lower row corresponding to the smaller constituents are weighted so as to be of the same weight as those for the larger constituents. The weight of each slide, with attached pulley-frame and pulley of the lower row, is about 60 grammes. This weight is exactly borne by the two straight portions of wire passing upward from the pulley (in vertical lines at equal distances on its two sides from its centre of gravity), and the weight of the ink-bottle marker is therefore to be made exactly equal to half this amount. Each slide, pulley-frame, and pulley of the upper row is pulled downward by the same amount (60 grammes), by the two straight portions of wire passing downward on each side. Hence the counterpoise is made to exactly balance this amount added to the weight of its own pulley, pulley-frame, and slide. Thus if the crank-pins were removed, and the slides and pulleys left perfectly free, with the ink-marker on one end, and the other end fixed, all are in equilibrium. Hence, if the machine turn infinitely slowly, the pressure on the guides is zero, and the pressure to be provided for in the actual motion is just what is needed to balance the couple constituted by the upward or downward pressure of the crank-pin in its slot, and the reaction against acceleration of the slide pulley-frame and pulley (and of the counterpoise and its revolving pulley in the case of the upper row of tidal pulleys) which is in a vertical through the centre of inertia of the whole.

In working at as slow a rate as one turn of the main shaft per four seconds (or one year's curves in twenty-four minutes) the reactions against acceleration in all parts of the machine are so small as to be scarcely perceptible in the main mechanism, however slight it may be made. A form and arrangement of guides and sliding pieces has therefore been chosen, which admits of the moving parts being very much lighter and less frictional than those of the harmonic analyser, or of the second tide predicter. Some details of the plan are shown in Fig. 11. The wire, fixed at one end, passed over and under the pulleys, and carrying the 30-gramme ink-bottle, is steel, of No. 50 B. W. G., weighing $\frac{1}{20}$ gramme per metre. Its whole length is 300 centimetres. Its elongation by a difference of pull of 1 gramme is $\frac{1}{40}$ millimetre: and it is strong enough to bear a weight of 500 grammes, or over fifteen times the weight of the ink-bottle.

The Paper is illustrated by several models and diagrams, from which Plate 1 and the woodcuts have been prepared.

Discussion.

Sir William Thomson commenced with an explanation of the models and drawings. He showed a model constructed to aid in the design for No. 2 Clyde tide gauge, in which was a pencil-marker on the second of the two plans described in the Paper and represented in Fig. 1. The cylinder carrying the paper was inclined to the vertical, and looked like the "leaning tower of Pisa." The cylinder could be taken off and re-applied in a moment. The bearing plate was slightly cupped, so that the weight of the cylinder pressed entirely on its rim. The spike in the centre bore no part of the weight; it was merely a guide to keep the cylinder in the middle. The cylinder was adapted either for a long roll of paper or for marking a week's or a fortnight's curve on one paper. It had been almost determined in the case of the Clyde tide gauges that instead of a very long slip of paper for a year's curves, the curve of a week or perhaps a fortnight should be traced on one piece of paper. There would be little trouble and no difficulty in the affair. The person who wound up the clock would carry the cylinder away and substitute another with paper properly placed on it. Or if he could be trusted to put on fresh paper, there would be a table in the tide-gauge house with papers ready ruled with the time and the datum lines. All that would be necessary was to lay the cylinder in the proper position and roll it over the paper so as, by aid of the hour-pins on the cylinder, to wind the paper upon it, and then fix it by two little adhesive overlaps. Referring to the geometrical slides of the two plans, and to the general principle of a geometrical slide, Sir W. Thomson remarked that good workmanship was too often put in requisition to overcome evils of a bad design. A good design in many cases required no fitting; and where it was possible it was better to manage with no fitting; for the finest fitting might be undone by a little warp in the material or by a piece of grit. In the pen- or pencil-markers exhibited there might be an inequality in the paper projecting as much as $\frac{1}{8}$ inch, but, if not too steep, it would not disturb the marker, which would be pressed out, and simply slip over it. He would test the instrument by shaking it roughly, and it would be seen that there was no error in the marking. With regard to the floater, he had a greenheart one shown in action. It weighed 50 grammes, and was a little heavier than water. Greenheart took kindly to the water, but india-rubber or gutta-percha, slightly weighted with metal, would do very well. Something was wanted which would just sink—not with too much force. If the motion of the wheelwork became arrested—if the pencil broke down or was caught in the paper—when the tide rose, the wire would become slack and therefore be liable to kink if the floater floated without any upward pull from the wire. The great safeguard was to beware of kinks. In fact the sinking floater might be called an anti-kink arrangement. In case of any arrested action of the wheelwork, the floater would sink, and keep the wire tight. The early specimen alluded to in the Paper, made by Mr Légé for the Author, was exhibited. In all the tide gauges since made for the Author a very light flat-rimmed wheel was substituted for that in the specimen, which was too heavy, and spirally grooved. There was no occasion for the grooves, because the height of riding of the thin platinum wire in the groove was infinitesimal in its effect on the reckoning of the water level; so that the riding of the wire was not a thing to be avoided. A tide gauge should always when possible be placed vertically over the vertical sea-tube. In cases in which this was not judged practicable, as in the tide gauge recently placed on the Admiralty pier, at Dover, a stouter wire, a larger floater, and a counterpoise besides the ink-bottle were necessary. Otherwise the simple arrangement was one shaft for the main wheel on which the wire was wound, and a second shaft carrying a drum on which the wire bearing the pencil was wound; then, when the tide rose, the pencil went down, and when the tide fell, the pencil went up. Thus in a tide-gauge curve on its drum in the instrument the high water was down, and the lower water up; the hours must therefore be marked from right to left, and the turning must be in the direction of the hands of a watch, so that when the paper was taken off and turned to make high water up and low water down, the part corresponding to past time would be to the left, and future to the right. In all the tide gauges he had made hitherto, he had used a long slip of paper. The long slip extending round the room contained a year's curve for Mauritius, drawn by Mr Roberts, by means of the South Kensington Tide Calculating Machine. At the rate of a foot a day the length of paper would be 365 feet. With seven curves on one paper the sum of the lengths would only be 52 feet. Fourteen curves would probably be a more convenient number to put upon one paper. He held in his hand a paper with thirty curves, drawn by a self-recording tide-gauge at Loanda (West Coast of Africa), and it would be seen that the appearance was very confusing. That number was too great, but he did not think that a fortnight's curves on one paper of twenty-four hours' length would be too much. He was certainly a convert to putting several curves on one paper. The Dover tide-curves, which he was able to exhibit through the kindness of Mr Druce, M. Inst. C.E., the harbour-engineer, were very interesting. It would be seen that the line was much thickened. There were oscillations of 2 or 3 feet in the tide gauge which had been recently put up there under the direction of Mr Druce, who had informed him that in all other respects its working was satisfactory. Mr Légé was now making a copper guard tube, in order to remedy that evil. It was to extend 6 feet below low-water mark, ending in a small hole about $\frac{1}{4}$ inch in diameter, so that it would be impossible for any pumping up and down to take place. If in two or three minutes the water rose by several inches, or even half a foot, the floater would show it. A rising and falling by wave-disturbance in five or ten seconds would not affect the floater to any sensible degree. Thus the quick oscillation up and down would be annulled, and he had no doubt that the tidal gauge would be in a perfectly satisfactory state.

With regard to the Tidal Harmonic Analyser, he had a rough model to exhibit. One of the twenty-four hours' tidal curves, or a paper containing seven or fourteen, was placed on the cylinder, which went round once in twelve hours. The circumference of the cylinder was half the twenty-four hours' length of the paper. The two ends of the paper were united, and it ran round (like an endless towel on a roller) when the cylinder was turned. The instrument had been explained in the paper, and he would only show the management of it. To move the pointer along to the right and left in tracing the curve he had to turn a little crank with his right hand alternately to right and to left. The left hand was always kept turning one way; the right hand alternately one way and the other. The manipulation might appear to be very puzzling, but he was informed that at the Meteorological Office the instrument worked satisfactorily, and that the manipulation became easy after a little practice. It would be seen that the disk oscillated alternately in one direction and the other, and caused the counter to turn alternately forwards and backwards if the centre of the globe remained in one position. But if, while that change of direction of motion of the disk took place, the globe was turned over to the other side by the motion of the fork bar, then the counter kept turning in one direction, and thus each of the eleven counters counted out the amount of one particular harmonic constituent of the complex variation represented by the curve presented for analysis.

With regard to the Tide Predicter he wished to explain an epicyclic mechanism for the combining of two simple harmonic motions, which he had described at a meeting of the British Association at Brighton in 1872, and which was the simplest possible way of producing and of combining two harmonic constituents, though essentially inapplicable to more than two. Fig. 12 represented a pinion fixed on the end of a stud. The large circle represented a wheel about twenty-eight and a half times the diameter of the pinion. There were seventeen teeth in the

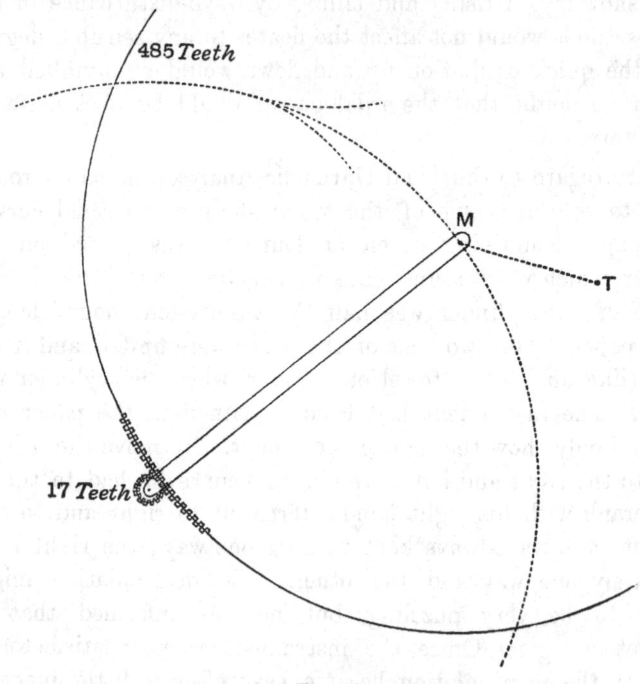

Fig. 12.

pinion, and four hundred and eighty-five in the wheel. Imagine first a fixed framework, the pinion rotating in one direction, and the wheel rotating in the opposite direction. The angular velocities of the wheel and pinion would be as 485 to 17. Take now the whole machine and turn it round the axis of the pinion with an angular velocity equal to the pinion's but in the contrary direction; that annulled the angular velocity of the pinion, and added the amount to the first-supposed angular velocity of the wheel; making the angular velocity of the wheel $485+17$, or

502, if 485 be called the angular velocity with which the bearing of the wheel was carried round. Now the "speeds" of the mean solar and mean lunar semi-diurnal tides were as 502 to 485, very exactly. (See M_2 in the Table of Speeds in Part III of the Paper.) Thus, by a crank-pin T, carried by the wheel in this mechanism, a solar tide was superimposed on a lunar tide. The point T (Fig. 12) changed its level according to the resultant effect of the mean lunar semi-diurnal tide and the mean solar semi-diurnal tide The semi-range of the first of these tides was the radius of the circle described by the wheel; the semi-range of the second was the distance of the point T from the centre of the wheel*. This epicyclic method was a mechanical realisation of the construction in kinematics corresponding to the "polygon of forces" well known in elementary statics, for the case of two constituents; and it was a very useful as well as simple method when there were only two to be combined.

With regard to the particular mode of combining the motion which he adopted, by a hair-spring passing under and over pulleys:—as stated in the Paper, Mr Tower had made the suggestion to him in a railway journey from Portsmouth to Brighton, at the commencement of the meeting of the British Association in 1872. Before the end of the meeting he wrote to Mr White, and gave him instructions, and before the end of the month he wrote to Mr Roberts, and told him that he had given instructions, for the construction of the wooden model now exhibited. Mr Roberts had been with him, before the meeting of the British Association, for a few days, assisting to complete the report of the Tidal Committee for the meeting of the Association. During that time he continued discussions that he had had with Mr Tower on the subject of his projected tide-predicting machine: and in the course of these discussions Mr Tower made several suggestions, and amongst them a suggestion as to floaters in tubes, according to which, on hydrostatic principle, the effect could be summed up. Pushing in one piston caused water to rise in a tube; pushing in another added its effect; pulling out another caused a corresponding subtraction from the whole; and so on, with any number of pistons. With that beautiful idea it could be seen

* A working model was exhibited at the meeting of the 15th of March. See conclusion of the Discussion.—W. T.

how the combination might be realised without mechanism. That was a very interesting suggestion of Mr Tower's inventive mind; but he did not himself regard it as a convenient practical solution of the problem, nor did Sir W. Thomson. Mr Roberts wrote to him: "I like Mr Tower's idea of floaters as well as any of the different things that have been contemplated; but at present I cannot see a good method of combination." In reply Sir W. Thomson commented thus, in a letter to Mr Roberts despatched before the end of August 1872: "The floats would not work well. He (Mr Tower) suggested also a plan with pulleys and a cord or chain which led me to a very simple plan with a long hair spring round pulleys centered on cranks. I have given White instructions to commence a partial model for trial." The result of those instructions was the wooden eight-component model (Fig. 8) now before the meeting. It had only yesterday come to his knowledge that in the report in the *Athenæum* of the meeting of the British Association at Bradford in 1873, the Tide Predicter had been described as "Mr Roberts' instrument." The origin of that misapprehension had been explained by Mr Roberts, in a letter to the Author of date October 23, 1873, informing him that a label describing the instrument as of his (Mr Roberts') design had been affixed to the instrument by mistake during his absence. No doubt the reporter had taken his information from the false label. When the machine was exhibited at Paris in 1878, in charge of Mr Légé, under direction of Mr Roberts, it was designated as Sir William Thomson's Tide-calculating Machine, and was accompanied with a description, extracted from the catalogue of the loan collection of South Kensington, which contained the following: "The general plan of the screw gearing for the motions of the different shafts is due to Mr Légé, the maker of the machine. The construction has been superintended throughout by Mr Roberts, and to him is due the whole arithmetical design of the gearing to give with sufficient approximation the proper periods to the several shafts." Mr Roberts' and Mr Légé's translation of that passage (giving less credit to Mr Roberts) was, "Tous les nombres ont été donnés par Mr Roberts." He should not have troubled the members with such a statement, but that he wished to make it clear that he had dealt in a perfectly fair manner with those who had worked for him on the tide-predicting machine.

[In the discussion which followed, Mr E. Roberts made some claims of priority in devices of construction to which Lord Kelvin in his final reply was unable to agree.

The discussion was continued by Mr G. F. Deacon, Mr H. Law, Mr T. Cushing, Mr G. J. Symons, Prof. J. D. Everett, Sir G. B. Airy, Mr J. B. Redman, Mr P. Adie, Mr E. A. Cowper, Mr W. Parkes, Mr H. Unwin.

In continuation of his reply, Sir W. Thomson spoke as follows:—]

Referring to the Astronomer Royal's remarks on the problem to be solved by the Harmonic Analyser, let

$$y = A_0 + A\cos nt + B\sin nt + A'\cos n't + B'\sin n't + A''\cos n''t + B''\sin n''t$$

be the theoretical equation of the curve given by the self-registering instrument, whatever it be (tide-gauge in the special case before the meeting), or given by plotting properly the observations (as in astronomical applications, pointed to by Sir George Airy). The nt, $n't$, $n''t$ corresponded to the θ, ϕ, ψ of Sir George Airy's verbal statement. The several counters of the machine, after it had been worked through a space corresponding to a whole time, T, gave the values of the integrals

$$\int_0^T y\,dt, \quad \int_0^T \cos nt\,y\,dt, \quad \int_0^T \sin nt\,y\,dt, \quad \&c.$$

Thus if C_0, C, D, C', D', C'', D'' denoted the readings of the several counters, there would be the following seven equations to determine the seven unknown quantities A_0, A, B, A', B', A'', B'':—

$$TA_0 + \frac{1}{n}\sin nT \,.\, A + \frac{1}{n}(1 - \cos nT)\,B + \&c. = C_0,$$

$$\frac{1}{n}\sin nT \,.\, A_0 + \frac{1}{2}\left(T + \frac{1}{2n}\sin 2nT\right)A + \frac{1}{2n}(1 - \cos 2nT)\,B + \left\{\frac{1}{2}\frac{\sin(n-n')\,T}{n-n'} + \frac{1}{2}\frac{\sin(n+n')\,T}{n+n'}\right\}A' + \&c. = C, \ \&c.$$

Here, provided T was large enough, the coefficient of A_0 was relatively very large, and all the others small, in the first equation; that of A was relatively large, and all the others small, in the second equation, and so on. Hence approximately (more and more approximately the larger was T),

$$A_0 = \frac{C_0}{T}, \qquad A = \frac{2C}{T + (2n)^{-1} \sin 2nT}, \text{ \&c.}$$

If T was not large enough to make these approximations sufficiently close, the first approximate values of A, B, A', B', A'', B'' were to be used in the equation for C_0 above, and a second approximate value for A_0 was then instantly calculated by it. Similarly the first approximate values of A_0, B, A_1, &c. were to be used in the equation for C, and a second approximate value for A calculated from it, and so on for B, A', B', A'', B''. These second approximate values were in most practical cases as accurate, practically speaking, as the available observations could give. Thus the Harmonic Analyser really did the work which seemed to the Astronomer Royal so complicated and difficult that no machine "could master it."

An interesting point had been brought forward by Mr Cushing in connection with engraving curves, and he had shown a beautiful specimen of the Ostend tide curves. He did not quite agree with him in his objection to the pencil. He had not the slightest difficulty in getting a sufficiently fine line with a pencil, and he therefore objected to introducing anything less simple. The diamond was valuable when they wished to engrave the result; but when that was not required it was an unnecessary expense. He knew of no case in which the pencil had been broken. With regard to the floater, he did not neglect the great power of the ocean, but he would not use a windmill 30 feet in diameter when a windmill 30 inches would suffice to do the work; and on that principle he preferred the smaller floater and a finer line connecting the floater with the wheel to the enormous floater shown by Mr Adie, involving large expenditure to provide a large enough tube under water to guard it. Another reason for not making the floater too heavy was that in case of accident it should not be able to break the wire. He liked to have the wire strong enough to bear the floater. It would not do to have it too strong, or it would not be flexible enough to go round the pulley; and the

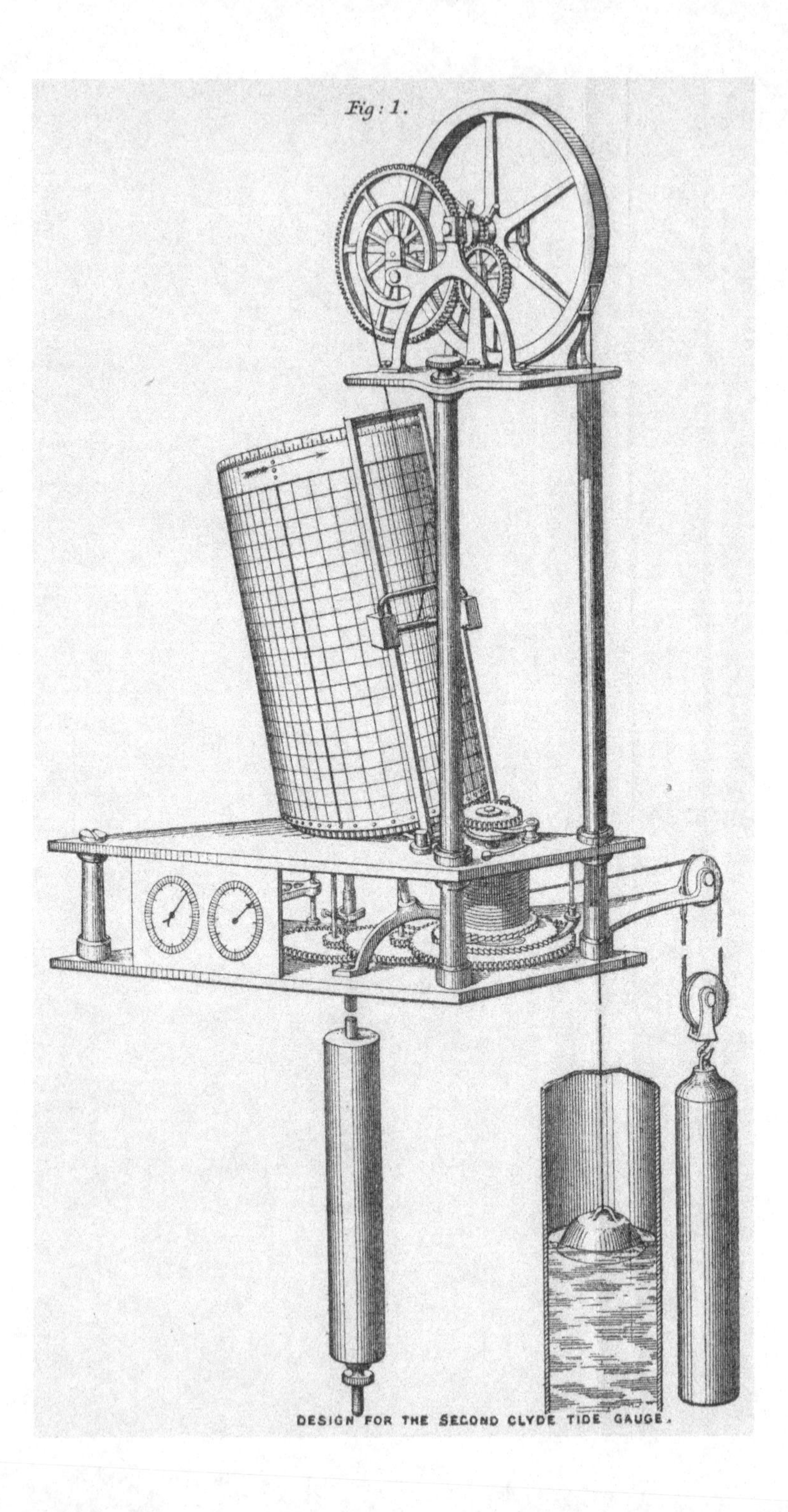

DESIGN FOR THE SECOND CLYDE TIDE GAUGE.

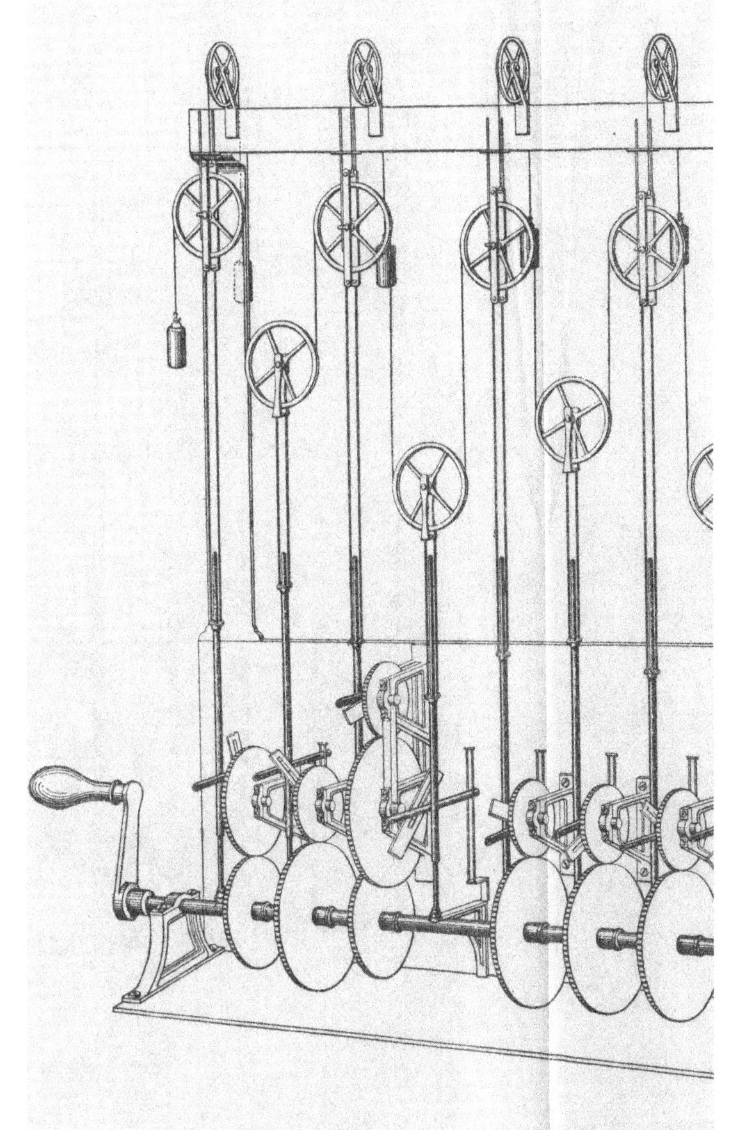

THE TIDE

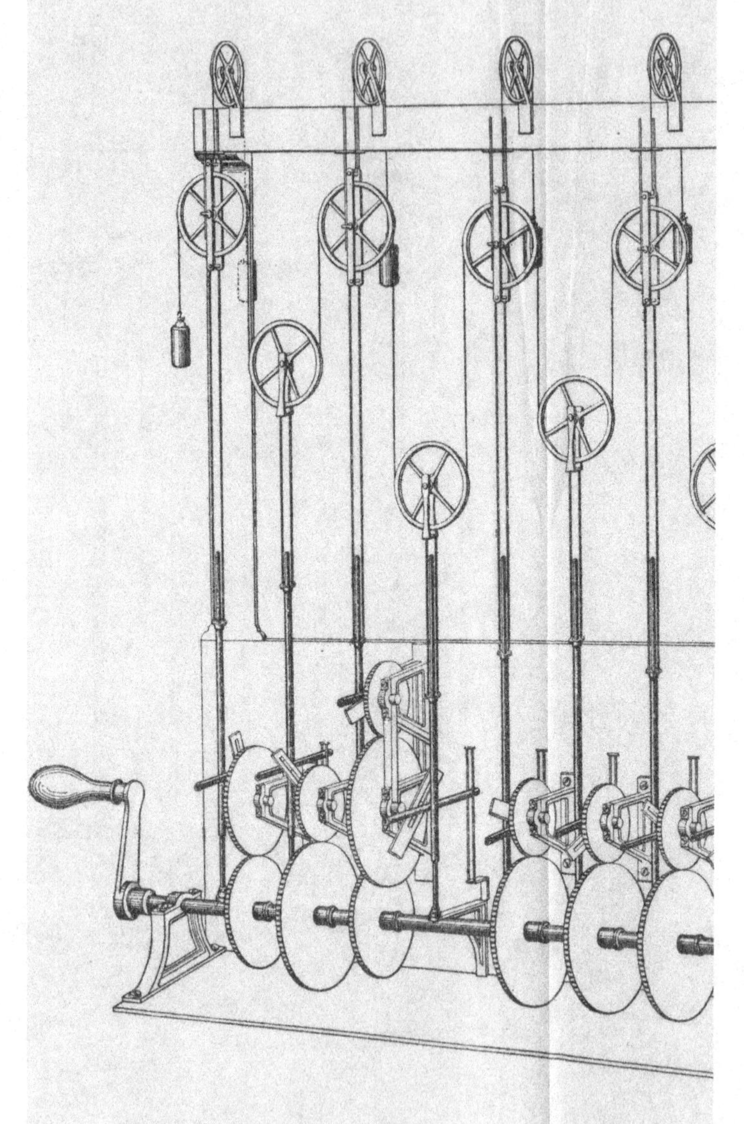

THE TIDE

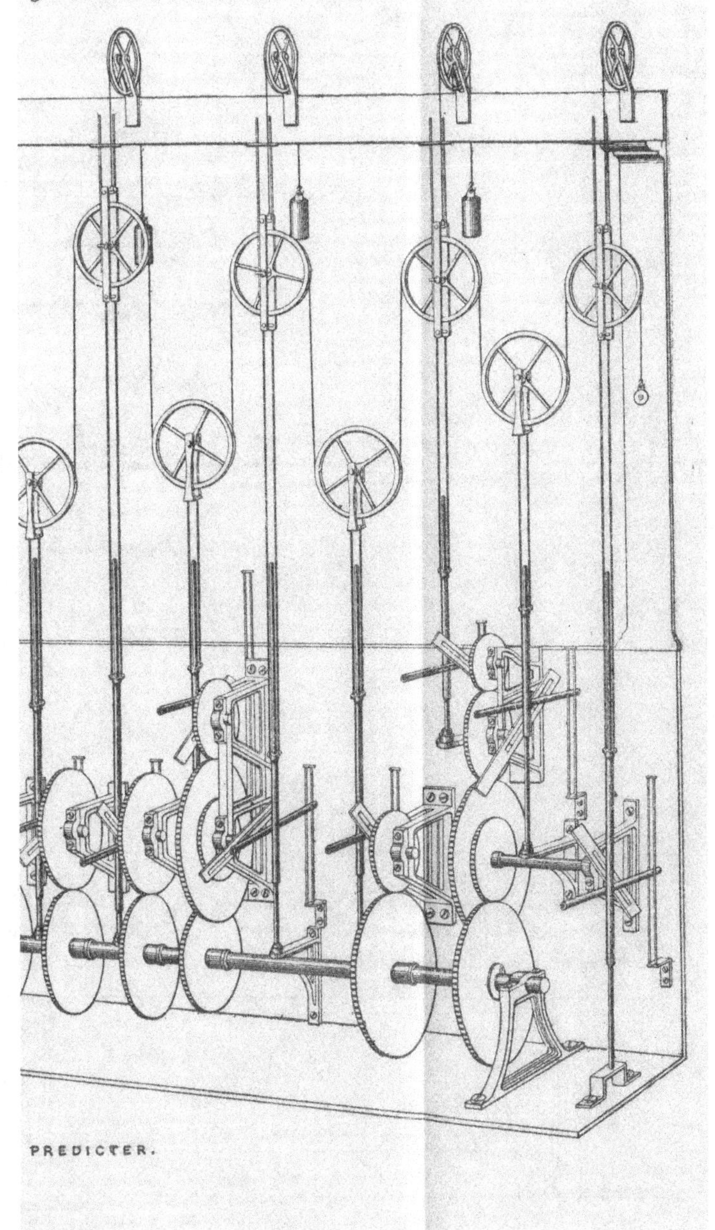
Fig: 10.
PREDICTER.

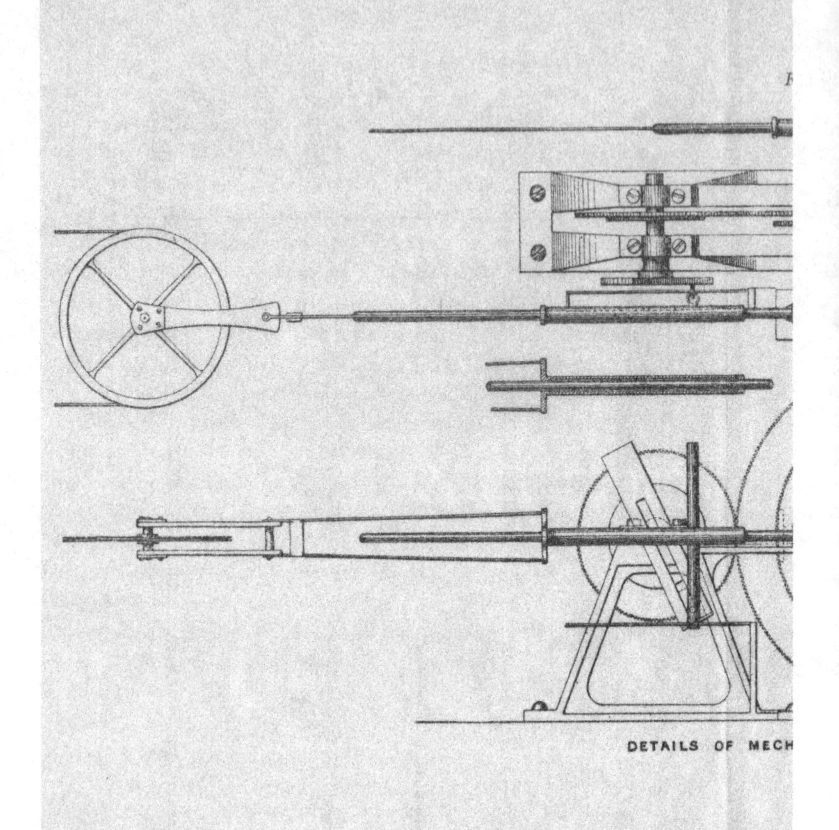

DETAILS OF MECH

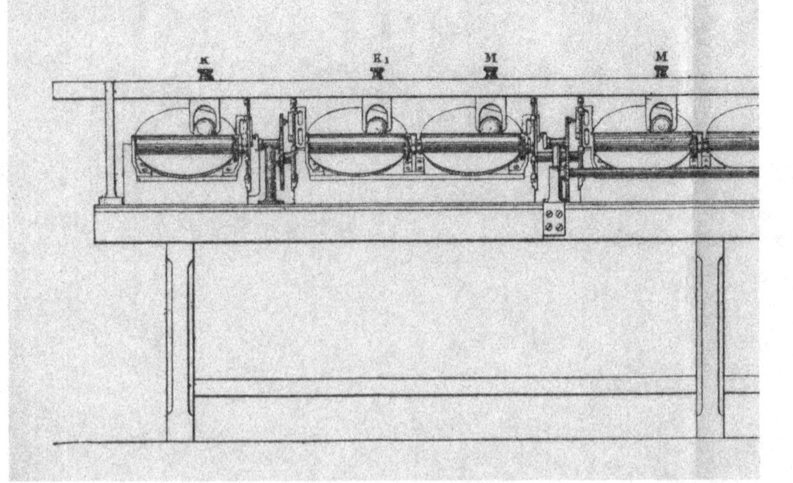

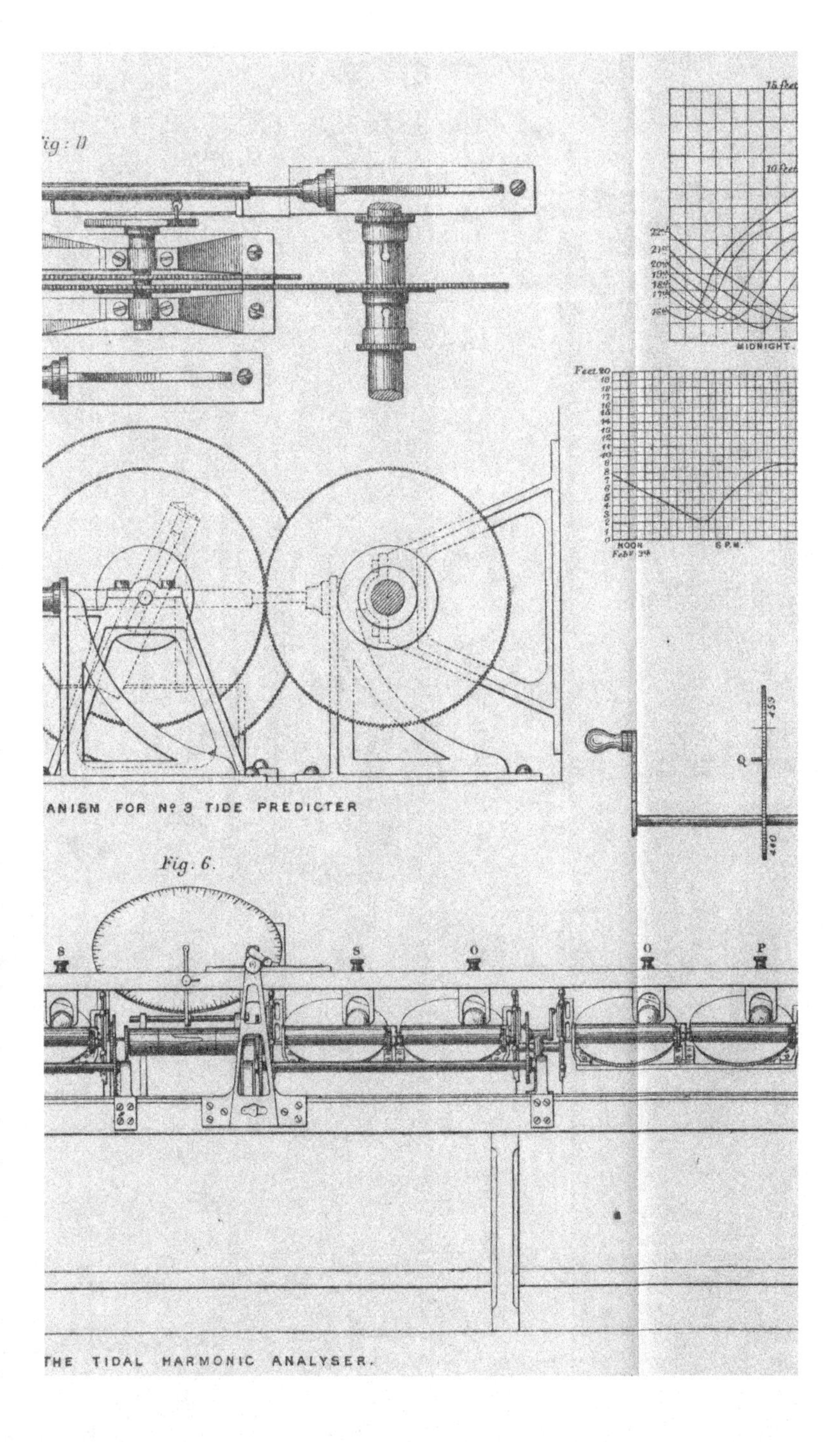
Fig: 11
ANISM FOR Nº 3 TIDE PREDICTER
Fig. 6.
S
S
O
O
P
Q
MIDNIGHT.
NOON
Feb. 3rd
6 P.M.
THE TIDAL HARMONIC ANALYSER.

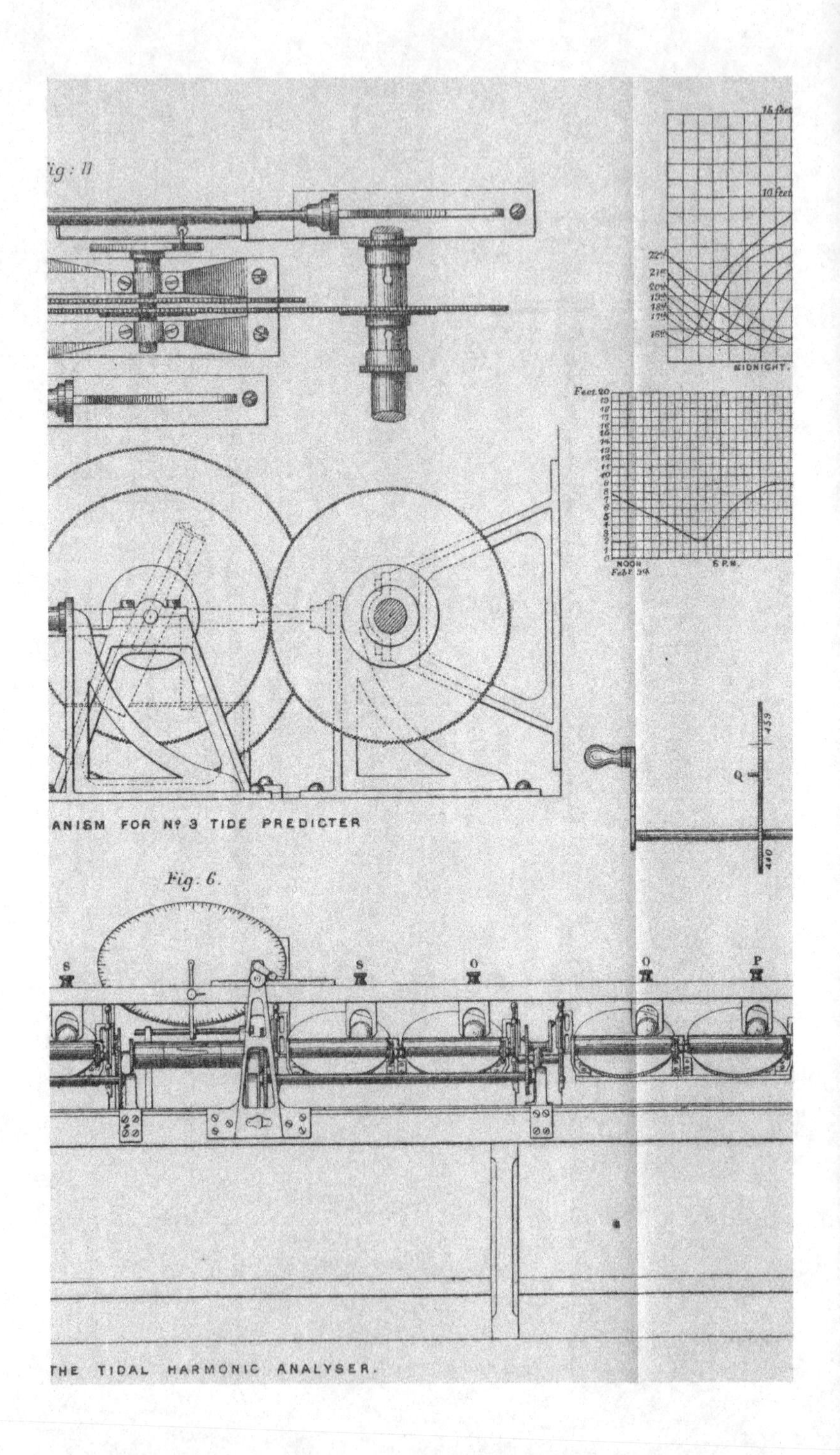
Fig: 11
ANISM FOR Nº 3 TIDE PREDICTER
Fig: 6.
S
S
O
O
P
Q
THE TIDAL HARMONIC ANALYSER.
NOON
6 P.M.
MIDNIGHT.

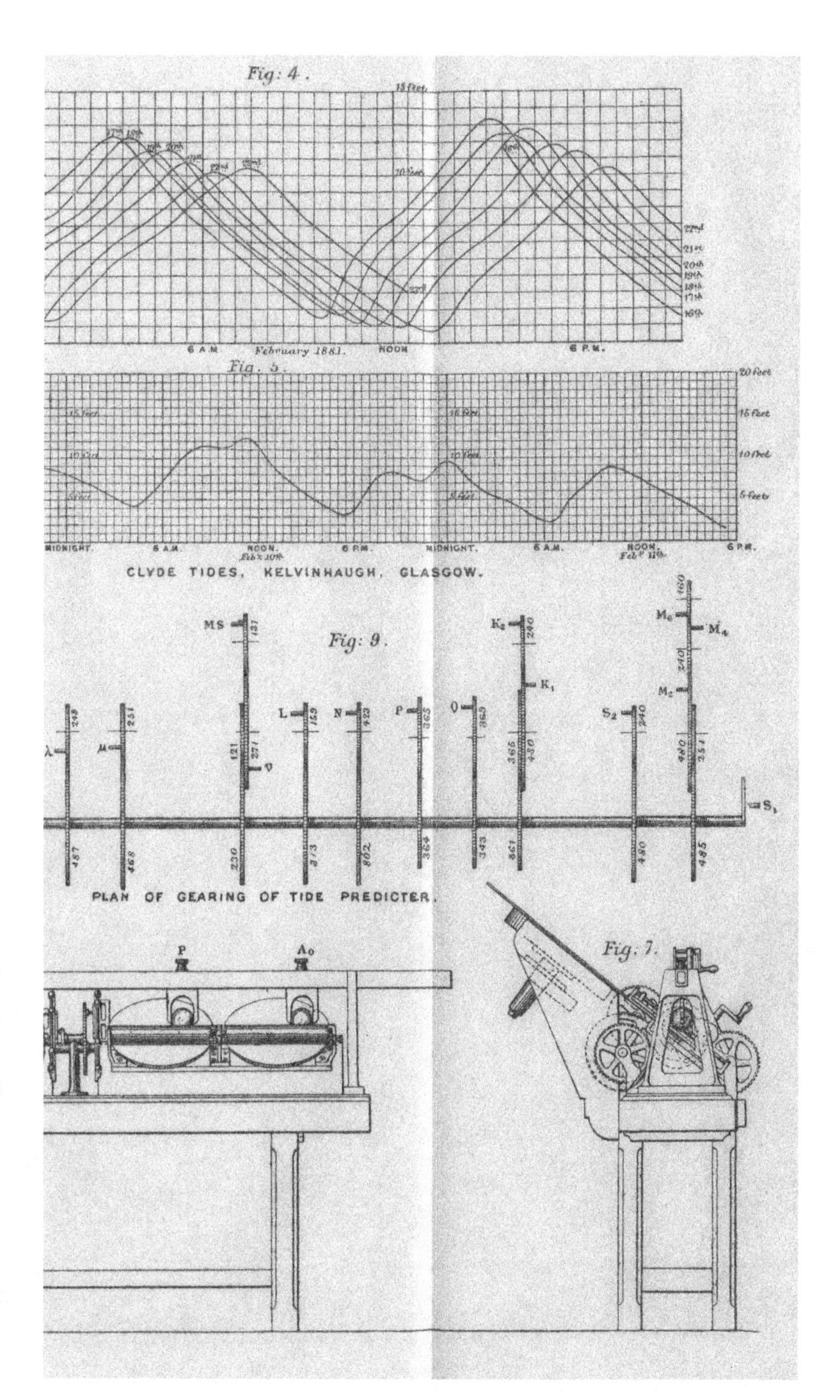
Fig: 4.
15 feet
10 feet
6 A.M.
February 1881.
NOON
6 P.M.
Fig. 5.
20 feet
15 feet
10 feet
5 feet
MIDNIGHT.
6 A.M.
NOON.
Feb 10th
6 P.M.
MIDNIGHT.
6 A.M.
NOON.
Feb 11th
6 P.M.
CLYDE TIDES, KELVINHAUGH, GLASGOW.
Fig: 9.
PLAN OF GEARING OF TIDE PREDICTER.
Fig: 7.

strain must not be too great or the ink-marker would have to be too heavily weighted. He desired to call attention to a model made since the last meeting, and placed before the present meeting, showing the epicyclic method of combination, which he had explained at the commencement of the discussion on the present Paper, and which he had first described more than nine years ago to the British Association at Brighton. It was the very simplest of all ways of combining two constituents. It was essentially limited to two, and for some time he contemplated introducing it for combining the two chief tidal constituents, the lunar semi-diurnal, and the solar semi-diurnal in the predicting machine. He had given M. Légé instructions for making the model only last Tuesday, and it was now exhibited to the members; having been made according to the drawing (Fig. 12), which they had seen at the last meeting. This plan had an interesting feature—it showed to the eye the priming and the lagging of the tides. Thus it showed that the times of tides were unaffected by the disturbing influence of the sun at spring and at neap tides, but that the amplitudes were affected. It showed exactly how at the quarters intermediate between neap and spring tides there was a lagging or a priming.

MISCELLANEOUS.

272. ARCHIBALD SMITH, AND THE MAGNETISM OF SHIPS*.

[From the 'Obituary Notices' in the *Proceedings of the Royal Society*, 1874.]

ARCHIBALD SMITH, only son of James Smith, of Jordanhill, Renfrewshire, was born on the 10th of August, 1813, at Greenhead, Glasgow, in the house where his mother's father lived. His father, who also was a Fellow of the Royal Society, had literary and scientific tastes with a strongly practical turn, fostered no doubt by his education in the University of Glasgow and his family connexion with some of the chief founders of the great commercial community which has grown up by its side. In published works on various subjects he left enduring monuments of a long life of actively employed leisure. His discovery of different species of Arctic shells, in the course of several years' dredging from his yacht, and his inference of a previously existing colder climate in the part of the world now occupied by the British Islands, constituted a remarkable and important advancement of Geological Science. In his *Voyage and Shipwreck of St Paul*, a masterly application of the principles of practical seamanship renders St Luke's narrative more thoroughly intelligible to us now than it can have been to contemporary readers not aided by nautical knowledge. Later he published a *Dissertation on the Origin and Connexion of the Gospels*; and he was engaged in the collection of further materials for the elucidation of the same subject up to the time of his death, at the age of eighty-five. Archibald Smith's mother was also of a family

* [It was the preparation of this obituary notice for the Royal Society which first directed Lord Kelvin's sustained attention to the problem of the compass in iron ships, in which he was to achieve so much.]

distinguished for intellectual activity. Her paternal grandfather was Dr Andrew Wilson, Professor of Astronomy in the University of Glasgow, whose speculations on the constitution of the sun are now generally accepted, especially since the discovery of spectrum-analysis and its application to solar physics. Her uncle, Dr Patrick Wilson, who succeeded to his father's Chair in the University, was author of papers in the *Philosophical Transactions* on Meteorology and on Aberration.

Archibald Smith's earliest years were chiefly passed in the old castle of Roseneath. In 1818 and 1819 he was taken by his father and mother to travel on the continent of Europe. Much of his early education was given him by his father, who read Virgil with him when he was about nine years old. He also had lessons from the Roseneath parish schoolmaster, Mr Dodds, who was very proud of his young pupil. In Edinburgh during the winters of 1820–22 he went to a day-school; and after that, living at home at Jordanhill, he attended the Grammar School of Glasgow for three years. As a boy he was extremely active, and fond of everything that demanded skill, strength, and daring. At Roseneath he was constantly in boats; and his favourite reading was anything about the sea, commencing no doubt with tales of adventurers and buccaneers, but going on to narratives of voyages of discovery, and to the best text-books of seamanship and navigation as he grew older. He had of course the ordinary ardent desire to become a sailor, incidental to boys of this island; but with him the passion remained through life, and largely influenced the scientific work by which he has conferred never-to-be-forgotten benefits on the marine service of the world, and made contributions to nautical science which have earned credit for England among maritime nations. He was early initiated into practical seamanship under his father's instruction, in yacht-sailing. He became an expert and bold pilot, exploring and marking passages and anchorages for himself among the intricate channels and rocks of the West Highlands, when charts did not supply the requisite information. His most loved recreation from the labours of Lincoln's Inn was always a cruise in the West Highlands. In the last summer of his life, after a naturally strong constitution had broken down under the stress of mathematical work on ships' magnetism by night, following days of hard work in his legal profession, he regained something of health and strength in sailing about with his boys in his yacht, between the beautiful coasts of the Frith of Clyde, but not enough. alas, to carry him through unfavourable influences in the winter that followed.

In 1826 he went to a school at Redland, near Bristol, for two years; and in 1828 he entered the University of Glasgow, where he not only began to show his remarkable capacity for mathematical science in the classes of Mathematics and Natural Philosophy, but also distinguished himself highly in Classics and Logic. Among his fellow students were Norman Macleod and Archibald Campbell Tait, with both of whom he retained a friendship throughout life. After completing his fourth session in Glasgow, he joined in the summer of 1832 a reading party, under Hopkins, at Barmouth in North Wales, and in the October following commenced residence in Trinity College, Cambridge.

While still an undergraduate he wrote and communicated to the Cambridge Philosophical Society a paper on Fresnel's wave-surface. The mathematical tact and power for which he afterwards became celebrated were shown to a remarkable degree in this his first published work. Fresnel, the discoverer of the theory, had determined analytically the principal sections of the wave-surface, and then guessed its algebraic equation. This he had verified, by calculating from it the perpendicular from the centre to the tangent plane; but the demonstration thus obtained was so long that he suppressed it in his published paper. Ampère by sheer labour had worked out a direct analytical demonstration, and published it in the *Annales de Chimie et de Physique**, where it occupies thirty-two pages, and presents so repulsive an aspect that few mathematicians would be pleased to face the task of going through it. With these antecedents, Archibald Smith's investigation, bringing out the desired result directly from Fresnel's postulates by a few short lines of beautifully symmetrical algebraic geometry, constitutes no small contribution to the elementary mathematics of the undulatory theory of light. It was one of the first applications in England, and it remains to this day a model example, of the symmetrical method of treating analytical geometry, which soon after (chiefly through the influence of the *Cambridge Mathematical Journal*) grew up in Cambridge,

* Volume for 1828.

and prevailed over the unsymmetrical and frequently cumbrous methods previously in use.

In 1836 he took his degree as Senior Wrangler and first Smith's Prizeman, and in the same year he was elected to a Fellowship in Trinity College.

Shortly after taking his degree, he proposed to his friend Duncan Farquharson Gregory, of the celebrated Edinburgh mathematical family, then an undergraduate of Trinity College, the establishment of an English periodical for the publication of short papers on mathematical subjects. Gregory answered in a letter of date December 4th, 1836, cordially entering into the scheme, and undertaking the office of editor. Being, however, on the eve of the Senate-House examination for his degree, he adds, "But all this must be done after the degree; for 'business before pleasure,' as Richard said when he went to kill the king before he murdered the babes." The result was, the *Cambridge Mathematical Journal*, of which the first number appeared in November 1837. It was carried on in numbers, appearing three times a year under the editorship of Gregory, until his death, and has been continued under various editors, and with several changes of name, till the present time, when it is represented by the *Quarterly Journal of Mathematics* and the *Messenger of Mathematics.* The original *Cambridge Mathematical Journal* of Smith and Gregory, containing as it did many admirable papers by Smith and Gregory themselves, and by other able contributors early attracted to it, among whom were Greatheed, Donkin, Walton, Sylvester, Ellis, Cayley, Boole, inaugurated a most fruitful revival of mathematics in England, of which Herschel, Peacock, Babbage, and Green had been the prophets and precursors.

It is much to be regretted that neither Cambridge, nor the university of his native city, could offer a position to Smith, enabling him to make the mathematical and physical science, for which he felt so strong an inclination, and for which he had so great capacity, the professional work of his life. Two years after taking his degree he commenced reading law in London; but his inclination was still for science. Relinquishing reluctantly a Trinity Lectureship offered to him by Whewell in 1838, and offered again and almost accepted in 1840, resisting a strong temptation to accompany Sir James Ross to the Antarctic regions on the scientific explorating expedition of the *Erebus* and *Terror* in 1840–41 and regretfully giving up the idea of a Scottish professorship, which, during his early years of residence in Lincoln's Inn, had many attractions for him, he finally made the bar his profession. But during all the long years of hard work through which he gradually attained to an important and extensive practice, and to a high reputation as a Chancery barrister, he never lost his interest in science, nor ceased to be actively engaged in scientific pursuits; and he always showed a lively and generous sympathy with others, to whom circumstances (considered in this respect enviable by him) had allotted a scientific profession.

About the year 1841 his attention was drawn to the problem of ships' magnetism by his friend Major Sabine, who was at that time occupied with the reduction of his own early magnetic observations made at sea on board the ships *Isabella* and *Alexander* on the Arctic Expedition of 1818, and of corresponding magnetic observations which had been then recently made on board the *Erebus* and *Terror* in Capt. Ross's Antarctic Expedition of 1840–41. The systematic character of the deviations, unprecedented in amount, experienced by the *Isabella* and *Alexander* in the course of their Arctic voyage, had attracted the attention of Poisson, who published in 1824, in the *Memoirs of the French Institute*, three papers containing a mathematical theory of magnetic induction, with application to ships' magnetism. The subsequent magnetic survey of the Antarctic regions, of which by far the greater part had to be executed by daily observations of terrestrial magnetism on ship-board, brought into permanent view the importance of Poisson's general theory; but at the same time demonstrated the necessity for replacing his practical formulæ by others, not limited by certain restrictions as to symmetry of the ship, which he had assumed for the sake of simplicity. This was the chief problem first put before Smith by Sabine; and his solution of it was the first great service which he rendered to the practical correction of the disturbance of the compass caused by the magnetism of ships. Twenty years later the work thus commenced was referred to in the following terms by Sir Edward Sabine*, in presenting, as President of the Royal Society,

* *Proceedings of the Royal Society*, Nov. 30, 1865, Vol. XIV. p. 499.

the Royal Medal which had been awarded to Archibald Smith for his investigations and discoveries in ships' magnetism:—"......Himself a mathematician of the first order, and possessing a remarkable facility (which is far from common) of so adapting truths of an abstract character as to render them available to less highly trained intellects, he derived at my request, from Poisson's fundamental equations, simple and practical formulæ, including the effects both of induced magnetism and of the more persistent magnetism produced in iron which has been hardened in any of the processes through which it has passed. The formulæ supplied the means of a sufficiently exact calculation when the results were finally brought together and coordinated. They were subsequently printed in the form of memoranda in the account of the survey in the *Philosophical Transactions* for 1843, 1844, and 1846.

"The assistance which, from motives of private friendship and scientific interest, Mr Smith had rendered to myself, was from like motives continued to the two able officers who had successively occupied the post of Superintendent of the Compass Department of the Navy; and the formulæ for correcting the deviation, which he had furnished to me, reduced to simple tabular forms, were published by the Admiralty in successive editions for the use of the Royal Navy.

"As, in the course of time, the use of steam machinery, the weight of the armament of ships of war, and generally the use of iron in vessels, increased more and more, the great and increasing inconveniences arising from compass irregularities were more and more strongly felt, and pressed themselves on the attention of the Admiralty and of naval officers.

"An entire revision of the Admiralty instructions became necessary; Mr Smith's assistance was again freely given; and the result was the publication of the *Admiralty Manual* for ascertaining and applying the deviations of the compass caused by the iron in a ship.

"The mathematical part of this work, which is due to Mr Smith, seems to exhaust the subject, and to reduce the processes by simple formulæ and tabular and graphic methods, to the greatest simplicity of which they are susceptible. Mr Smith also joined with his fellow-labourer, Captain Evans, F.R.S., the present Superintendent of the Compass Department of the Navy, in laying before the Society several valuable papers containing the results of the mathematical theory applied to observations made on board the iron-built and iron-plated ships of the Royal Navy."

This is not an occasion for explaining in detail the elaborate investigations sketched in the preceding statement by Sir Edward Sabine; but the writer of the present notice, having enjoyed the friendship of Archibald Smith since the year 1841, and having had many opportunities, both in personal intercourse and by letters, of following the progress through thirty years of his work on ships' magnetism, may be permitted a brief reference to some of the points which have struck him as most remarkable:—

1. Harmonic reduction of observations.

2. Practical expression of the full mathematical theory.

3. Heeling error.

4. Dygograms.

5. Rules for positions of needles on compass card, with dynamical and magnetic reasons.

1. *Harmonic reduction of observations.*—The disturbance of the compass produced by the magnetism of a ship is found by observation to be the same, to a very close degree of approximation, when the ship's head is again and again brought to the same bearing, no great interval of time having intervened, and no extraordinary disturbance by heavy sea or otherwise having been experienced in the interval. Overlooking these restrictions for the present, we may therefore say, in Fourier's language, that the disturbance of the compass is a periodic function of the angle between the vertical plane of any line fixed relatively to the ship, and any fixed vertical plane, when the ship, on "even keel" or with any constant inclination, is turned into different azimuths—the period of this function being four right angles. Hence also the disturbance of the compass is a periodic function of the angle between the vertical plane of the chosen line moving with the ship, and the vertical plane through the magnetic axis of the compass. The line moving with the ship being taken as a longitudinal line drawn horizontally from the stern towards the bow, and the fixed vertical plane being taken as the magnetic meridian, the angle first mentioned is called for brevity "the ship's *magnetic course*," and the other "the ship's *compass course*."

One of Smith's earliest contributions to the compass problem was the application of Fourier's grand and fertile theory of the expansion of a periodic function in series of sines and cosines of the argument and its multiples, now commonly called the harmonic analysis of a periodic function. To facilitate the practical working out of this analysis, he gave tables of the products of the multiplication of the sines of the "rhumbs" by numbers, and by arcs in degrees and minutes; also tabular forms and simple practical rules for performing the requisite arithmetical operations. These tables, tabular forms, and rules, just as Smith gave them about thirty years ago, are in use in the Compass Department of the Admiralty up to the present time. From every ship in Her Majesty's Navy, in whatever part of the world, a table of observed deviations of the compass, at least once a year is sent to the Admiralty, and is there subjected to the harmonic analysis. The observations having been accurately and faithfully made, a full history of the magnetic condition of the ship is thus obtained, and want of accuracy, or want of faithfulness, if there has been any, is surely detected. The rigorous carrying out of this system, with all the method and business-like regularity characteristic of the scientific departments of our Admiralty, has undoubtedly done more than anything else to promote the usefulness of the compass, and to render its use safe throughout the British Navy. Smith's tables and forms for harmonic analysis have proved exceedingly valuable in many other departments of practical physics besides ships' magnetism. The writer of this article found them most useful fifteen years ago in reducing for the Royal Society of Edinburgh Forbes's observations of the underground temperature of Calton Hill, the Experimental Gardens, and Craigleith Quarry, in the neighbourhood of Edinburgh; and the forms, with a suitable modification of the tables, have proved equally useful in the harmonic analysis of tidal observations for various parts of the world, carried out by the Tidal Committee of the British Association, with the assistance of sums of money granted in successive years from 1868 to 1872.

2. *Practical expression of the full mathematical theory.*— Poisson himself, in making practical application of his theory, had simplified it by assuming particular conditions as to symmetry of the iron in the ship, and even with these restrictions had left it in a form which seemed to require further simplification before

it could be rendered available for general use. Airy, in taking up the problem with this object, at the request of the Admiralty in the year 1839, founded his calculations on a supposition that, "by the action of terrestrial magnetism every particle of iron is converted into a magnet whose direction is parallel to that of the dipping needle, and whose intensity is proportional to the intensity of terrestrial magnetism." This supposition, which is approximately true only for the ideal case of the iron of the ship being all in the shape of globes placed at such considerable distances from one another as not to exercise mutual influence to any sensible degree, leads to a law of dependence between the ship's force on the compass needle, and the angular coordinates of the ship, which differs from that of the complete theory, as shown afterwards by Smith, only in the want of his constant term A of the harmonic development,—a difference which, in ordinary cases, does not vitiate sensibly the practical application. In introducing the supposition, Airy correctly anticipated that it would in general lead to results sufficiently accurate and complete for practical purposes. But he said "it would have been desirable to make the calculations on Poisson's theory, which undoubtedly possesses greater claims on our attention (as a theory representing accurately the facts of some very peculiar cases) than any other. The difficulties, however, in the application of this theory to complicated cases are great, perhaps insuperable." These difficulties were wholly overcome by the happy mathematical tact of Archibald Smith, who reduced the full expression of Poisson's theory, including the effect of permanent magnetism, the great practical importance of which had been discovered by Airy, to a few simple and easily applied formulæ. These formulæ are now in regular use in the Compass Department of the Admiralty, for the practical deduction of rigorous results from the harmonic analysis already referred to. In fact the full expression of the unrestricted theory, as given by Archibald Smith in Part III. of the *Admiralty Manual*, is even simpler and more ready for ordinary use than the partial and restricted expressions which Poisson and Airy had given for practical application of the theory.

3. *Heeling error.*—Poisson's general formulæ express three rectangular components of the resultant force at the point where the compass is placed, due to the magnetism induced in the ship by the terrestrial magnetic force. To these Airy added the

components of force due to permanent magnetism of the ship's iron, which, though not ignored by Poisson, had been omitted by him, because, considering the probability of scattered directions of the magnetic axes of permanent magnetism in the isolated masses of iron existing in wooden ships and their armaments, he justly judged that permanent magnetism could not seriously disturb a properly placed compass in a wooden ship; and iron ships were scarcely contemplated in those days. This general theory of Poisson and Airy expresses the resultant force in terms of three angular coordinates, specifying the position of the ship. In the practical application these coordinates are most conveniently taken as:—(1) the ship's "magnetic course," defined above; (2) the inclination of the longitudinal axis of the ship to the horizon; (3) the inclination to the horizon of a plane drawn through this line perpendicular to the deck. The second coordinate has no name and is of no importance in the compass problem; for under steam, or even under sail, the average inclination of the longitudinal axis (chosen as horizontal for the ship in still water) is never so great as to produce any sensible effect on the compass disturbance, and the magnetic effects of pitching in the heaviest sea are not probably ever so great as to produce any seriously inconvenient degrees of oscillation in the compass card. The third coordinate is called the "heel"; and its magnetic effect on the compass is called "the heeling error." The heeling error was investigated by Airy in his earliest work on the compass disturbance; but at that time, when iron sailing ships were comparatively rare, he confined his ordinary practical correction of compass error to the case of a ship in different azimuths on even keel. Since that time the heeling error has come to be of very serious practical importance, on account of the great number of iron sailing ships, and of screw steamers admitting of being pressed by sail to very considerable degrees of "heel." Archibald Smith took up the question with characteristic mathematical tact and practical ability, and gave the method for correcting the heeling error—which is now, I believe, universally adopted in the Navy, and too frequently omitted (without the substitution of any other method) in the mercantile marine.

4. *Dygograms.*—This is the name given by Smith to diagrams exhibiting the magnitude and direction of the resultant of the

terrestrial magnetic force and the force of the ship's magnetism at the point occupied by the compass. The solution of the problem of finding for a ship in all azimuths on even keel the dygogram of the whole resultant force is given by him in the chapter headed "Ellipse and Circle," of the *Admiralty Manual*, Appendix 2 (3rd edition, 1869, page 169—171). But it is only for horizontal components of force that he has put dygograms into a practical form; and for this case, which includes the whole compass problem of ordinary navigation, his dygograms are admirable both for their beauty and for their utility. "Dygogram Number I" is the curve locus of the extremity of a line drawn from a fixed point O, in the direction, and to a length numerically equal to the magnitude, of the horizontal component of the resultant force experienced by the needle when the ship is turned through all azimuths. This curve (however great the deviations of the compass) he proves to be the Limaçon of Pascal—that is to say, the curve (belonging to the family of epitrochoids) described by the end of an arm rotating in a plane round a point, which itself is carried with half angular velocity round the circumference of a fixed circle in the same plane. The length of the first-mentioned arm is equal to the maximum amount of what is called (after Airy) the quadrantal deviation; the radius of the circle last mentioned is the maximum amount of what Airy called the polar magnet deviation, and Smith the semicircular deviation. (When, as the writer of this article trusts before long will be universally the case*, the quadrantal deviation is perfectly corrected by Airy's method of soft iron correctors, the dygogram Number I will be reduced to a circle.) Besides the form of the curve in any particular case, which depends on the ratio of the first-mentioned radius to the second, to complete the diagram and use it we must know the position of the fixed point through which the resultant radius-vector is to be drawn, and must show in the diagram the magnetic bearing of the ship's head, for which any particular point of the curve gives the resultant force. Smith gave all these elements by simple and easily executed constructions, in the first and second editions of the *Admiralty Manual*. In the third edition he substituted, for his first method of construction of the dygogram curve, a modification of it due

* The barrier against this being done hitherto has been the perniciously great length of the compass needles used at sea, the shortest being about six inches.

to Lieut. Colongue of the Russian Imperial Navy and of the Imperial Compass Observatory, Cronstadt, and added several elegant constructions, also due to Lieut. Colongue, for the geometrical solution of various compass problems, by aid of the dygogram Number I.

The annexed diagram is the dygogram Number I for the *Warrior*, drawn accurately (by aid of a circular board rolling

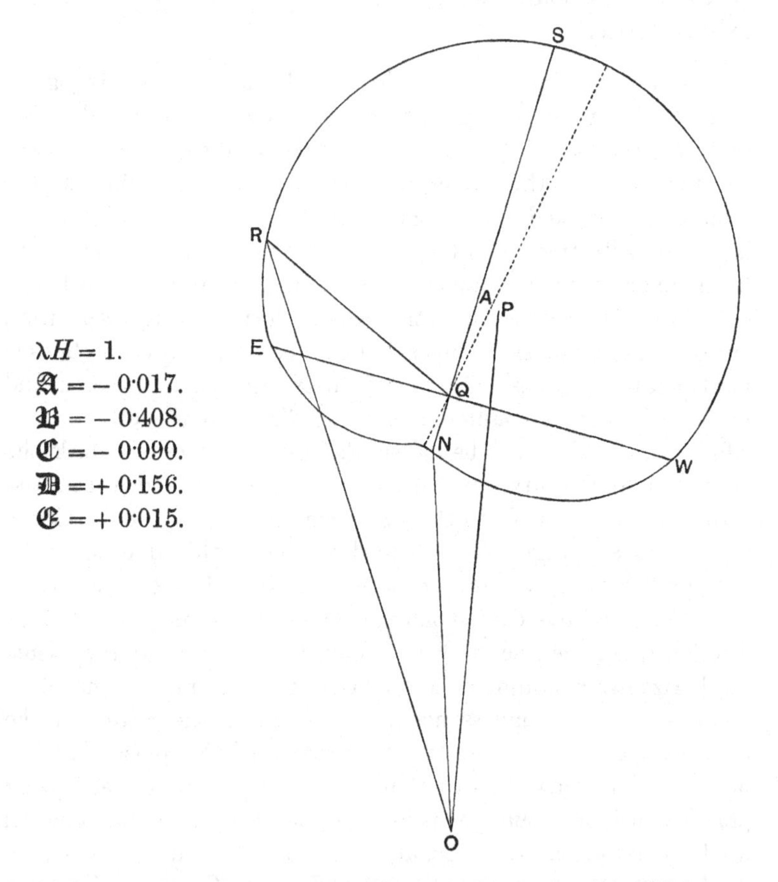

***Rule for using Dygogram No.* I.—In the diagram Q is a fixed point of the "Limaçon," called "the pole of the dygogram." It lies in the axis of symmetry, which is indicated by a dotted line. NQS, EQW are two lines through Q at right angles to one another; and O, P are two points, in positions fixed by the ship's magnetic elements. The length OP represents "mean force on compass to north" (λH). Take any point R, on the curve, such that NQR is equal to the ship's "magnetic course," then is ROP the "deviation" of the compass, and OR represents the horizontal component of the force on it.**

upon a fixed circular board of equal diameter, in the manner described by Smith in the *Admiralty Manual*, Appendix II., under the heading "Mechanical Construction of Dygogram, No. I"), according to the following data deduced from observations made at Spithead in October 1861. The notation $\mathfrak{A}, \mathfrak{B}, \mathfrak{C}, \mathfrak{D}, \mathfrak{E}$, is that which was introduced by Smith when he first substituted the rigorous formulæ for the approximate harmonic formulæ which had previously sufficed: it is explained in the Appendix to this notice.

Dygogram Number II may be deduced from dygogram Number I by attaching a piece of paper to the half-speed revolving arm, and letting the tracing-point of the limaçon leave its trace also on this paper, which will be a circle, while at the same time the fixed point from which the resultant radius-vector is drawn will trace another circle on the moving paper. The fresh diagram thus obtained consists of two circles. Mark one of these circles (the second in the order of the preceding description) with the points of the compass*, like a compass card; or (better) mark simply degrees all round from North taken as zero; and mark with degrees counted in reverse direction the other circle, which, for brevity, will be called the auxiliary circle. Mark the ship's magnetic course on the circumference of the ideal compass card. From this point to the corresponding point on the auxiliary circle draw a straight line. The direction of this line shows by the parallel to it, through the centre of the ideal compass card, the compass course corresponding to any chosen magnetic course. The length of the line, drawn in the manner described, represents the horizontal resultant force of the earth and ship, at the point occupied by the compass needle, in terms of the radius of the ideal compass card, as unity. The writer of the present article believes that this construction will yet prove of very great practical utility, although hitherto it has not come into general use†. Its geometrical beauty attracted the notice even of Cayley, who has contributed to the Admiralty Compass Manual a

* The ancient system of marking 32 points on the compass card, and specifying courses in terms of them, has always been very inconvenient, and is now beginning to be generally perceived to be so.

† A short demonstration of it, deduced directly from Smith's fundamental formulæ, is appended to the present article for the sake of mathematical readers who may not have the Admiralty Compass Manual at hand.

second method of solving, by means of it, one of Smith's compass problems.

Construction from ship's and earth's magnetic elements. With O as centre and OH equal to "mean force on compass to north" (λH) describe a circle. Make

$$NL = \mathfrak{A},\ OB = -\mathfrak{B};\quad BC = -\mathfrak{C};\quad CD = -\mathfrak{D};\quad Dh = -\mathfrak{E};$$

with C as centre describe a circle through h.

The following diagram shows (for an ideal case, as possibly a

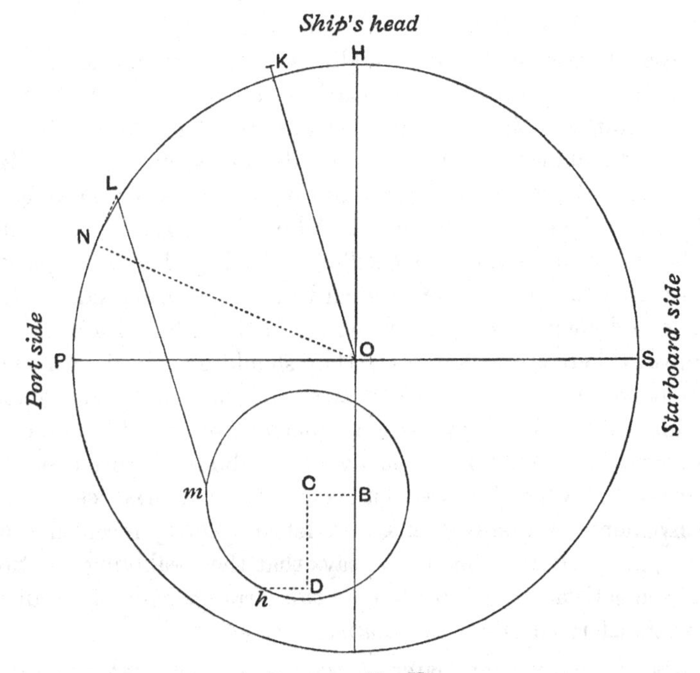

Dygogram No. II.

Use.—Make hm equal in angular value to $\widehat{NH}$; then OK, parallel and equal to mL, shows direction of needle and magnitude of horizontal component force on it when correct magnetic north is in direction ON, and ship's head OH : NOH being ship's "magnetic course," KOH is the corresponding "compass course."

turret ship of the future, with very large values of the usually small magnetic elements $\mathfrak{A}$ and $\mathfrak{E}$) the Dygogram of two circles, modified to suit the Chinese compass (or needle unloaded with compass card). This modification is convenient for the theoretical explanation and proof appended to the present notice.

5. *Rule for positions of needles on compass card, with dynamical and magnetic reasons.*—In 1837 a Committee, consisting of Captain Beaufort, Hydrographer to the Admiralty, Captain Sir J. C. Ross, R.N., Captain Johnson, R.N., Major Sabine, R.A., and Mr S. H. Christie, was appointed to remedy defects of the compasses at that time in use in Her Majesty's fleet, and to organize a system of compass management ashore and afloat. The labours of that Committee have conferred signal benefit, not only on the British Navy, but on the navies and mercantile marine services of all nations—in the 'Admiralty Standard Compass,' and in the establishment in 1843 of the British Admiralty Compass Department. The qualities of the magnetic needles and their arrangement on the card occupied much attention of the Committee. Smith's attention was called to the subject by his friend Sabine; and he gave a rule for placing the needles, which was adopted by the Committee, and has ever since been followed in the construction of the Admiralty compass. The rule is, that when there are two needles used they should be placed with their ends on the compass card at 30° on each side of the ends of a diameter; and that when (as in the Admiralty Standard Compass) there are four needles, they should be placed with their ends at 15° and 45° from the ends of the diameter. The object of this rule was to give equal moments of inertia round all horizontal axes, and so to remedy the "wabbling" motion of the compass card when balanced on its pivot, which has been found inconvenient. Captain Evans, in a letter recently received from him by the writer of this notice, says that the "wabbling" motion has been satisfactorily corrected by this arrangement of needles; "it is transformed into a '*swimming*' motion."

About twenty years later it was discovered that the same arrangement gives, by a happy coincidence, a very important magnetic merit to the Admiralty compass, which had not been contemplated by Smith when he first gave his rule. To explain this, it must be premised that practical compass-adjusters had experienced difficulties in correcting the compass deviation of certain ships by Airy's method (which consists in using soft iron to correct the quadrantal deviation, and permanent magnets to correct the semicircular), and had reported that in such cases they had found it advantageous to substitute compasses with two needles for a single-needle compass. The attention of Captain

Evans was drawn to this subject by the observations made in the *Great Eastern* on her experimental voyage from the Thames to Portland, and afterwards when she was lying at Holyhead and Southampton, from which he found that although the deviations had been carefully corrected by Mr Gray, of Liverpool, with magnets and soft iron, and were in fact nearly correct on the cardinal and quadrantal points, there were errors of between 5° and 6° on some of the intermediate points. These observations indicated the existence of a considerable error, which was neither semicircular nor quadrantal, and thus apparently some source of error which had not been taken into account by Airy in his plan for correction. To explain the cause of these and similar results in other ships, previously considered to be anomalous, Captain Evans instituted a series of experiments with compasses, and magnets and soft iron placed in different positions with respect to them. He soon found that the greatness of the supposed anomaly in the *Great Eastern* depended on the unusually great length of the needles of her standard compass (two needles* of $11\frac{1}{2}$ inches in length, placed near each other on the card). The results of the observations and experiments, reduced by aid of Napier's graphic method, and subjected to a thorough harmonic analysis, are described in a joint paper by Smith and Evans, published in the Transactions of the Royal Society for 1861. They show, in the expression for the deviation, sextantal and octantal terms† very large in the case of the *Great Eastern*, and comparatively small when the Admiralty standard compass was tested in circumstances otherwise similar. Whether single needles or double needles were used, it was found that the smaller the needle the smaller were the sextantal and octantal terms. Single needles gave greater terms of this class than double needles of the same magnitude, arranged as in the Admiralty compass.

The merit of giving almost evanescent sextantal and octantal terms, discovered in the Admiralty standard compass, "suggested

* Compass needles becoming larger with the ships, by a process of "Artificial Selection" unguided by intelligence, have sometimes attained to the monstrous length of 15 inches, or even more, in some of the great modern passenger-steamers fitted out by the owners regardless of expense, and only desiring efficiency, trusting to instrument-makers of the highest name.

† That is to say, terms consisting of coefficients multiplying the sines or cosines of six times and of eight times the ship's magnetic azimuth.

the idea, that the arrangement of the needles in that compass might produce, in the case of deviations caused by a magnet or mass of soft iron in close proximity to it, a compensation of the sextantal and octantal deviations; and this, on the subject being mathematically investigated [on the approximate hypothesis that the intensity of magnetization is uniform through the length of each needle, and equal in the different needles], proved to be the case, this particular arrangement of needles reducing to zero the coefficients of the terms involving the square of the ratio of the length of the needle to the distance of the disturbing iron; so that this remarkable result was obtained that the arrangement of needles which produces the equality in the moments of inertia is, by a happy coincidence, the same as that which prevents the sextantal deviation in the case of correcting magnets, and the octantal deviation in the case of soft iron correctors. The consequence is, that with the Admiralty compass cards, or with cards with two needles each 30° from the central line, correcting magnets and soft iron correctors may be placed much nearer the compass than can safely be done with a single-needle compass card, and that the large deviations found in iron ships may be thus far more accurately corrected."

It will be understood that the preceding statement, even as an index of subjects, gives but a very incomplete idea of Smith's thirty years' work on magnetism. Further information is to be found in his papers in the Transactions and Proceedings of the Royal Society, some of them contributed in conjunction with Sabine or with Evans, others in his own name alone. In 1850 he published separately * an account of his theoretical and practical investigations on the correction of the deviations of a ship's compass, which was afterwards given as a supplement to the Admiralty "Practical Rules" in 1855. The large deviations found in iron-plated ships of war having rendered necessary the use of the exact instead of "the approximate formulæ," this article was rewritten by Smith for the Compass Department of the Admiralty. It now forms Part III of the 'Admiralty Manual for the deviations of the Compass,' edited by Evans and Smith, to which are added appendices containing a complete mathematical statement of the general theory, proofs of the practical formulæ, and constructions

* *Instructions for Computation of Tables of Deviations*, by Archibald Smith. Published for the Hydrographic Office of the Admiralty.

and practical methods of a more mathematical character than those given in the body of the work for ordinary use. A separate publication, of 'Instructions for correcting the Deviation of the Compass,' by Smith, was made by the Board of Trade in 1857.

It is satisfactory to find that the British Admiralty *Compass Manual*, embodying as it does the result of so vast an amount of labour, guided by the highest mathematical ability and the most consummate practical skill, has been appreciated as a gift to the commonwealth of nations by other countries than our own. It is adopted by the United States Navy Department, and it has been translated into Russian, German, Portuguese, and French. Smith's mathematical work, and particularly his beautiful and ingenious geometrical constructions, have attracted great interest, and have called forth fresh investigations in the same direction, among the well-instructed and able mathematicians of the American, Russian, French, and German Navy Departments.

The laborious and persevering devotion to the compass problem, which has been shown by British mathematicians and practical men, by Sabine, Scoresby, Airy, Archibald Smith, by Captains Johnson and Evans of the Compass Department of the Admiralty, and by Towson and Rundell, who acted as secretaries to the Liverpool Compass Committee, has been an honour to the British nation in the eyes of the world. Referring particularly to the Liverpool Compass Committee, Lieut. Collet, of the French Navy, the French translator of the *Admiralty Manual*, in a history of the subject which he prefixes to his translation, says:—"Aidé par des libéralités particulières, soutenu surtout par cette sorte de ténacité passionnée, tout particulière à la nation anglaise, qui, en inspirant les longues et patientes recherches conduit sûrement au succès et sans laquelle tous les moyens d'action sont impuissants à surmonter les obstacles, ce Comité fit paraître successivement trois rapports qui fixèrent d'une manière définitive la plupart des questions controversées, et qui indiquèrent nettement, pour celles qui restaient à résoudre, la marche qu'il fallait suivre et les véritables inconnues du problème." And in an official publication by the American Navy Department, containing an English translation of Poisson's memoir, followed by the whole series of papers, theoretical and practical, on ships' magnetism, which have appeared in this country, we find the following passage, which must be gratifying to all who feel British scientific work and appreciation of it by other nations, to be a proper subject for national pride:—"...With the complex conditions thus introduced, and the more exacting requirements of experience in their practical treatment, came the necessity for constantly aiming at *that complete analysis of the magnetic phenomena of the ship* which has been so prominent and characteristic a feature of the English researches."

The constancy to the compass problem in which Smith persevered with a rare extreme of disinterestedness, from the time when Sabine first asked him to work out practical methods from Poisson's mathematical theory, until his health broke down two years before his death, was characteristic of the man. It was pervaded by that "ténacité passionnée" which the generous French appreciation, quoted above, describes as a peculiarity of the English nation; but there was in it also a noble single-mindedness and a purity of unselfishness to be found in few men of any nation, but simply natural in Archibald Smith.

Honourable marks of appreciation reached him from various quarters, and gave him the more pleasure from being altogether unsought and unexpected. The Admiralty, in 1862, gave him a watch. In 1864 he received the honorary degree of LL.D. from the University of Glasgow. The Royal Society awarded to him the Royal Medal in the year 1865. The Emperor of Russia gave him, in 1866, a gold Compass emblazoned with the Imperial Arms and set with thirty-two diamonds, marking the thirty-two points. Six months before his death Her Majesty's Government requested his acceptance of a gift of £2000, as a mark of their appreciation of "the long and valuable services which he had gratuitously rendered to the naval Service in connexion with the magnetism of iron ships, and the deviations of their Compasses." The official letter intimating this, dated Admiralty, July 1st, 1872, contains the following statement, communicated to Smith by command of the Lords of the Admiralty:—"To the zeal and ability with which for many years you have applied yourself to this difficult and most important subject, My Lords attribute in a great degree the accurate information they possess in regard to the influence of magnetism, which has so far conduced to the safe navigation of iron ships, not only of the Royal and Mercantile Navies of this country, but of all nations."

The writer of this notice has obtained leave to quote also the following from a letter from the First Lord of the Admiralty, Mr Goschen, of date February 23rd, 1872, announcing to Mr Smith that the Government had determined to propose to Parliament that the sum of £2000 should be awarded to him "as a mark of recognition of the great and successful labours" which he had "bestowed on several branches of scientific enquiry of deep importance to Her Majesty's Navy."

"I am aware that you have treated your arduous work in this direction as a labour of love; and therefore I do not consider that the grant which Parliament will be requested to sanction is in any way to be looked upon as a remuneration of your services.... I trust you will regard it as a mark of recognition on the part of the country, of your great devotion to enquiries of eminent utility to the public, conducted in the leisure hours which remained to you in a laborious profession."

The following letter, which was addressed to the Editor of the *Glasgow Herald*, and published in that paper last January, will be read with interest by others as well as those for whom it was originally written:—"As an intimate friend of the late Archibald Smith of Jordanhill, I desire to call your attention to a passage in your article of the 30th December upon him, which might perhaps convey a wrong impression to the minds of your readers.

"You say that 'mathematics...in its application to practical navigation was the amusement of his lighter hours.' The truth is, that the profession of a Chancery barrister, which the claims of a large family forbade him to abandon, occupied his best energies from early morning till late in the evening—in other words, what would in the case of most people, be called 'his whole time'; and compass investigation of the most minute and severe nature, undertaken after midnight, and carried on far into the morning hours by a man whose brain had been working all day, and must work again the next day, can hardly be called 'the amusement of lighter hours.' The same remark applies to vacations, during which his magnetic papers were constantly with him—on railway journeys, on board the yacht, the last subject of his thoughts at night, the first in the morning, at one time depriving him, to an alarming extent, of the power of sleep; for, unlike the labours of law, these abstruse subjects cannot be dismissed at will.

"The fact is that, in addition to the love of science for her own sake, he was penetrated by the conviction of the usefulness of his work. His splendid abilities, supported by a constitution of ususual vigour, were freely and heartily devoted to the service of his country, and the good of his fellow-creatures. 'Think how many lives it will save,' was his answer to an anxious friend who begged him to relinquish labours so exhausting, and to give himself ordinary rest. But the inevitable result followed; and though in earlier days it had seemed as if nothing could hurt his constitution, and his friends might anticipate for him the length of days for which many of his family had been remarkable, yet the continued mental strain did its work too surely, and in 1870 he was compelled to give up his profession with shattered health, to spend two short years with those he loved, and then sink into a premature grave. You observe that 'from the very commencement of his career Her Majesty's Government (to their credit be it said) appreciated the supreme importance of his researches.' In justice to the Government, it ought also to be mentioned, that they asked [twelve years ago] what acknowledgement should be made to him for work undertaken at their request, and when Smith named *a watch*, it was presented to him by the Admiralty. The testimonial presented to him during the past year, 'not as representative of the money value of his researches, but as a mark of their appreciation of their worth,' and still more, the graceful letter in which Mr Goschen intimated to him that it was awarded, gave him pleasure, and his friends must always be glad that it did not come too late.

"The truth is, Sir—and it is for this reason that I address you—that services such as his, rendered at such heavy cost to himself and his sorrowing friends, deserve the highest reward which can be given, namely, the gratitude of the nation."

One more extract in conclusion. The following from the *Solicitors' Journal and Reporter* of January 11th, 1873, contains a brief statement regarding the estimation in which Smith was held in relation to his legal profession:—

"When Mr James Parker was made Vice-Chancellor he appointed Mr Smith his Secretary; and he was also Secretary to the Decimal Coinage Commission, which made its final report in 1859. In that report there is a *résumé* of the subject by Mr Smith; and one may see there not only the special knowledge which he had collected on the matter in hand, but an example of his thorough and exhaustive style, close, compressed, and rich with fruits which it had cost him long labours and careful thought to mature. Ungrudgingly and without parade he used to offer the products of his toil: 'This,' he said to the writer, pointing to one half page of figures in his book, 'cost me six weeks of hard work.' It was thus he ever worked: no pains seemed to be too much; and consequently a marvellous neatness and elegance adorned all that he did. In his profession, although he did not attain the same exceptional eminence as in science, there was much that deserves notice. His mental characteristics were of course more or less apparent here. As a draughtsman few could compare with him for conciseness and perspicuity. His opinions were much esteemed; and his arguments, though far from brilliant in manner, had in them so much sound law and careful and subtle analysis that they were always of interest and value, and commanded the respect and attention of the judges. The important change which substituted figures for words as to dates and sums occurring in bills in Chancery was made, it is believed, at his suggestion. The well-known case of *Jenner* v. *Morris* (on appeal 3 D. F. & J. 45, 9 W. R. 391), is an instance of one of his successful arguments; and the case of *Deare* v. *Soutten* (9 L. R. Eq. 151, 18 W. R. 203), in which the former case was reconsidered and confirmed, illustrates the research and industry which he was wont to use in all matters which came before him. A judgeship in Queensland was offered to him about the year 1864; but he declined it."

In private life those who knew Archibald Smith best loved him most; for behind a reserve which is perhaps incident to engrossing thought, especially when it is concerned with scientific subjects, he kept ever a warm and true heart; and the affectionate regrets of his friends testify to the guileless simplicity and sweetness of his disposition, which nothing could spoil or affect. Simplicity indeed was remarkable through his whole character, in his "child's pure delight in little things," in the pleasure with which he shared his children's pursuits, and in the childlike Christian faith unvexed by doubt, which leavened all his life, and showed itself in his high principles, his truth, his singleness of aim. His moral and intellectual life were one. The scientific accuracy of his mind was in harmony with a moral nature which could not bear the slightest deviation from truth: and the ungrudging pains he bestowed on all that he undertook, whether great or small, sprung not only from a workmanlike love of finished work, but from conscientious endeavour in all things to do his best. About the close of 1870 he was compelled by illness to give up work; but two years later he had wonderfully rallied, and though he was not strong enough to resume his legal or scientific work, he was able to take his old interest in his boys' mathematical and other studies. A few weeks before his death he revised the instructions for compass observations to be made on board the *Challenger*, then about to sail on the great voyage of scientific investigation now in progress, and he spoke several times of the satisfaction it gave him to feel able again to do such work without fatigue. The attack of illness which closed his life was unexpected and of but a few hours' duration. In 1853 he married a daughter of Vice-Chancellor Sir James Parker, then deceased; and he leaves six sons and two daughters. He died on the 26th of December, 1872.

Appendix.

1. *Smith's Deduction of Practical Formulæ from Poisson's Mathematical Theory.*

Let the components of the terrestrial magnetic force*, parallel to three rectangular lines of reference fixed with reference to the ship, be denoted by X, Y, Z; the components at the point occupied by the compass† of combined magnetic force of earth and ship by X', Y', Z'; the components of that part of the ship's action depending on "permanent" or "subpermanent" magnetism, by P, Q, R, quantities which mathematically must be regarded as slowly varying parameters, their variations to be determined for

* That is to say, the force experienced by a unit magnetic pole. The direction of the force is taken as that of the force experienced by a south pole, or (according to Gilbert's original nomenclature) the pole of a magnet which is repelled from the southern regions of the earth. British instrument-makers unhappily mark the north pole with S and the south with N.

† The length of the needle is supposed infinitely small in comparison with the distance of the nearest iron of the ship from the centre of the compass.

each ship by observation; and the components of that part of the ship's action which depends on transiently induced magnetism by p, q, r; so that we have

$$X' = X + p + P, \quad Y' = Y + q + Q, \quad Z' = Z + r + R \quad \ldots(1).$$

Lastly, let (p, x), (q, x), (r, x) be the values which p, q, r would have if the earth's force were of unit intensity, and in the direction of x; (p, y), (q, y), (r, y) the same for y; and (p, z), (q, z), (r, z) the same for z. By the elementary law of superposition of magnetic inductions the actual value of p will be

$$(p, x)\, X + (p, y)\, Y + (p, z)\, Z;$$

and corresponding expressions will give q and r. Hence, and by (1), we have

$$\left.\begin{aligned} X' &= X + (p, x)\, X + (p, y)\, Y + (p, z)\, Z + P \\ Y' &= Y + (q, x)\, X + (q, y)\, Y + (q, z)\, Z + Q \\ Z' &= Z + (r, x)\, X + (r, y)\, Y + (r, z)\, Z + R \end{aligned}\right\} \quad \ldots\ldots(2).$$

These equations were first given by Poisson in 1824, in the fifth volume of the Memoirs of the French Institute, p. 533. From these Smith worked out practical formulæ for the main case of application, that of a ship on even keel, thus: let

H be the earth's horizontal force;

H' the resultant of the earth's and ship's horizontal forces;

θ the dip;

ζ the ship's "magnetic course";

ζ' the ship's "compass course";

$\delta = \zeta - \zeta'$ the deviation of the compass.

Then, if the directions of x be longitudinal from stern to head, y transverse to starboard, z vertically downwards, we have

$$X = H \cos \zeta, \quad Y = -H \sin \zeta, \quad Z = H \tan \theta,$$

$$X' = H' \cos \zeta', \quad Y' = -H' \sin \zeta'.$$

Resolving along and perpendicular to the direction of H we find, after some reductions,

$$\left.\begin{aligned} \frac{H'}{\lambda H} \sin \delta &= \mathfrak{A} + \mathfrak{B} \sin \zeta + \mathfrak{C} \cos \zeta + \mathfrak{D} \sin 2\zeta + \mathfrak{E} \cos 2\zeta \\ \frac{H'}{\lambda H} \cos \delta &= 1 + \mathfrak{B} \cos \zeta - \mathfrak{C} \sin \zeta + \mathfrak{D} \cos 2\zeta - \mathfrak{E} \sin 2\zeta \end{aligned}\right\} \ldots(3),$$

where

$$\left.\begin{aligned} \lambda &= 1 + \frac{(p,x)+(q,y)}{2}, & \mathfrak{A} &= \frac{1}{\lambda}\frac{(q,x)-(p,y)}{2} \\ \mathfrak{B} &= \frac{1}{\lambda}\left[(p,z)\tan\theta + \frac{P}{H}\right], & \mathfrak{C} &= \frac{1}{\lambda}\left[(q,z)\tan\theta + \frac{Q}{H}\right] \\ \mathfrak{D} &= \frac{1}{\lambda}\frac{(p,x)-(q,y)}{2}, & \mathfrak{E} &= \frac{1}{\lambda}\frac{(q,x)+(p,y)}{2} \end{aligned}\right\} \ldots(4).$$

Dividing the first by the second, of (3) we find

$$\tan\delta = \frac{\mathfrak{A} + \mathfrak{B}\sin\zeta + \mathfrak{C}\cos\zeta + \mathfrak{D}\sin 2\zeta + \mathfrak{E}\cos 2\zeta}{1 + \mathfrak{B}\cos\zeta - \mathfrak{C}\sin\zeta + \mathfrak{D}\cos 2\zeta - \mathfrak{E}\sin 2\zeta} \ldots(5),$$

which gives the deviation on any given magnetic course, ζ, when the five coefficients $\mathfrak{A}$, $\mathfrak{B}$, $\mathfrak{C}$, $\mathfrak{D}$, $\mathfrak{E}$ are known. Multiplying both members by the denominator of the second member, and by $\cos\delta$, and reducing, we find

$$\sin\delta = \mathfrak{A}\cos\delta + \mathfrak{B}\sin\zeta' + \mathfrak{C}\cos\zeta' + \mathfrak{D}\sin(\zeta+\zeta') + \mathfrak{E}\cos(\zeta+\zeta) \quad \ldots\ldots\ldots(6),$$

or

$$\sin\delta = \mathfrak{A}\cos\delta + \mathfrak{B}\sin\zeta' + \mathfrak{C}\cos\zeta' + \mathfrak{D}\sin(2\zeta'+\delta) + \mathfrak{E}\cos(2\zeta+\delta) \quad \ldots\ldots\ldots(7).$$

These give the deviations expressed nearly, though not wholly, in terms of the compass courses.

When the deviations are of moderate amount, say not exceeding 20°, equation (6) or (7) may be put under the comparatively simple and convenient form

$$\delta = A + B\sin\zeta' + C\cos\zeta' + D\sin 2\zeta' + E\cos 2\zeta' \quad \ldots(8),$$

in which the deviation is expressed wholly in terms of the compass courses; and this will be sufficiently exact for practical purposes.

It will be seen that the $\mathfrak{A}$, $\mathfrak{B}$, $\mathfrak{C}$, $\mathfrak{D}$, $\mathfrak{E}$ are nearly the natural sines of the angles A, B, C, D, E.

2. *Dygograms of Class II.*

Take lengths numerically equal to X, Y, Z and X', Y', Z' for the coordinates of two points. The axes of coordinates being fixed relatively to the ship, conceive the ship to be turned into all positions round a fixed point taken as the origin of coordinates; or for simplicity imagine the ship to be fixed and the direction of the earth's resultant force to take all positions, its magnitude remaining constant: the point (X, Y, Z) will always lie on a spherical surface, [(9) below]; and the point (X', Y', Z') will always lie on an ellipsoid fixed relatively to the ship. For we have

$$X^2 + Y^2 + Z^2 = I^2 \quad \text{..................(9)},$$

where I denotes the earth's resultant force. Now by (2) solved for X, Y, Z, we express these quantities as linear functions of

$$X' - P, \quad Y' - Q, \quad Z' - R.$$

Substituting these expressions for X, Y, Z, in (9) we obtain a homogeneous quadratic function of $X' - P$, $Y' - Q$, $Z' - R$, equated to I^2, which is the equation of an ellipsoid having P, Q, R, for the coordinates of its centre.

It is noteworthy that the point (X', Y', Z') is the position into which the point (XYZ) of an elastic solid is brought by a translation (P, Q, R), compounded with a homogeneous strain and rotation represented by the matrix

$$\left|\begin{matrix} 1+(p,x), & (p,y), & (p,z) \\ (q,x), & 1+(q,y), & (q,z) \\ (r,x), & (r,y), & 1+(r,z) \end{matrix}\right| \quad \text{......(10)}.$$

Instead of drawing at once the dygogram surface for the resultant of the force of earth and ship (X', Y', Z'), draw according to precisely the same rule, the dygogram surfaces for (X, Y, Z), the earth's force, and $(X' - X, Y' - Y, Z' - Z)$, the force of the ship. The first of these will be a sphere of radius I. The second will be an ellipsoid having its centre at the point (P, Q, R). Let ON and OM be corresponding radius vectors of these two surfaces. On OM, ON describe a parallelogram $MONK$, OK is the resultant force of earth and ship at the point occupied by the ship's compass. Vary the construction by taking a "triangle of forces" instead of the parallelogram, thus:—Produce MO through O to m, making Om equal to MO; in other words, draw the dygogram surface representing $(X-X', Y-Y', Z-Z')$; and of it let Om be the radius vector corresponding to OM of the spherical-surface dygogram of the earth's force. Join Nm; through O draw OK equal and parallel to Nm. OK (the same line as before) is the radius vector of the resultant dygogram surface, corresponding to ON of the spherical dygogram. The law of correspondence between N on the spherical surface and m on the ellipsoid is, according to (2) above, that m is the position to which M is brought—translation $(-P, -Q, -R)$ and strain* with rotation, represented by the matrix

$$\begin{vmatrix} (p,x), & (p,y), & (p,z) \\ (q,x), & (q,y), & (q,z) \\ (r,x), & (r,y), & (r,z) \end{vmatrix} \quad ....(11).$$

Take any plane section (large or small circle) of the spherical surface. The corresponding line on the ellipsoid is also a plane section, but generally in a different plane from the other. For example, let the ship revolve round a vertical axis OZ; in other words, relatively to the ship let ON revolve round OZ in a cone whose semi-vertical angle is θ, the dip. The locus of N is a horizontal circle whose radius is H, the horizontal component of the earth's magnetic force. The corresponding locus of m is an ellipse, not generally in the plane perpendicular to OZ—that is to say, not generally horizontal. This ellipse and that circle are Smith's "Ellipse and Circle" (*Admiralty Manual*, 3rd edition, 1869, App. 2, p. 168). The projection of the ellipse on the plane of the circle is the dygogram of what is wanted for the practical problem, namely the horizontal component of the ship's force.

By a curious and interesting construction (*Admiralty Manual*, page 175) Smith showed that, when $\mathfrak{A}$ and $\mathfrak{E}$ are zero, the ellipse and circle are susceptible of a remarkable modification, by which, instead of them, an altered circle and another circle (generally smaller) are found, with a perfectly simple law of corresponding

* This strain must include reflexion in a plane mirror so as not to exclude negative values exceeding certain limits in the constituents of the matrix. It is to be borne in mind that, imaginary values of the elements being excluded, strain and reflection can only alter spheres or ellipsoids to spheres or ellipsoids, not to hyperboloids.

points, to give, in accordance with the general rule stated above, the magnitude and direction of the resultant of horizontal force on the ship's compass. But in point of fact the comparison with Dygogram No. I, by which (pages 168, 169) Smith introduced Dygogram No. II, taken along with his previous mechanical construction of Dygogram No. I (pages 166, 167), proves that Dygogram No. II, simplified to two circles, is not confined to cases in which $\mathfrak{A}$ and $\mathfrak{E}$ vanish, and so gives to this beautiful construction a greatly enhanced theoretical interest. It is to be also remarked that, although the necessity for supposing $\mathfrak{A}$ and $\mathfrak{E}$ zero has been hitherto of little practical moment, as their values are very small for ordinary positions of the compass in all or nearly all ships at present in existence, the greatly increased quantity of iron in the new turret ships, and its unsymmetrical disposition in the newest projected type (the *Inflexible*), may be expected to give unprecedently great values to $\mathfrak{E}$ and $\mathfrak{A}$. The happy artifice by which Smith found two circles to serve for the "ellipse and circle" consisted in altering the radius of the first circle from H to λH. If, further, we alter it in magnitude and direction, and made it represent the resultant of λH to north and $\mathfrak{A}$ to east, thus including part of the ship's force, namely $(\lambda - 1) H$ to north and $\mathfrak{A}$ to east, along with the earth's horizontal force in one circular dygogram, the residue of the horizontal component of the ship's force has also a circular dygogram. The construction thus obtained is fully described and illustrated by a diagram under the heading "Dygogram No. II," above. The proof of this is very simple. The following is the analytical problem of which it is the solution:—In the general equations (2) suppose Z to be constant, and put

$$X' - (p, z) Z - P = X'', \quad Y' - (q, z) Z - Q = Y'' \quad \ldots(12).$$

We have

$$\left.\begin{aligned} X'' &= [1 + (p, x)] X + (p, y) Y \\ Y'' &= \qquad (q, x) X + [1 + (q, y)] Y \end{aligned}\right\} \quad \ldots\ldots\ldots(13).$$

Now imagine two dygogram curves (ellipses or circles) to be constructed as the locus of points (x, y), (x', y') given by the equations

$$\left.\begin{aligned} &X^2 + Y^2 = H^2, \\ &x = X + (\alpha X + \beta Y); \qquad y = Y + (\gamma X + \delta Y); \\ &x' = X'' - X - (\alpha X + \beta Y); \quad y' = Y'' - Y - (\gamma X + \delta Y) \end{aligned}\right\} \ldots(14);$$

and let it be required to find $\alpha, \beta, \gamma, \delta$ so that these two curves may be circles; we have four equations for these four unknown quantities. Then, as

$$x' + x = X'', \quad y' + y = Y''$$

the resultant of the radius vectors of the two concentric circles thus obtained is the resultant of the constituent (X'', Y'') of the force on the compass; and by (12) we have only to shift the centre of one of them to the point whose coordinates are $(p, z)\,Z + P$, $(q, z)\,Z + Q$, to find two circles such that the resultant of corresponding radius vectors through the centre of one of them shall be the whole horizontal component of the force on the compass. Thus we have Smith's beautiful and most useful Dygogram of two Circles.

273. H. C. Fleeming Jenkin.

[From the 'Obituary Notices' in the *Proceedings of the Royal Society*, 1885.]

Amongst the men who have laboured earnestly and successfully to place on a sound scientific basis the practice of engineering, the late accomplished occupant of the Chair of Engineering at the University of Edinburgh, Henry Charles Fleeming Jenkin, will hold a distinguished place. Born in Kent on the 25th March, 1833, the only son of Captain Charles Jenkin, R.N., he was sent to school in Scotland at the early age of seven years, where, under Dr Burnett, of Jedburgh, for three years, and after that for three years in the Edinburgh Academy, the first six years of his school life were spent. In 1846 he was placed at a school in Frankfort; in 1847 he was for a time in Paris; and, finally, in 1850, he graduated as a Master of Arts at the University of Genoa.

He began his training as an engineer in a locomotive workshop at Genoa, under Philip Taylor, of Marseilles, where he remained for about a year. He returned to England in 1851, and served a three years' apprenticeship in the works of the Fairbairns, of Manchester. After a varied experience of practical work, Mr Jenkin, in 1857, entered the service of Messrs Newall, in their submarine cable factory at Birkenhead, where they were engaged in the manufacture of a part of the first Atlantic cable, and afterwards of cables for the Mediterranean and the Red Sea. His energy and talents very soon obtained for him the position of chief of the engineering and electrical staff. In this connexion Jenkin was brought into close relation with the able engineers and electricians who were then working out to a practical result the great problem of submarine telegraphy. These circumstances determined the direction in which his energies were more especially to be applied, and he became early known as an electrical engineer of high standing.

At the beginning of 1859 he became known to Sir William Thomson, and entered into constant correspondence with him in connexion with the testing of conductivity and insulation of submarine cables, and the speed of signalling through them. After Faraday's discovery of the *existence of specific inductive capacity*, and his now celebrated, though then ignored, determinations of it for flint glass, shell-lac, and sulphur, the first correct determination of the specific inductive capacity of any substance was made by Jenkin by means of observations arranged for the purpose on some of the submarine cables in the factory at Birkenhead.

In 1861 Mr Jenkin joined Mr H. C. Ford as partner, and with him for seven years he carried on an extensive practice in telegraphic and general engineering. During the last two years of this partnership Jenkin held the post of Professor of Engineering at University College, London, and in 1868 the partnership was dissolved on account of his appointment to fill the Chair of Engineering in the University of Edinburgh, which he occupied till his death, teaching with much success.

In 1859 he began to write upon scientific subjects, encouraged to do so, as he has himself remarked, by Sir William Thomson. His published papers are in all about forty in number. Of these a large proportion deal with questions arising from the science and practice of submarine electrical engineering, and were published within the ten years 1859 to 1869—a period of the great progress in submarine telegraphy.

Professor Fleeming Jenkin took a very important part in the work of the British Association Committee on Electrical Standards, appointed on the suggestion of Sir William Thomson at the Manchester meeting of 1861, for the purpose of promoting the practical use of Gauss and Weber's system of absolute measurement, by which lasting benefit has been conferred on electric and magnetic science. Jenkin was made Secretary of this Committee; and, in conjunction with Professor Clerk Maxwell, carried out the most important of the experiments instituted by the Committee.

Through having been so intimately concerned in the beginnings of ocean telegraphy, Jenkin became associated with Sir William Thomson and Mr C. F. Varley in the development of the instruments by which the transmission of messages over long submarine cables was for the first time made practicable. During later years he and Sir William Thomson acted as joint engineers for various cable companies, their latest work in that capacity being the Atlantic and other cables of the Commercial Cable Company.

For the last two years he was much occupied with a new mode of electric locomotion, a very remarkable invention of his own, to which he gave the name of "Telpherage." He persevered with endless ingenuity in carrying out the numerous and difficult mechanical arrangements essential to the project, up to the very last days of his work in life. He had completed almost every detail of the realisation of the system which was recently opened for practical working at Glynde, in Sussex, four months after his death.

His book on *Magnetism and Electricity*, published as one of Longmans' elementary series in 1873, marked a new departure in the exposition of electricity, as the first text-book containing a systematic application of the quantitative methods inaugurated by the British Association Committee on Electrical Standards. In 1883 the seventh edition was published, after there had already appeared two foreign editions, one in Italian and the other in German.

His papers on purely engineering subjects, though not numerous, are interesting and valuable. Amongst these may be mentioned the article "Bridges," written by him for the ninth edition of the *Encyclopædia Britannica*, and afterwards republished as a separate treatise in 1876; and a paper "On the Practical Application of Reciprocal Figures to the Calculation of Strains in Framework," read before the Royal Society of Edinburgh, and published in the *Transactions* of that Society in 1869. But perhaps the most important of all is his paper "On the Application of Graphic Methods to the Determination of the Efficiency of Machinery," read before the Royal Society of Edinburgh, and published in the *Transactions*, Vol. XXVIII. (1876—78), for which he was awarded the Keith Gold Medal. This paper was a continuation of the subject treated in Reuleaux's *Mechanism*, and, recognising the value of that work, supplied the elements required to constitute from Reuleaux's kinematic system a full machine receiving energy and doing work.

Professor Jenkin's activity was not, however, confined to purely scientific pursuits. The very important practical subject of healthy houses largely engaged his attention during the last eight or ten years of his life, and he succeeded so well in impressing its importance on public opinion, that he obtained the establishment in many large towns of Sanitary Protection Associations. He also took great interest in technical education, and was always ready in word and deed to aid in its promotion. His literary abilities were of no mean order, and as a critic he made several marked successes, among which his reviews of Darwin's *Origin of Species* and of Munro's *Lucretius* (the atomic theory) may be referred to as of high scientific merit.

He was elected a Fellow of [the Royal] Society in 1865; he was also a Vice-President of the Royal Society of Edinburgh, a Member of the Institution of Civil Engineers, and of the Institution of Mechanical Engineers, and in 1883 he received the honorary degree of LL.D. from the University of Glasgow. He died on the 12th of June, 1885, after a few days' illness, due to a slight surgical operation.

274. THE SCIENTIFIC WORK OF SIR GEORGE STOKES [OBITUARY NOTICE].

[From *Nature*, February 12, 1903, pp. 337, 338.]

STOKES ranged over the whole domain of natural philosophy in his work and thought; just one field—electricity—he looked upon from outside, scarcely entering it. Hydrodynamics, elasticity of solids and fluids, wave-motion in elastic solids and fluids, were all exhaustively treated by his powerful and unerring mathematics.

Even pure mathematics of a highly transcendental kind has been enriched by his penetrating genius; witness his paper "On the Numerical Calculation of a Class of Definite Integrals and Infinite Series*," called forth by Airy's admirable paper on the intensity of light in the neighbourhood of a caustic, practically the theory of the rainbow. Prof. Miller had succeeded in observing thirty out of an endless series of dark bands in a series of spurious rainbows for the determination of which Airy had given a transcendental equation, and had calculated, of necessity most laboriously, by aid of ten-figure logarithms, results giving only two of those black bands. Stokes, by mathematical supersubtlety, transformed Airy's integrals into a form by which the light at any point of any of those thirty bands, and any desired greater number of them, could be calculated with but little labour and with greater and greater ease for the more and more distant places where Airy's direct formula became more and more impracticably laborious. He actually calculated fifty of the roots, giving the positions of twenty black bands beyond the thirty seen by Miller.

With Stokes, mathematics was the servant and assistant, not the master. His guiding star in science was natural philosophy.

* *Collected Mathematical and Physical Papers*, Vol. I. pp. 329—357. From Cambridge Philosophical Society, March 11, 1850.

Sound, light, radiant heat, chemistry, were his fields of labour, which he cultivated by studying properties of matter with the aid of experimental and mathematical investigation.

His earliest published papers (Cambridge Philosophical Society, April 25, 1842, and May 29, 1843, followed, November 3, 1846, by a supplement) were on fluid motion; the second of these and its supplement contained a beautiful mathematical solution of the problem of finding the motion of an incompressible fluid in the interior of a rectangular box to which is given any motion whatever, starting from rest with the contained liquid at rest. This solution, as shown in Thomson and Tait's *Natural Philosophy*, §§ 704 and 707, is also applicable to the very practical problem of finding the torsional rigidity of a rectangular bar of metal or glass. For every oblong rectangular section, the solution may be put in one or other of two interestingly different forms, which are identical when the cross-section is square and are always both convergent. One of them converges much more rapidly than the other when one of the diameters of cross-section is more than two or three times the other. Regarding these two solutions, Thomson and Tait (§ 707) say*:—

"The comparison of the results gives astonishing theorems of pure mathematics, such as rarely fall to the lot of those mathematicians who confine themselves to pure analysis or geometry instead of allowing themselves to be led into the rich and beautiful fields of mathematical truth which lie in the way of physical research."

The 1843 paper contained his theory of the viscosity of fluids, and his definite mathematical equations for its influence in fluid motion, which constituted the complete foundation of the hydrokinetics of the present day. In the same paper, by reference to known facts, with reference to natural and artificial solids, glass, iron, india-rubber, jelly, and results of experimental investigations, he relieved the theory of elastic solids from what is now known as the Navier-Poisson doctrine of a constant proportion between the moduluses of resistance to compression and of rigidity (resistance to change of shape), and, following Green, gave us the

* [The theorem referred to is, however, a natural result in the expansion of the Jacobian elliptic functions, in terms of which the solution for the rectangular area can now of course be simply expressed.]

equations of equilibrium and motion of isotropic elastic solids, with their two distinct moduluses, which constitute the whole theory of equilibrium and motion of elastic solids as we have it at this day.

Seven years later, building on the foundation he had laid, he communicated another great paper to the Cambridge Philosophical Society*, "On the Effect of the Internal Friction of Fluids on the Motion of Pendulums." In this paper he solved three very difficult problems, taxing severely the mathematical power of anyone trying to attack them.

(1) The oscillations of a rigid globe in a mass of viscous fluid contained in a spherical envelope having for its centre the mean position of the globe.

(2) The oscillations of an infinite circular cylinder in an unlimited mass of viscous fluid.

(3) Determination of the motion of a viscous fluid about a globe moving uniformly with small velocity through it.

(4) The effect of fluid friction in causing the rapid subsidence of ripples in a puddle and the slow subsidence from day to day of ocean waves when the storm which produced them is followed by a calm.

Of solution (3) he makes a most interesting application to explain the suspension of clouds by determining from the known viscosity of air the terminal velocity of an exceedingly minute rigid globule of water falling through air. His formula for this has been used with excellent effect in the Cavendish Laboratory by Prof. J. J. Thomson and his research corps; first, I believe, by Townsend in determining approximately the diameter of the globules in a mist produced by electrolysis, by observing its rate of subsidence when left to itself in a glass bell.

In the interval between the two great papers of 1843 and 1850, Stokes gave another magnificent hydrokinetic paper†, "Theory of Oscillatory Waves," containing a thoroughly original and masterly investigation of a most difficult problem, the determination of the motion of *steep* deep-sea waves. As an illustration of his

* December 9, 1850, *Math. and Phys. Papers*, Vol. III. pp. 1—144.

† Camb. Phil. Soc. March, 1847, *Math. and Phys. Papers*, Vol. I. pp. 197—229, with supplement first published in the reprint *Math. and Phys. Papers*, Vol. I. pp. 316—326.

results, he gives a diagram (*M. and P. P.* Vol. I. p. 212) showing the shape of a deep-sea wave in which the difference of level between crest and hollow is seven-fortieths of the wave-length—an admirable triumph of mathematical power.

He proved (Vol. I. p. 227) that the steepest possible wave has a crest of 120°, with slope of 30° down from it before and behind. He *hoped* to work out fully its shape, and would no doubt have succeeded had time permitted.

Four short papers of July, 1845, February, 1846, May, 1846, and July, 1846*, show that in those early times Stokes had taken to heart the wave theory of light. His later splendid work on light has given such great results that even in the scientific world Stokes is often thought of only as a worker in optics and the wave theory of light. Truly his work in this province is more than enough for the whole life-time of a hard-working searcher in science.

A short paper of great value†, "On the Formation of the Central Spot of Newton's Rays beyond the Critical Angle," touches in its title a physical question of fundamental importance—*What motion takes place in the ether close behind the perfect mirror presented by total internal reflection?* And the answer to it given in the paper is admirably clear and satisfactory.

A little later, we find one of the most important of all of Stokes's papers on light‡, "The Dynamical Theory of Diffraction." This paper contains the full mathematical theory of the propagation of motion in a homogeneous elastic medium. It contains, also, application of the theory to the disturbance produced in ether by a Fraunhofer grating for the two cases of incident light, (1) with its vibrations *in* the plane of incidence, and (2) with its vibrations *perpendicular to that plane* (therefore parallel to the lines of the grating). Lastly, it contains a description of an elaborate experimental investigation by himself, and a comparison of the results with theory, from which he concluded that the plane of polarisation is the plane perpendicular to the direction of vibrations in plane polarised light. This conclusion, notwith-

* *Math. and Phys. Papers*, Vol. I. pp. 141—157.

† Camb. Phil. Soc. December 11, 1848, *Math. and Phys. Papers*, Vol. II. pp. 56—81.

‡ Camb. Phil. Soc. November 26, 1849, *Math. and Phys. Papers*, Vol. II. pp. 243—328.

standing adverse criticism by Holtzmann*, was confirmed by Lorenz, of Copenhagen†. The same conclusion was arrived at from the dynamics of the blue sky by Stokes and Rayleigh, and from the dynamics of reflection at the surface of a transparent substance by Lorenz and Rayleigh. We may now consider it one of the surest truths of physical science.

The greatest and most important of all the optical papers of Stokes was communicated to the Royal Society on May 27, 1852, under the title "On the Change of the Refrangibility of Light‡." In this paper, his now well-known discovery of fluorescence is described, according to which a fluorescent substance emits in all directions from the course through it of a beam of homogeneous light. The periods of analysed constituents of this fluorescent light, in all Stokes's experiments, were found to be longer than the period of the exciting incident light. But I believe fluorescent light of shorter periods than the exciting light has been discovered in later times.

Stokes found that the fluorescence vanished very quickly after cessation of the incident light. A beautiful supplement to his investigation was made by Edmond Becquerel showing a persistence of the fluorescent light for short times, to be measured in thousandths of a second, after the cessation of the exciting light.

Stokes's fundamental discovery of fluorescence is manifestly of the deepest significance in respect to the dynamics of waves, and of intermolecular vibrations of ether excited by waves, and causing fresh trains of waves to travel through the fluorescent substance. The prismatic analysis of the fluorescent light for any given period of incident light was investigated by Stokes for a large number of substances in his first great paper on the subject, and was followed up by further investigations by Stokes himself in later years, of which some of the results are given in his paper "On the Long Spectrum of the Electric Light" (*Phil. Trans.* June 19, 1862).

Stokes's great paper on the refrangibility of light is the last paper of the last volume (Vol. III.) hitherto published of his

* *Poggendorff's Annalen*, Vol. XCIX. 1856, or *Phil. Mag.* Vol. XIII. p. 135.

† *Poggendorff's Annalen*, Vol. III. 1860, or *Phil. Mag.* Vol. XXI. p. 321.

‡ *Phil. Trans.* and *Math. and Phys. Papers*, Vol. III. pp. 259—407.

mathematical and physical papers. It is to be hoped that with the least possible delay we shall have a complete collected re-publication of *all* his other papers*. Every one of them, however small, will in all probability be found to be a valuable contribution to science; witness, for example, his paper of twenty-one lines in the *Phil. Mag.* for October, 1872. Let us hope that manuscript may be found for the communication to the Royal Society promised at the end of that paper.

Stokes's scientific work and scientific thought is but partially represented by his published writings. He gave generously and freely of his treasures to all who were fortunate enough to have opportunity of receiving from him. His teaching me the principles of solar and stellar chemistry when we were walking about among the colleges some time prior to 1852 (when I vacated my Peterhouse fellowship to be no more in Cambridge for many years) is but one example. Many authors of communications to the Royal Society during the thirty years of his secretaryship remember, I am sure gratefully, the helpful and inspiring influence of his conversations with them. I wish some of the students who have followed his Lucasian lectures could publish to the world his *Opticae Lectiones*; it would be a fitting sequel to the "Opticae Lectiones" of his predecessor in the Lucasian Chair, Newton.

The world is poorer through his death, and we who knew him feel the sorrow of bereavement.

* [The re-publication of Stokes's papers has been completed by the addition of Vols. IV. and V.; and two important volumes of scientific correspondence have been added, at Lord Kelvin's urgent suggestion, for the reasons stated in the text.]

275. JAMES WATT.

An Oration delivered in the University of Glasgow on the Commemoration of its Ninth Jubilee.

THE name of James Watt is famous throughout the whole world, in every part of which his great work has conferred benefits on mankind in continually increasing volume up to the present day.

It is fitting that the University of Glasgow, in this celebration of its ninth jubilee, should recollect with pride the privilege it happily exercised a hundred and forty-five years ago of lending a helping hand, and extending the beneficent solace of personal friendly intercourse of professors of mathematics, philosophy, and classical literature, and giving a workshop within its walls, to a young man of no university education, struggling to commence earning a livelihood as a mathematical-instrument-maker, in whom they discovered something of the genius, destined for such great things in future.

James Watt's paternal grandfather, Thomas Watt, was the son of an Aberdeenshire farmer, who died in battle in one of the wars of Montrose early in the seventeenth century. As a poor orphan, he was rescued from destitution by benevolent relatives of whom no records are known. He settled in Carsedyke as a teacher of navigation, or "Professor of the Mathematicks," as he was styled on his tombstone.

Carsedyke, on the site of the present Port-Glasgow, was a borough of barony under a charter of Charles II (1669), a mile or two from Greenock, to which a charter had been granted by Charles I thirty-four years earlier (1635). The two boroughs, one hundred years after Thomas Watt's settlement in Carsedyke, had together a population reckoned at 4100. But as early as 1700 they possessed between them four ships and two barques besides probably a somewhat large fleet of open or half-decked fishing-boats, then a nursery of excellent seamen soon to be employed in the rapidly growing over-sea trade of Glasgow, Port-Glasgow, and Greenock. It is difficult to imagine how Thomas Watt could have supported himself on a professorship of mathematics in the latter part of the seventeenth century, or even in teaching mathematical navigation; and it seems probable that he may have taken active part in the creation of the four ships and two barques, furnished with which Clyde navigation entered on the eighteenth century. However this may have been, he certainly was a public-spirited man, working for the good of his fellow-citizens as elder of the parish and presbytery, and chief magistrate of the borough of Greenock, anxiously caring for the minds and morals of the little community, and applying his capacity for scientific engineering "to repairing the church, widening the bridge, and trying by mathematical standards the weights and measures used in the borough*." He died at the good old age of ninety-five (or ninety-two according to another reckoning judged less probable by Muirhead) leaving two sons, John and James, who both inherited from him mathematical and engineering capacity. The elder practised as a surveyor in Glasgow, and died in 1737 at the age of fifty, leaving behind him a survey-chart, made in 1734, of the Clyde River and Frith from Rutherglen above Glasgow, to Loch Ryan and the coast of Ireland, and including the islands of Islay, Colonsay, and part of Mull; which was engraved and published twenty-five years later (1759) by his brother James assisted by *his* two sons, James and John. Of these the younger brother, John, died on board one of his father's ships on a voyage to America two years later. The elder brother, James, was *the* James Watt; and was twenty-four years old, occupied in his workshop in the University of Glasgow, when he assisted his father and brother and uncle in the production of the now celebrated chart.

James Watt's father was an energetic, practical man. After serving an apprenticeship to a shipbuilder in Carsedyke, he settled at Greenock at the age of thirty, and lived a busy life of work as a shipwright; a ship-chandler supplying vessels with nautical apparatus, stores, and instruments; a builder; and a

* This and all other statements distinguished by quotation marks are from Muirhead's *Life of James Watt* except when some other origin is indicated.

merchant. For upwards of twenty years he was a member of the Town Council of Greenock, and, during great part of that time, its Treasurer; a magistrate; and always a zealous and enlightened promoter of the improvements of the town of which he was an inhabitant. Above all, it is recorded by one who knew him well, that "he was an intelligent, upright, and benevolent man."

About 1729 he married Agnes Muirhead, "a fine-looking woman, with pleasing, graceful manners, a cultivated mind, an excellent understanding, and an equal cheerful temper...descended from an old Scottish family of Muirheads settled in the shire of Clydesdale time immemorial, and certainly before the reign of David the First of Scotland, anno 1120." Six children were born, of whom the three eldest died in early childhood. The fourth was the great James Watt, and the fifth was his brother John, who died in 1762.

James Watt was very delicate as a child and unable to take much part in the healthy sports and school work of other boys of his age, and early, like many other men of genius, manifested a very contemplative disposition. "His parents were indulgent, yet judicious in their kindness; and their child was docile, grateful, and affectionate. From an early age he was remarkable for manly spirit, a retentive memory, and strict adherence to truth; he might be wilful or wayward, but never was insincere. He received from his mother his first lessons in reading, his father taught him writing and arithmetic. Owing to variable health, his attendance on public classes at Greenock was irregular; his parents were proud of his talents; and encouraged him to prosecute his studies at home. His father gave him a small set of carpenter's tools, and one of James' favourite amusements was to take his little toys to pieces, reconstruct them, and invent new playthings."

From a paper entitled "Memoranda of the early years of Mr Watt, by his cousin, Mrs Marion Campbell," his biographer, Mr Muirhead, quotes the following interesting statement, "That his powers of imagination and composition were early displayed, appears from the following incident. He was not fourteen when his mother brought him to Glasgow to visit a friend; his brother John accompanied them; on Mrs Watt's return to Glasgow some weeks after, her friend said, 'You must take your son James home; I cannot stand the state of excitement he keeps me in; I am worn out with want of sleep; every evening before ten o'clock, our usual hour of retiring to rest, he contrives to engage me in conversation, then begins some striking tale, and, whether humorous or pathetic, the interest is so overpowering, that all the family listen to him with breathless attention; hour after hour strikes unheeded; in vain his brother John scolds him and pulls him by the arm, Come to bed, James; you are inventing story after story to keep us up with you till after midnight, because you love company, and your severe fits of toothache prevent your sleeping at an earlier hour.'

"Sitting one evening with his aunt, Mrs Muirhead, at the tea-table, she said: 'James Watt, I never saw such an idle boy; take a book or employ yourself usefully; for the last hour you have not spoken one word, but taken off the lid of that kettle and put it on again, holding now a cup and now a silver spoon over the steam, watching how it rises from the spout, and catching and connecting the drops of hot water. Are you not ashamed of spending your time in this way?'

"It appears that when thus blamed for idleness, his active mind was employed in investigating the properties of steam; he was then fifteen; and once in conversation he informed me that before he was that age he had read twice with great attention 'S'Gravesande's *Elements of Natural Philosophy*,' adding that it was the first book on that subject put into his hands and that he still thought it one of the best. When health permitted, his young ardent mind was constantly occupied, not with one but many pursuits. Every new acquisition in science, languages, or general literature, seemed made without an effort. While under his father's roof, he went on with various chemical experiments, repeating them again and again until satisfied of their accuracy from his own observations. He had made for himself a small electrical machine, and sometimes startled his young friends by giving them sudden shocks from it."

After the age of thirteen he was often in Glasgow with his uncle, Mr Muirhead, taking opportunity to learn something of anatomy and chemistry. While at home with his parents he attained to considerable proficiency in Latin, and learned something of Greek, at the grammar school of Greenock; but he studied mathematics with much greater zest under Mr John Marr, a relative of his family. He also got great benefit in seeing his father's business affairs, and so making the acquaintance of optical instruments of various kinds for astronomy and navigation, and learning the highly scientific and interesting mechanics of sailing ships. He had a small forge set up for his own use, at which he worked in making and repairing instruments of all kinds. Thus while his delicate health prevented him from being an athlete with other boys of his age, he early became a skilled mechanic; and a skilled mechanic he remained, taking pleasure in the exercise of his handicraft, to the very end of his life.

In June, 1754, Watt came to live in Glasgow under care of relations of his mother; and was introduced by one of them, Prof. Geo. Muirhead, to Prof. James Moore (his colleague in the editorship of the magnificent Glasgow edition of Homer in four folio volumes) and to Adam Smith, Robert Simson, and other professors in the University, whose friendship he enjoyed as long as they lived. Looking forward to earning his livelihood as a mathematical-instrument-maker, Watt was advised by the professor of natural philosophy, Dr Dick, to go to London for better instruction in the art than he could get in Glasgow. Accordingly, on the 7th June, 1755, young Watt rode out of Glasgow in charge of his old mathematical master, John Marr, who was going south to act as naval instructor on board the *Hampton Court*, a seventy-gun ship then lying at anchor in the Thames. They travelled by Coldstream, Newcastle, Durham, York, Doncaster, Newark, and Biggleswade, the whole way to London on horseback in twelve days, on two of which not more than a Sabbath day's journey was performed.

Touching letters to young Watt's father from himself and Mr Marr showed the great difficulty they had to find in London a competent instrument-maker who would consent to give the required instruction, and the great anxiety of the son to avoid being a burden on his father, whose means had been seriously straitened through want of prosperity of his Greenock business. However, with the assistance of Mr Marr and the good offices of Dr Dick, an arrangement was at last happily concluded with a very good man, John Morgan, mathematical-instrument-maker in Finch Lane, Cornhill,—young Watt to receive a year's instruction in instrument-making, for which he was in return to pay twenty guineas and give his labour for the year. In Muirhead's book we have an interesting account of the young pupil's work and life during the year. He lodged under the roof of his master, but had to find his own food, which cost him eight shillings a week, "lower than that he could not reduce it." To diminish the expense to his father, he earned some money on his own account by rising early and gaining something by work done before the shop-time. At night he was, as he wrote to his father, "thankful enough to go to bed with his body wearied and his hand shaking from ten hours' hard work." "We work to nine o'clock every night, except Saturdays."

In his letters he regrets the charge his living must be to his father, and says he is striving all he can to improve himself that he may be sooner able to assist him and to assure his own maintenance.

Of young Watt's time in London Muirhead tells us, "An unexpected danger at that time hung over his destiny, which might have cut short, at least for a season, his projects of further improvement in natural science and postponed *sine die* his return to Glasgow College, with all its interesting consequences. This sword of Damocles was the chance of being impressed for the navy. He writes in the spring of 1756 that he avoids 'a very hot press just now by seldom going out.' And on a later day he adds 'they now press anybody they can get, landsmen as well as seamen, except it be in the liberties of the city, where they are obliged to carry them before my Lord Mayor first, and unless one be either a 'prentice or a creditable tradesman, there is scarce any getting off again. And if I was carried before my Lord Mayor, I durst not avow that I wrought in the city, it being against their laws for any unfreeman to work, even as a journeyman, within the Liberties.'"

Our country is happier and freer now than it was a hundred and fifty years ago. Volunteer sailors and soldiers compete enthusiastically for the honour of fighting their country's battles. Every employer is free by law to give work as he pleases; and every worker, old or young, is free by law to take work where he can find it.

Watt might probably have got good work in London after his year of pupilage had he decided to try for it. But the hard struggle had told upon his health. With violent rheumatic pain and "weariness all over his body" he found himself compelled to seek the benefit he expected to derive from the "ride homeward" and from his native air. So at the end of August, 1756, he took leave of London and of Mr Morgan, who, dying in 1758, was not destined to witness the future success of his pupil. But before leaving Watt made a small investment of twenty guineas in "half a hundred additional tools" and the materials necessary for "a great many more that he knew he must make himself."

Soon after his arrival in Glasgow, an occasion for good employment of that little stock-in-trade came to him through the good offices of his friend the Professor of Natural Philosophy, Dr Dick, who asked him to assist him in unpacking a valuable collection of astronomical instruments just arrived from Jamaica. These instruments had been constructed at great cost by the best makers in London for their late proprietor, Mr Alex. Macfarlane, a merchant and amateur astronomer, long resident in Jamaica, who died in 1755, having bequeathed the contents of his observatory to the University in which he had received his education. I doubt whether any of you here present may remember the old Macfarlane Observatory in the upper eastern part of the college green of the old Glasgow College in High Street. I remember it well, and remember being taught to take transits of the sun and stars about 1838 or 1839 on Alex. Macfarlane's own old transit instrument by my father's colleague, Dr Nicol, afterwards my own colleague, and the father of my late colleague, Prof. John Nicol. That transit instrument and, I believe, other instruments from Mr Macfarlane's old observatory in Jamaica are still doing good work for the University of Glasgow in its present observatory on Dowanhill. A minute of a University meeting held on the 26th October, 1756, regarding them is interesting—the Professor of Greek and the Professor of Natural Philosophy appointed as a deputation to call on the youthful mechanic James Watt! "Several of the instruments from Jamaica having suffered by the sea-air, especially those of iron, Mr Watt, who is well skilled in what relates to the cleaning and preserving of them, being accidentally in town, Mr Moor and Mr Dick are appointed to desire him to stay some time in town to clean them, and put them in the best order for preserving them from being spoiled." A record of a few weeks later tells us that "a precept was signed to pay James Watt five pounds sterling for cleaning and refitting the instruments lately come from Jamaica." This was probably the first money he earned since the termination of his pupilage.

He was then within a few weeks of twenty-one, and wished to commence as soon as possible the regular exercise of the trade for which he had been preparing. But he was not allowed by city and trade rules to work as an instrument-maker in the City of Glasgow, because he was neither the son of a burgess, nor married to the daughter of a burgess, nor a passed apprentice to any trade. He was forbidden to set up even a humble workshop with himself as solitary tenant within the limits of the borough. The University is now happily within the borough of Glasgow. Happily it was not in the borough in 1757, and it was able to give James Watt protection from tyrannical usages outside its bounds. By midsummer of that year he received permission to occupy an apartment and open a shop within the precincts of the College, and to use the designation of "Mathematical-instrument-maker to the University." In the autumn of the same year the foundation-stone of an astronomical observatory, to receive the collection of the Jamaica instruments which he had refitted and set up, and to be called the Macfarlane Observatory, was laid. Probably the completion of that undertaking gave some of the earliest employment to Watt in his University workshop.

In work for outside the University he seems early to have made some progress, as we may judge from the following interesting letter to his father of date 15th September, 1758: "As I have now had a year's trial here, I am able to form a judgment of what may be made of this business, and find that unless it be the Hadley's instruments, there is little to be got by it, as at most other jobs I am obliged to do the most of them myself; and as it is impossible for one person to be expert at everything, they very often cost me more time than they should do. However, if there could be a ready sale procured for Hadley's quadrants, I could do very well, as I and one lad can finish three in a week easily; and selling them at 28*s.* 6*d.*, which is vastly below what they were ever sold at before, I have 40*s.* clear on the three. So it will be absolutely necessary that I take a trip to Liverpool to look for customers, and hope that upon the profits of what I shall be able to sell there, I can go to London in the spring, when I make no doubt of selling more than I can get made; all which I want your advice on. And if that does not succeed I must fall into some other way of business, as this will not do in its present situation." The sale, however, of the profitable Hadley's quadrants in Glasgow appears to have increased so much as to have rendered the proposed speculative trading voyage to Liverpool unnecessary.

A year later, it is interesting to find an advertisement (dated October 22, 1759) of an engraved map of the Frith of Clyde "to be sold by James Watt at his shop in the College of Glasgow." This was the final outcome of the survey made two years before he was born by his uncle, John Watt, of which I have already told you.

While still continuing to make mathematical and nautical instruments in his University workshop, we find him also making organs, guitars, flutes, and violins, and making or repairing harps, guitars, mandolines, viol-de-gambas, and double-basses, in 1761 and 1762. Of this excursion from mere mathematical-instrument-making Robison, then a post-graduate theological student (afterwards successor of Black as Lecturer on Chemistry) in the University of Glasgow, wrote, "We imagined that Mr Watt could do anything; and, though we all knew that he did not know one musical note from another, he was asked if he could build this organ (an organ wanted for a Masonic Lodge in Glasgow). He said 'Yes,' but he began by building a very small one for his friend, Dr Black, which is now in my possession. In doing this a thousand things occurred to him which no organ-builder ever dreamed of—nice indicators of the strength of the blast, regulators of it, etc. He then began to study the philosophical theory of music. Fortunately for me, no book was at hand but the most refined of all, and the only one that can be said to contain any theory at all—Smith's *Harmonics*. Before Mr Watt had half-finished this organ, he and I were completely masters of that most refined and beautiful theory of the beats of imperfect consonances. He found that by these beats it would be possible for him, totally ignorant of music, to tune this organ according to any system of temperament; and he did so, to the delight and astonishment of our best performers."

While thus interestedly occupied in the fascinating study of musical instruments, Watt entered on his life-long work on steam-power. In a note by himself appended to Professor Robison's dissertation on steam-engines, he says, "My attention was first directed in the year 1759 to the subject of steam-engines, by the late Dr Robison, then a student in the University of Glasgow, and nearly of my own age. He at that time threw out an idea of applying the power of the steam-engine to the moving of wheel-carriages, and to other purposes, but the scheme was not matured, and was soon abandoned on his going abroad.

"About the year 1761, or 1762, I tried some experiments on the force of steam in a Papin's digester, and formed a species of steam-engine by fixing upon it a syringe, one-third of an inch diameter, with a solid piston, and furnished also with a cock to admit the steam from the digester, or shut it off at pleasure, as well as to open a communication from the inside of the syringe to the open air, by which the steam contained in the syringe might escape. When the communication between the digester and syringe was opened, the steam entered the syringe, and by its action on the piston raised a considerable weight (15 lbs.) with which it was loaded," which shows that he had steam at 170 lbs. per square inch to deal with. "When this was raised as high as was thought proper, the communication with the digester was shut, and communication with the atmosphere opened, the steam then made its escape, and the weight descended. The operations were repeated, and, though in this experiment the cock was turned by hand, it was easy to see how it could be done by the machine itself, and to make it work with perfect regularity. But I soon relinquished the idea of constructing an engine upon this principle, from being sensible it would be liable to some of the objections against Savery's engine, viz., the danger of bursting the boiler, and the difficulty of making the joints tight, and also that a great part of the power of the steam would be lost, because no vacuum was formed to assist the descent of the piston. I described this engine in the fourth article of the specification of my patent of 1769, and again in the specification of another patent in the year 1784, together with a mode of applying it to the moving of wheel-carriages."

Precisely that single-acting, high-pressure, syringe-engine, made and experimented on by James Watt one hundred and forty years ago in his Glasgow College workshop, now in 1901, with the addition of a surface-condenser cooled by air to receive the waste-steam, and a pump to return the water thence to the boiler, constitutes the common road motor, which, in the opinion of many good judges, is the most successful of all the different motors which have been made and tried within the last few years. Without a condenser, Watt's high-pressure, single-acting engine of 1761 only needs the cylinder-cover with piston-rod passing steam-tight through it (as introduced by Watt himself in subsequent developments), and the valves proper for admitting steam on both sides of the piston and for working expansively, to make it the very engine which, during the whole of the past century, has done practically all the steam work of the world, and is doing it still, except on the sea or lakes or rivers, where there is plenty of condensing water. Even the double and triple and quadruple expansion engines, by which the highest modern economy for power and steam engines has been obtained, are splendid mechanical developments of the principle of expansion, discovered and published by Watt, and used, though to a comparatively limited extent, in his own engines. One thing James Watt did not know—the thermodynamic value of high temperature without high pressure. This was absolutely unknown, and nothing towards it was thought of by engineers or philosophers, till it was discovered by Sadi Carnot and published in his *Puissance Motrice du Feu* in 1824. Thus James Watt did not see merit in superheated steam. Its use, introduced thirty years ago by John Elder, and only largely coming into practice within these last two or three years, gives the finishing touch of Science to obtain the highest economy in the modern steam engine.

With all the essential ideas of the finally successful engine in his mind, a long and arduous struggle to realise them for practical usefulness lay before Watt. He soon relinquished the idea of constructing a high-pressure, non-condensing engine, and, by being employed to repair a model of Newcomen's engine a year or two later, he was brought back to steam power as developed in Newcomen's engine, which essentially involved condensation. Having been for fifty-three years official guardian of the model with which Watt's practical work on the steam-engine thus commenced, I may be pardoned for asking your sympathy in recalling some trivial details of its history. In the records of the University of Glasgow we find two minutes, with six years interval:

"University meeting, 25th June, 1760. Mr Anderson is allowed to lay out a sum not exceeding two pounds sterling to recover the steam-engine from Mr Sisson, instrument-maker at London."

"University meeting, 10th June, 1766. An account was given in by James Watt for repairing and altering the steam-engine with copper pipes and cisterns, amounting to £5. 11*s*. The said machine being the property of the College, and having been in such a situation that it did not answer the end for which it was made, the Principal is appointed to grant a precept for payment of the said account, which is to be stated upon the fund for buying instruments to the College."

Sisson was a highly-skilled maker of astronomical instruments in London. The great French astronomer, Delambre, tells us that he made a mural quadrant for the Greenwich Observatory, and another for the private observatory of the King of England, and adds the remark: "Thus Sisson maintained the honour and pre-eminence of England." Sisson soutint à cet égard l'honneur *et la pre-eminence* de l'Angleterre*. Yet it seems that he did not succeed in making the Newcomen model work.

Mr John Anderson was my official great-great-grandfather as Professor of Natural Philosophy in the University of Glasgow, having been appointed to the Chair on the death, in 1759, of Watt's appreciative and devoted friend, Dr Dick, and having been himself for five years previously Professor of Hebrew in the University. He occupied the Chair for thirty-nine years; and unhappily, somewhat out of temper with the College or University in the later years of his encumbency, made a will founding a rival institution, to be called Anderson's University, with a condition that in it not a lecturer, nor teacher, nor a porter, not even an instrument-maker, was to be employed who had worked for the old University. I don't believe this condition of Anderson's will was ever fulfilled. The Andersonian Institution has, from its foundation to the present day, worked in perfect harmony with

* Delambre, *Histoire de l'Astronomie au dixhuitième Siècle*, p. 237, 1827.

the University—perhaps even more perfect harmony than if it had been founded as an officially incorporated College of the University.

Watt has told us that it was in the winter of 1763—4 that he was engaged repairing the model, and we see that his account for the work done was not given in till the 10th June, 1766; so we may fairly conclude that he had it in hand for more than two years, and made a great many experiments with it. In the course of these experiments he noticed with surprise the large quantity of water required to condense the steam—five or six times as much as the water primarily evaporated. In conference with Joseph Black, Lecturer on Chemistry in Glasgow College, it was found that this was a splendid and previously unthought-of example of the doctrine of latent heat, then fresh from Black's original discovery of it. With very primitive and imperfect instrumental appliances, Watt measured the amount of the latent heat of condensation of steam at different temperatures and pressures, and found for its variations a roughly approximate law. When, eighty-one years later, a student under Regnault in his laboratory in the College of France, I used to hear him speaking of "la loi de Watt," and telling us that it was the nearest approach to the truth which he found among the results of previous experimenters, I felt some pride in thinking that the experiments on which it was founded had been made in Glasgow College.

In working on the Newcomen model, Watt found that it essentially involved great waste of heat by performing the condensation in the cylinder by the injection of cold water, which not only cooled the steam but the whole metal of the cylinder. To remedy this fault, he invented the separate condenser, and established the principle of working with the cylinder always hot and dry Thus during the five years from 1761—6 Watt had worked out all the principles and invented all that was essential in the details for realising them in the most perfect steam-engines of the present day.

In 1763 Watt ceased to live in his College rooms, and took a small house in the town; and in 1764 he married a cousin of his own, Margaret Miller, who for nine years did everything possible to support him and to brighten his life through the severe trials which were before him in his great work, rendered harder by continued ill-health. Of this time of his life we find an interesting statement in Miss Campbell's Memoranda: "Even his powerful mind sank occasionally into misanthropic gloom, from the pressure of long-continued nervous headaches and repeated disappointments in his hopes of success in life. Mrs Watt, from her sweetness of temper and lively, cheerful disposition, had power to win him from every wayward fancy—to rouse and animate him to active exertions. She drew out all his gentle virtues, his native benevolence, and warm affections."

I wish I could tell you of his early trials and failures to realise the steam-engine for practical purposes with the co-operation and assistance of his enthusiastic friend, Dr Roebuck, the founder of the Carron Ironworks. In 1770, deeply depressed by hope deferred and almost constant bad health, he writes: "I am resolved, unless those things I have brought to some perfection reward me for the time and money I have lost on them, if I can resist it, to invent no more. Indeed, I am not near so capable as I was once. I find that I am not the same person I was four years ago, when I invented the fire-engine, and foresaw, even before I made a model, almost every circumstance that has since occurred. I was at that time spurred on by the alluring hope of placing myself above want, without being obliged to have much dealing with mankind, to whom I have always been a dupe. The necessary experience in great was wanting; in acquiring it I have met with many disappointments. I must have sunk under the burthen of them if I had not been supported by the friendship of Dr Roebuck. I have now brought the engine near a conclusion, yet I am not in idea nearer that rest I wished for than I was four years ago. However, I am resolved to do all I can to carry on this business, and if that does not thrive with me, I will lay aside the burthen I cannot carry."

With a family of three children the necessity to earn money gradually led him, as his biographer Muirhead tells us, "more frequently to forsake the solitary vigils of his workshop in the city for the active labours of his profession of a civil engineer. 'Somehow or other,' as he modestly expresses it—or, as we cannot doubt, from his ability and integrity having now become well known—the magistrates of Glasgow had for two or three years past employed him in various engineering works of importance." In 1767 he was employed, in conjunction with Mr Robert Mackell, to make a survey for a small canal intended to unite the rivers Forth and Clyde, by a line known as the Loch Lomond passage. He attended Parliament on the part of the subscribers to this scheme, and it appears from some of his letters to Mrs Watt that he was not much enamoured of the public life of which he thus obtained a glimpse; "close confined attending this confounded Committee of Parliament," he says, "I think I shall not long to have anything to do with the House of Commons again: I never saw so many wrong-headed people on all sides gathered together."

It seems that on his journey from London on that occasion he made the acquaintance of Dr Erasmus Darwin (grandfather of *the* Charles Darwin), who writes to him from Lichfield, in August, 1767: "Now, my dear new friend, I first hope you are well and less hypochondriacal, and that Mrs Watt and your child are well. The plan of your steam improvements I have religiously kept secret, but begin myself to see some difficulties in your execution which did not strike me when you were here. I have got another and another new hobby-horse since I saw you. I wish the Lord would send you to pass a week with me, and Mrs Watt along with you—a week, a month, a year. You promised to send me an instrument to draw landscapes with. If you ever move your place of residence for any long time from Glasgow, pray acquaint me.—Adieu. Your friend, E. Darwin." The dear new friend did leave Glasgow seven years later to live in Darwin's neighbourhood, and formed with him "the Lunar Society," an association of kindred spirits all devoted to the pursuit of natural knowledge and filled with mutual esteem and affection—Erasmus Darwin, Watt, Boulton, Dr Small, Wedgwood, Day (author of the delightful *Sandford and Merton* of our childhood), Dr Withering, Keir, Galton, Edgeworth, and Dr Priestley. The Lunar Society dined together every month at two o'clock on the day of full moon, in order to have the benefit of its light in returning to their homes at night! Our scientific and friendly symposiums, alas! are shorter in these degenerate days.

In 1769 he made a survey and estimate for a navigable canal from the collieries at Monkland in Lanarkshire to the City of

Glasgow, which, as Muirhead tells us, "was carried out under his own directions and superintendence, to the great advantage of the public as well as of the parties to the undertaking." His civil engineering work came to a melancholy close in 1773 while he was engaged in a survey of the Caledonian Canal. In the autumn of that year he was suddenly summoned home by the intelligence of the dangerous illness of his wife, but arrived too late. She had died after having given birth to a still-born child. They had had four children, of whom two died in infancy, one daughter, who married in Glasgow but died early, and a son, James Watt, of Aston Hall, who long survived his father, and died unmarried in 1848.

The death of his first wife in Glasgow was the turning point in Watt's life. For thirty-eight years, except the one year of trade apprenticeship in London, his home had been in Scotland. During seventeen happy years in the University and City of Glasgow, chequered with much of painfully anxious care, he had laid a secure foundation for future ease and prosperity. He had emerged from the feeble and unstable health of his early life. He had taken in 1769 his first patent for engines realising steam-power, for which a twenty-five years' extension from 1775 was afterwards granted by Act of Parliament. He had entered into partnership with Mr Boulton. He had in April, 1773, removed to Soho, Birmingham, his first practical steam-engine from Kinneil, a highland glen near Carron with sufficient water supply for condensation, where, after primary trials, it had been lying useless for some years perishing from long exposure to the weather. In terms of his partnership with Boulton he was to make his home in the neighbourhood of Soho, but this was not done before the death of his wife. A few months later he left Scotland, and thenceforward to the end of his life his home was in England.

I wish we had an hour to devote in imagination to James Watt in England for the remaining forty-five years of a beautiful and hard-working and useful and happy life. All I can say just now is—read of it in Muirhead, and in Arago's Éloge of Watt.

Greenock and the University and City of Glasgow never lost James Watt though he ceased to live among them in 1774. The University conferred the honorary degree of LL.D. upon him in

1806. In 1808 he founded the Watt prize in Glasgow College by a letter to Dr Wm. Taylor, the Principal of the University, in which he said: "Entertaining a due sense of the many favours conferred upon me by the University of Glasgow, I wish to leave them some memorial of my gratitude, and, at the same time, to excite a spirit of inquiry and exertion among the students of Natural Philosophy and Chemistry attending the College, which appears to me the more useful, as the very existence of Britain, as a nation, seems to me in great measure to depend upon her exertions in science and in the arts." In 1816 he made a donation to the town of Greenock for the purchase of scientific books, stating as his intention "to form the beginning of a scientific library for the instruction of the youth of Greenock, in the hope of prompting others to add to it, and of rendering his townsmen as eminent for their knowledge as they are for their spirit of enterprise."

Watt became

Fellow of the Royal Society of Edinburgh in	1784
Fellow of the Royal Society of London in ...	1785
Member of the "Société Batave" in ...	1787
Correspondent of the French Academy of Sciences in	1808
One of the eight "Associés Étrangers" of the French Academy of Sciences in ...	1814

I do not know if any University in the world ever had a tradesman's workshop and saleshop within its walls even for the making and selling of mathematical instruments prior to 1757. But whether the University of Glasgow is or is not unique in its beneficent infraction of usage in this respect, I believe it is certainly unique in being the first British University, perhaps the first University in the world, to have an engineering school and professorship of engineering (commenced under Prof. Lewis Gordon about 1843).

Glasgow was, I believe, certainly the first University to have a chemical teaching laboratory for students, started by its first professor of chemistry, Thomas Thomson, some time between 1818 and 1830. Glasgow was, I believe, also certainly the first University to have a physical laboratory for the exercise and instruction of students in experimental work, which grew up with very imperfect appliances between 1846 and 1856. Pioneer though it was in those three departments, it has been outstripped within the last ten or fifteen years by other Universities and Colleges in the elaborate buildings and instruments now needed to work them effectively for the increase of knowledge by experimental research and the practical instruction of students. But there is no lagging to-day in the resolution to improve to the utmost in affairs of practical importance; and we almost see attainment of the further aspiration to excel over all others in the James Watt Engineering Laboratory of the University of Glasgow, to be ready for work before the expected meeting of the Engineering Congress next September.

And now, through the magnificently generous kindness of Mr Andrew Carnegie to the people among whom he has made for himself a summer home in the land of his birth, all the four Scottish Universities can look forward to a largely increased power of benefiting the world by scientific research, and by extending their teaching to young people chosen from every class of society as likely to be made better, and happier, and more useful to our country by University Education.

276. Obituary Notice of Professor Tait*.

[From *Proceedings of the Royal Society of Edinburgh*; read December 2, 1901.]

When Professor Tait last February resigned the chair of Natural Philosophy in the University of Edinburgh, we hoped that the immediate relief from strain and anxiety regarding his duty might conduce to a speedy recovery from the severe illness under which he was then suffering. I was indeed myself sanguine in looking forward to an unbroken continuation of the friendly intercourse with him which I had enjoyed through forty-one years of my life. A slight abatement of the graver symptoms, and a cheering return to some mathematical work left off six months before, gave hope that a change from George Square to Challenger Lodge in June, on the invitation of his friend and former pupil Sir John Murray, might be the beginning of a recovery. But it was not to be. Death came suddenly on the 4th of July, and our friend is gone from us.

Peter Guthrie Tait was born at Dalkeith on 28th April 1831. After early education at Dalkeith Grammar School, and Circus Place School, Edinburgh, he entered the celebrated Edinburgh Academy, of which he remained a pupil till 1847, when he entered the University of Edinburgh. After a session there under Kelland and Forbes, he entered Cambridge in 1848 as an undergraduate of Peterhouse, and in 1852 he took his degree as Senior Wrangler and First Smith's Prizeman, and was elected to a Fellowship of his College. He remained officially in Peterhouse as mathematical lecturer till 1854, when he was called to Queen's College, Belfast, as Professor of Mathematics. This was a most happy appointment for Tait. It made him a colleague of, and co-worker on the electrolytic condensation of mixed oxygen and hydrogen and on ozone, with Andrews, the discoverer of a procedure producing

* [See *Memoir* of Prof. Tait by C. G. Knott, 1910, uniform with his *Collected Papers*.]

continuous change in a homogeneous substance, from liquid to gaseous and from gaseous to liquid condition. Through Andrews it introduced him to William Rowan Hamilton, the discoverer of the principle of varying action in dynamics, and the inventor of the captivatingly ingenious and beautiful method of quaternions in Mathematics. It gave him six years of good duty in Queen's College, well done, in teaching Mathematics; and for some time also Natural Philosophy, in aid of his colleague Stevelly. During those bright years in Belfast he found his wife, and laid the foundation of a happiness which lasted as long as his life.

In 1860 he was elected to succeed Forbes as Professor of Natural Philosophy in the University of Edinburgh. It was then that I became acquainted with him, and we quickly resolved to join in writing a book on Natural Philosophy, beginning with a purely geometrical preliminary chapter on kinematics, and going on thence instantly to dynamics, the science of Force, as foundation of all that was to follow. I found him full of reverence for Andrews and Hamilton, and enthusiasm for science. Nothing else worth living for, he said; with heart-felt sincerity I believe, though his life belied the saying, as no one ever was more thorough in public duty or more devoted to family and friends. His two years as "don" of Peterhouse and six of professorial gravity in Belfast had not wholly polished down the rough gaiety, nor dulled in the slightest degree the cheerful humour, of his student days; and this was a large factor in the success of our alliance for heavy work, in which we persevered for eighteen years. "A merry heart goes all the day, Your sad, tires in a mile-a." The making of the first part of "T and T′" was treated as a perpetual joke, in respect to the irksome details of interchange of drafts for "copy," amendments in type, and final corrections of proofs. It was lightened by interchange of visits between Greenhill Gardens, or Drummond Place, or George Square, and Largs, or Arran, or the old or new College of Glasgow; but of necessity it was largely carried on by post. Even the postman laughed when he delivered one of our missives, about the size of a postage stamp, out of a pocket handkerchief, in which he had tied it to make sure of not dropping it on the way.

One of Tait's humours was writing in charcoal on the bare plaster wall of his study in Greenhill Gardens a great table of living scientific worthies *in order of merit.* Hamilton, Faraday, Andrews, Stokes, and Joule headed the column, if I remember right; Clerk Maxwell, then a rising star of the first magnitude in our eyes, was too young to appear on the list.

About 1878 we got to the end of our "Division II." on "Abstract Dynamics"; and, according to our initial programme, should then have gone on to "properties of matter," "heat," "light," "electricity," "magnetism." Instead of this we agreed that for the future we could each work more conveniently and on more varied subjects, without the constraint of joint effort to produce as much as we could of an all-comprehensive text-book of Natural Philosophy. Thus our book came to an end with only a foundation laid for our originally intended structure.

Tait's first published work was undertaken in conjunction with a Peterhouse friend, Steele, who was his second in the University both as Wrangler and Smith's Prizeman. They commenced their work together immediately after taking their degrees; but Steele died before more than two or three chapters had been written, and Tait finished it alone, and published it four years later under the title *Tait and Steele's Dynamics of a Particle* (1856). It has gone through many editions, and still holds its place as a text-book.

Tait's second published book, *Elements of Quaternions,* was commenced under the auspices of Hamilton; but, in deference to his wish, not published till 1867. It has gone through three editions, and is, I believe, the text-book for all those who wish to learn the subject.

Tait also produced several valuable Treatises, short, *readable,* interesting, and useful, on various subjects in physical science:—

Sketch of Thermodynamics (1867).

Recent Advances in Physical Science (1876).

Heat (1884, 2nd edition 1892).

Light (1884, 3rd edition 1900), based on article in *Encyclopœdia Britannica.*

Properties of Matter (1885, 4th edition 1899).

Dynamics (1895), based on article "Mechanics" in *Ency. Brit.*

Among smaller articles contributed to the *Ency. Brit.* are "Quaternions," "Radiation and Convection," and "Thermodynamics," all reprinted in the collected papers. A small 50-page book on *Newton's Laws of Motion* is a remarkably concise statement of the foundations of dynamical science. It is Tait's last published work, primarily intended as a help to medical students attending his special three months' course of lectures for them on Natural Philosophy.

In the Royal Society of Edinburgh we all know something of how Tait has enriched its Proceedings and Transactions by his interesting and varied papers on mathematical and physical subjects from year to year, since 1860 when he came to Edinburgh to succeed Forbes as Professor of Natural Philosophy in the University. Nearly all of these are now collected, along with a considerable number of other scientific papers which he brought out through other channels, arranged in order of time, from 1859 to 1898; one hundred and thirty-three articles in all; republished by the Cambridge University Press in two splendid quarto volumes of 500 pages each; a worthy memorial of a life of laborious whole-hearted devotion to science.

The "Scientific Papers" collected in these two volumes abound in matter of permanent scientific interest; and literary interest too, as witness the short articles on "Hamilton," "Macquorne Rankine," "Balfour Stewart," "Clerk Maxwell," and "The Teaching of Natural Philosophy." Of all the mathematical papers in the collection, one of those which seem to me most fundamentally important is Part IV. of "Foundations of the Kinetic Theory of Gases," in which we find the first proof (and, I believe, the only proof hitherto given) of the theorem enunciated first by Waterston and twelve years later independently by Clerk Maxwell, asserting equal average partition of energy between two sets of masses larger and smaller, taken as hard globes, to represent the molecules of two different gases thoroughly mixed together. The collection contains also papers describing valuable experimental researches made by Tait through many years on various subjects: Thermo-electricity; Thermal Conductivity of Metals; Impact and Duration of Impact; Pressure Errors of the 'Challenger' thermometers; Compressibility of Water, Glass, and Mercury (contributed originally to the "Physics and Chemistry" of H.M.S. 'Challenger'). His work for the 'Challenger' Report was a splendid series of very difficult experimental researches carried on for about nine years (1879 to 1888), with admirable scientific inventiveness, and no less admirable zeal and perseverance. One little scientific bye-product of extreme interest I cannot refrain from quoting. Referring to a hermetically sealed glass tube under tests for strength to resist great water pressure, "I enclosed the glass tube in a tube of stout brass, closed at the bottom only, but was surprised to find that it was crushed almost flat on the first trial [when the glass tube broke]. This was evidently due to the fact that water is compressible, and therefore the relaxation of pressure (produced by the breaking of the glass tube) takes time to travel from the inside to the outside of the brass tube; so that for about 1/10000th of a second that tube was exposed to a pressure of four or five tons weight per square inch on its outer surface, and no pressure on the inner. The impulsive pressure on the bottom of the tube projected it upwards so that it stuck in the tallow which fills the hollow of the steel plug. Even a piece of gun-barrel, which I substituted for the brass tube, was cracked, and an iron disc, tightly screwed into the bottom of it to close it, was blown in. I have since used a portion of a thicker gun-barrel, and have had the end welded in. But I feel sure that an impulsive pressure of ten or twelve tons weight would seriously damage even this. These remarks seem to be of interest on several grounds; for they not only explain the crushing of the open copper cases of those of the 'Challenger' thermometers which gave way at the bottom of the sea, but they also give a hint explanatory of the very remarkable effects of dynamite and other explosives when fired in the open air. (It is easy to see that, *ceteris paribus*, the effects of this impulsive pressure will be greater in a large apparatus than in a small one.)"

In a communication on "Charcoal Vacua" to the Royal Society of Edinburgh of July 5, 1875, imperfectly reported in *Nature* of July 15 of that year, the true dynamical explanation of one of the most interesting and suggestive of all the scientific wonders of the nineteeth century, Crookes' radiometer, was clearly given. The phenomenon to be explained is that in highly rarefied air a disc of pith or cork or other substance of small thermal conductivity, blackened on one side, and illuminated by light on all sides, even the cool light of a wholly clouded sky, experiences a steady measurable pressure on the blackened side. Many naturalists, I believe, had truly attributed this fact to the blackened side being rendered somewhat warmer by the light; but none before Tait and Dewar had ever imagined the dynamical cause,—the largeness of the free path of the molecule of the highly rarefied air, and the greater average velocity of rebound of the molecules from the warmer side. *Long free path* was the open sesame to the mystery.

The Keith Medal of the Royal Society of Edinburgh was awarded to Professor Tait in the year 1869, and again in 1874; and one of the Royal Medals of the Royal Society of London was awarded to him in the year 1886. The Gunning Victoria Jubilee Prize of the Royal Society of Edinburgh was awarded to him in 1890.

Enthusiast as he was in experimental and mathematical work, he never allowed this to interfere with his University teaching, to which, from beginning to end of the forty years of his Professorship, he devoted himself with ever fresh vigour, and with unremitting faithfulness, as his primary public duty. How happily and usefully and inspiringly he performed it, has been remembered with gratitude by all who have ever had the privilege of being students in his class.

With not less devotion and faithfulness during all these years he has worked for the Royal Society, of which he was elected a Fellow when he came to Edinburgh as Professor. At the commencement of the following session he was elected a Member of Council; and in 1864 he became one of the Secretaries to the ordinary meetings. In 1879, in succession to Professor Balfour, he was elected to the General Secretaryship; and he held this office till the end of his life.

His loss will be felt in the Society, not only as an active participator in its scientific work, but also as a wise counsellor and guide. It has been put on record that "The Council always felt that in his hands the affairs of the Society were safe, that nothing would be forgotten, and that everything that ought to be done would be brought before it at the right time and in the right way." In words that have already been used by the Council, I desire now to say on the part, not only of the Council, but of all who have known Tait personally, and of a largely wider circle of scientific men who know his works,—"We all feel that a great man has been removed; a man great in intellect, and in the power of using it, and in clearness of vision and purity of purpose and therefore great in his influence, always for good, on his fellow-men; we feel that we have lost a strong and true friend."

After enjoying eighteen years' joint work with Tait on our book, twenty-three years without this tie have given me undiminished pleasure in all my intercourse with him. I cannot say that our meetings were never unruffled. We had keen differences (much more frequent agreements) on every conceivable subject,—quaternions, energy, the daily news, politics, *quicquid agunt homines*, etc., etc. We never agreed to differ, always fought it out. But it was almost as great a pleasure to fight with Tait as to agree with him. His death is a loss to me which cannot, as long as I live, be replaced.

The cheerful brightness which I found on our first acquaintance forty-one years ago remained fresh during all these years, till first clouded when news came of the death in battle of his son Freddie in South Africa, on the day of his return to duty after recovery from wounds received at Magersfontein. The cheerfulness never quite returned. The sad and final break-down in health came after a few weeks of his University lectures in October and November of last year. His last lecture was given on December 11, 1900.

277. ADDRESS ON INSTALLATION AS CHANCELLOR OF THE UNIVERSITY OF GLASGOW.

[November 29th, 1904.]

MR PRINCIPAL, Your Royal Highness, My Lord Duke, My Lords, Ladies, and Gentlemen,—I thank you all for your kindness in coming here to-day. Members of the General Council, I thank you for the great honour you have conferred on me in electing me to be Chancellor of the University of Glasgow, as successor to Lord Stair, by whose death we have lost a steadfast friend, and a wise ruler in all cases in which his judgment was required. Recent constitutional changes have diminished the executive power and responsibility of the Chancellor. Nevertheless I shall always endeavour to be as heartily devoted to the good of the University as any of my predecessors in the office can have been.

It has been a great pleasure to me to-day, as the first duty of my office, to have the privilege of conferring the honorary degree of Doctor of Laws on a Royal Lady, daughter of our late beloved Queen Victoria, and sister to his Majesty the King: a lady who has endeared herself to all in Glasgow, and throughout our West Country, by her kind sympathy and active co-operation in all good works. Her Royal Highness is gratefully remembered by us all as having been President of Queen Margaret College from its foundation to the time of its absorption into the University.

It has also been for myself a very agreeable duty to confer the degree on each of the nine distinguished men presented by the Senate of the University as worthy of this honour.

To be Chancellor of one of the Universities of our country is indeed a distinguished honour. For *me*, to be Chancellor of this my beloved University of Glasgow, is *more* than an honour. I am a child of the University of Glasgow. I lived in it 67 years

(1832 to 1899). But my veneration for the ancient Scottish University, then practically *the* University for Ulster, began earlier than that happy part of my life. My father, born in County Down, was for four years (1810 to 1814) a student of the University of Glasgow; and in his Irish home as first Professor of Mathematics in the newly-founded Royal Belfast Academical Institution, his children were taught to venerate the University of Glasgow.

One of my earliest memories of those old Belfast days is of 1829, when the joyful intelligence came that the Senate of the University of Glasgow had conferred the honorary degree of Doctor of Laws on my father. Two years later came the announcement that the Faculty of Glasgow College had elected him to the Professorship of Mathematics.

My father's experiences as a Glasgow student are naturally of supreme interest to myself. May I briefly speak of them, not because your kindness to me tempts me to be egotistical, but because the difficulties overcome and the precious life-long benefits won, by the struggles of an Irish student of the University of Glasgow in the beginning of the nineteenth century, illustrate the vitality and efficiency of the University in that primitive time.

There were no steamers, nor railways, nor motor cars in those days. Can young persons of the present time imagine life to be possible under such conditions? My father and his comrade-students, chiefly aspirants for the ministry of the Presbyterian Synod of Ulster and for the medical profession in the North of Ireland, had to cross the Channel twice a year in whatever sailing craft they could find to take them. *Once* my father was fortunate enough to get a passage in a Revenue cutter, which took him from Belfast to Greenock in ten hours. Another of his crossings was in an old smack whose regular duty was to carry lime, not students, from Ireland to Scotland. The passage took three or four days; in the course of which the little vessel, becalmed, was carried three times round Ailsa Craig, by flow and ebb of the tide.

At the beginning of his fourth and last University session, 1813—14, my father and a party of fellow-students, after landing at Greenock, walked thence to Glasgow. On their way they saw a prodigy—a black chimney moving rapidly beyond a field on the left-hand side of their road. They jumped the fence, ran across the field, and saw, to their astonishment, Henry Bell's 'Comet' (then not a year old) travelling on the river Clyde between Glasgow and Greenock. Their successors, five years later, found in David Napier's steamer 'Rob Roy' (which in 1818 commenced plying regularly between Belfast and Glasgow) an easier, if a less picturesque and adventurous, way between the College of Glasgow and their homes in Ireland. Those students who had experience of cross-channel passages before and after the advent of the 'Rob Roy' may well have been grateful to their University, not only for what it did for themselves, but for what, sixty years before, it did for steam navigation, in giving to James Watt a scientific home, and congenial friends, *and a workshop*, in the old University territory adjoining to the High Street of Glasgow.

In the course of his four student-years, my father attended the classes of Humanity, Moral Philosophy, Mathematics, Natural Philosophy, Anatomy, Divinity. Though his passion was for Science, and especially Mathematics and Natural Philosophy, he attended during his first three sessions, and won prizes, in the Latin Class, then happily as now, called Humanity.

It is scarcely possible to overestimate the life-long good gift presented to a scientific student a hundred years ago as now, by Universities, in giving something of the *Literae Humaniores* to all who can and will take it.

In 1834, two years after my father was promoted from Belfast to the Glasgow Professorship of Mathematics, I became a matriculated member of the University of Glasgow. And now, forgive me if I speak to you of memories, trivial and not trivial, of old Glasgow College, which crowd into my mind. The little tinkling bell in the top of the College tower, calling College servants and workmen to work at six in the morning: the majestic tolling of the great bell wakening at seven o'clock Professors (and students too I believe in the older times when students lived in College): then, again, the lively little tinkling bell calling the professors and students of Moral Philosophy, and Senior Greek, and Junior Latin, at half-past seven, to work in their class-rooms. Woe to the student of Latin who reached the class-room door ten seconds after the quick little bell's last stroke. He was shut out by the doorkeeper, unfailingly ruthless, by inexorable order; and had to wend his way through the darkness to his lodging, sorrowfully losing the happy hour's reading of Virgil, or Horace, or Livy, with his comrades, under their bright young Professor, William Ramsay; and knowing, poor fellow, that he had got an indelible black mark against his name. Rarely did even a single student of a large class experience this disaster. It was a sharp, healthy, beneficial discipline, rigorously maintained by one of the kindest and most considerate of all the Professors who have ever guided students in the Scottish Universities.

But I must not tax your patience by lingering over happy recollections of my earliest student life in Glasgow College. As to Latin, I followed my father's example and attended divisions of that Class during three sessions. To this day I look back to William Ramsay's lectures on Roman Antiquities, and readings of Juvenal and Plautus, as more interesting than many a good stage-play that I have seen in the theatre. Happy it is for our University, and happy for myself, that his name, and a kindred spirit, are with us still in my old friend and colleague, our senior Professor, George Ramsay.

Greek, under Sir Daniel Sandford, and Lushington; Logic under Robert Buchanan; Moral Philosophy under William Fleming; Natural Philosophy and Astronomy under John Pringle Nichol; Chemistry under Thomas Thomson, a very advanced teacher and investigator; Natural History under William Couper; were, as I can testify by my own experience, all made interesting and valuable to the students of Glasgow University, in the thirties and forties of the nineteenth century. Sandford, in teaching his Junior Class the Greek alphabet, and a few characteristic Greek words, and the Scottish pronunciation of Greek, gave ideas, and something touching on Philology, to very young students, which remains on their minds after the heavier grammar and syntax which followed have vanished from their knowledge. Logic was delightfully unlike the *Collegium Logicum* described by Goethe to the young German student through the lips of Mephistopheles. Even the dry bones of predicate and syllogism were made by Professor Buchanan very lively for six weeks among the students of Logic and Rhetoric in Glasgow College sixty-seven years ago; and the delicious scholastic gibberish of *Barbara celarent* remains with them an amusing recollection. A happy and instructive illustration of the Inductive Logic was taken from Wells' *Theory of Dew,* then twenty years old.

My predecessor in the Natural Philosophy Chair, Dr Meikleham, taught his students reverence for the great French mathematicians Legendre, Lagrange, Laplace. His immediate successor in the teaching of the Natural Philosophy Class, Dr Nichol, added Fresnel and Fourier to this list of scientific nobles: and by his own inspiring enthusiasm for the great French School of mathematical physics, continually manifested in his experimental and theoretical teaching of the Wave Theory of Light, and of practical astronomy, he largely promoted scientific study and thorough appreciation of science, in the University of Glasgow. In this Hall you see side by side two memorial windows presented to the University to permanently mark its admiration of *three* men of genius, John Caird; John Pringle Nichol; and his son, John Nichol; who lived in it, and worked for *it* and *for the world,* in the two departments of activity for which Universities exist, the Humanities and Science.

As far back as 1818 to 1830 Thomas Thomson, the first Professor of Chemistry in the University of Glasgow, began the systematic teaching of practical chemistry to students; and aided by the Faculty of Glasgow College, which gave the site and the money for the building, realised a well-equipped laboratory, which preceded, I believe, by some years Liebig's famous Laboratory of Giessen, and was, I believe, the first established of all the laboratories in the world for chemical research and the practical instruction of University students in chemistry. *That* was at a time when an imperfectly-informed public used to regard the University of Glasgow as a stagnant survival of medievalism and used to call its Professors the Monks of the Molendinar!

The University of Adam Smith, James Watt, Thomas Reid, was never stagnant. For two centuries and a quarter it has been very progressive. Nearly two centuries ago it had a laboratory of human anatomy. Seventy-five years ago it had the first chemical students' laboratory. Sixty-five years ago it had the first Professorship of engineering of the British Empire. Fifty years ago it had the first physical students' laboratory—a deserted wine cellar of an old professorial house, enlarged a few years later by the annexation of a deserted examination room. Thirty-four years ago, when it migrated from its four-hundred-years-old site off the High Street of Glasgow to this brighter and airier hill-top, it acquired laboratories of physiology and zoology; but too small and too meagrely equipped.

And now, every University in the world has, or desires to have, laboratories of Human Anatomy, of Chemistry, of Physics, of Physiology, of Zoology. Within the last thirty years laboratories of Engineering, of Botany, and of Public Health, have been added to some of the Universities of the British Empire, with highly beneficent results for our country and the world. All these the University of Glasgow now has.

During the last fifty years our University has grown in material greatness, and in working power, to an extent that its most ardent well wishers in the first half of the nineteenth century could scarcely have imagined possible. Two successive legislative commissions (1858 and 1889) have reformed its constitution; and broadened its foundations; and added to its financial resources; and admitted women to its membership, with all the privileges of students and graduates. Splendidly liberal subscriptions by the people of Glasgow, and by a world-wide public outside, backed by powerful aid from the National Treasury, enabled the University on leaving its ancient site, to enter into the grand group of buildings on Gilmore Hill, in which it has happily lived ever since. A few years later, the generous gift of £45,000 by the late Marquis of Bute built the Hall called after his name, in which we are now met. At the same time the adjoining Randolph Hall and staircase were built by a portion of the Legacy left to the University by the late Mr Randolph. The Queen Margaret College and grounds were presented to the University by Mrs Elder for the education of women; who also added largely to the endowment of the Engineering Professorship, and founded the Professorship of Naval Architecture.

Other generous donors have given an Engineering Laboratory with Lecture-rooms; and Botanical Buildings; and great and much-needed extensions in the Anatomical Department. The Carnegie Trust, and Principal Story's University Equipment Scheme, are at present providing two new buildings: one of these is for extensions in the Medical School. The other, in which I naturally take the most personal interest, is for the Natural Philosophy Department, including Lecture-rooms and a Physical Laboratory; all designed, and at present being realized, under the able direction of my successor in the Natural Philosophy Chair, Professor Andrew Gray.

In the Province of the Humanities, the working power of the University for instruction and research has been largely augmented during the last fifty years by the foundation of new Professorships: Conveyancing, and English Language and Literature, and Biblical Criticism, and Clinical Surgery, and Clinical Medicine, and History (in my opinion the most important of all in the literary department), and Pathology, and Political Economy now worthily cultivated in the University of Adam Smith.

In Mathematics, and in the Science of dead Matter, Professorships of Naval Architecture and Geology; Lectureships of Electricity, of Physics, and of Physical Chemistry; and Demonstratorships and official Assistantships in all departments; have most usefully extended the range of study, and largely strengthened the working corps for Research and Instruction.

My Lord Provost, may I venture to congratulate the City of Glasgow on having for her god-daughter a University so splendidly equipped, and so admirably provided with workers. I repeat my thanks to the General Council for having elected me to be Chancellor; and I heartily wish success and happiness to all Professors, and Lecturers, and Teachers, and Students, in their united activity in every department of the Humanities, and every Field of Science.

INDEX

(The numbers refer to pages)

CAMBRIDGE: PRINTED BY JOHN CLAY, M.A. AT THE UNIVERSITY PRESS.

Printed in the United States
By Bookmasters